THE MILITARY GEOGRAPHY OF THE SOLAR SYSTEM

Walter J Gomez

ISBN 978-1482568066

To my Telecha, and the Fabulous Five (Alexa, Benjamin, Blake, Corrina, and Preston James)

CONTENTS

PREFACE

This book is a synthesis of my experiences as a U.S. Air Force Medic, intelligence officer, and professor of geography at the U.S. Air Force Academy, as well as a decade of work as a computer professional, and a stint with a multi-national corporation in Central America and the Caribbean. I draw on these experiences to present what military geography might look like in the 21st century. The idea for this book was born in 1971, during my assignment to the Laos theatre of the Vietnam War. In the course of 99 missions as an airborne air intelligence officer over Laos, I came to realize how much the electromagnetic spectrum was being utilized throughout the whole process of "flying and fighting" in that conflict. More generally, the central theme in this book revolves around the fact that all human activities occur within a physical environment, or geography. As a military geographer, I focus on the geographic elements that enable or hinder military aerospace operations, within the Earth's atmosphere and now in the solar system that lies beyond it. A corollary of this is that all human activities occur in a spatial construct, which geographers call a region.

So, this book essentially presents a geographer's view of the solar system and the conduct of military operations therein. Consider that the science of geography has grown and evolved at an accelerated pace during the past century, both with respect to the scope of subject matter and the methods and tools that are used geographic studies. The range of subject matter now includes phenomena of all kinds, and which occurs throughout the relevant portions of the Universe. Hence, I am proposing that all matter and energy, beginning with the sub-particles of the atom, and extending to the stars that gave them birth, are a part of a fabric of space-time that can be analyzed in spatial terms.

The geographer also looks for the controlling variable(s) that underlie the spatial nature of any region or field. This geographical concept of the spatial region is a powerful key to understanding and analyzing all manner of human activity, on Earth, and now the rest of the solar system. One particular type of human activity can be generally referred to as military operations. Military geography has traditionally been concerned with the natural and cultural environment in which these military operations are carried out. So, an underlying theme of this book is the nature of the interaction between the various "environments" and the conduct of military operations throughout the entire solar system.

INTRODUCTION

My main purpose in writing this book is to propose a model for a military geography which can be useful in the modern era of the space age; a time-period which has seen a technological revolution that has affected the conduct of military aerospace activities since the middle of the 20th century. If I had written this book in 1963, when I began my undergraduate studies in geography, much of what I describe today (both with respect to the science of geography itself and the subject-matter) would have been alien to me then. At that time in the history of the science of geography, for example, the theory of continental drift and plate tectonics had not yet been generally accepted as a comprehensive theory to explain the spatial distribution of the continents on Earth. At the same time, the geographers mostly kept their eyes on the surface of our planet; only looking up to the sky to determine the effects of the atmosphere on surface activities. Meanwhile, the oceans essentially were relegated to the science of oceanography.

The technology for inquiry and analysis that is now available to the geographer-scientist in the 21st century was still only in the earliest stages of development as late as the 1960s. It was still the era of the camera and the notepad for acquiring data, and the art of cartography to "write the spatial book" about a given portion of the surface of the Earth. Since then, however, remote sensing technologies have been developed to enable geographers to "see" in electromagnetic wavelengths that occur outside the realm of visible light. The instruments for capturing geographic data also have become more powerful and sophisticated; therefore, we are now able to see farther and with greater clarity. In the centuries, prior to the 19th century, the main devices for acquiring data were the human eye or one of the other natural senses. It was only in the 17th century that we began to develop such data-acquisition systems as the telescope and its attendant camera for studying the skies. On the other end of the scale, the microscope and such specialized devices as the radiometer and the spectrometer have become available only during the last two centuries. All together, these devices now enable scientists to study data at variable scales, to "see" more data at a time (bandwidth), and to see them more clearly, throughout the range of electromagnetic wavelengths.

Also, the captured data now can be now stored in digital form instead of on notebooks. The sketches and the photographs of the 19th and 20th centuries have now been largely superseded by the computer-generated imagery, which is itself stored on electronic-magnetic media that are exceeding the gigabyte scale in the capacity to store both bytes and pixels. Along with this almost unlimited storage capacity, data is being converted into usable information by evermore sophisticated and powerful computer software. So, the information can now be presented to the "user" in an almost "real-time" time-frame, and on a multitude of presentation media, including many new hand-held devices. More importantly, information is now available as a finished product to a greater variety of "end users" in the field. And, the information can be displayed in a virtual dimension, where the mind's-eye can be portrayed in a three-dimensional and animated "reality."

In my chosen the specialty of military geography, I have witnessed similar trend lines in the development of the science that attempts to put military operations into a spatial context. Once again, there has been an explosion of applicable technology in the areas of remote sensors and the computerized systems; thus enabling the processing of inputted data within a "real-time" cycle that produces usable information to the geographer in the field. But the most significant trend has been manifested in the ascension of military operations into the highest altitudes of the terrestrial atmosphere, and then into outer space itself. This last phenomenon has made it necessary to expand the boundaries of military geography to deal with the expanded military area of operations into the regions of the solar system that lie outside Earth. In this sense, the celestial sphere, which encompasses the entire Universe, has superseded the geosphere as the outer limits of geographic inquiry.

During my professional life, I have been fortunate to have worked and studied in many other fields outside geography, which I believe has given me a somewhat more "universal" outlook and perspective, which I now bring to the writing of this book. Moreover, the solid foundation that I have been given as a geographer enables me to always view data through a "spatial" lens, regardless of the particular field of science in which I find them. Some of my mentors have described this perspective as the "geographer's eye." Through my formal education as a geographer and the cumulative experience of my colleagues, I also have gained an innate "tool box" of concepts and models with which to make sense of the natural and cultural phenomena around me. It is this "geographic education" that I have applied to my work in the areas of military intelligence and operations, computer science, and in international business, during the past six decades. Thus, it seems that without even any apparent conscious effort, I invariably have framed each situation which I encounter in a spatial context, where most phenomena have both an absolute and relative location, and which occur within a conceptual "region" of my own construction.

Therefore, one of the main theses in this book is that the very definition of "geography" has effectively been altered – and expanded - by many events which have occurred since the 1960s. Even though the term "geography," in the strictest sense, applies to the study of the Earth, I believe that one must now expand the meaning to include virtually all matter and energy in the Universe – and at the levels of scale that range between the subatomic and the planetary, to the cosmic. For these reasons, I will refer to the "geography of the solar system" in the same vein as I would the geography of the northern hemisphere of the Earth. I also will employ the geographic skills that I have learned on Earth to describe and analyze the "geography" of the other planets and bodies of the solar region. With that, I now venture into a presentation of what I understand as a study of the military geography of the solar system.

———

One of the main arguments in this book is that the definition of the term, "military aviator" must be expanded, especially in the space age, where the term "astronaut" is applied to all the crew members of a space vehicle. That is to say, the term "military aviator" is a rubric which should encompass, in my opinion, all those who participate in any military aerospace activities. Indeed, the concept of "military aviator" already has been changed and expanded since the days of World War I, through the period of the "Cold War," and now, into the 21st century with the advent of the unmanned aerial vehicle (UAV). In the first days of military aviation, this term practically applied only to the pilot who flew the airplane, and the one who navigated by visual references on the ground, and who dropped the bombs on the ground targets. However, its meaning has expanded along with the increasing complexity in military aerospace operations since those early days.

The expansion began with the realization that the pilot alone could not always perform all the tasks of a mission; there had to be a division of labor. So a second "passenger" had to be brought along to perform such things as photo-reconnaissance or dropping bombs, at first. Then there emerged recognized need for a full-time navigator. And so it went, until WWII, when the concept of a team of military aviators working as a team to accomplish the mission became formalized. Then during the Cold War, the Strategic Aerospace Command further institutionalized the widened concept of "military aviator" with the team of highly-trained crewmembers of the B-52s. And, there was a further expansion of the rubric with the deployment of missile crews for the Minute Man ICBMs during the Cold War era.

The Vietnam War introduced an even more diverse population of "military aviators." There were, for example, the "EWOs" (Electronic Warfare Officers) who flew with certain F-4 Phantom configurations. Their primary function was to detect and defend against any electronically-controlled enemy air defense radar sites that might menace their mission. They utilized onboard sensors to detect enemy threats from SAMs and AAA sites. Based on their intelligence findings, these EWOs then utilized flares, chaff, and other electronic countermeasures to protect their aircraft from the radar-controlled anti-aircraft weaponry. Then there were the gunships, such as the AC-130 Specters, whose crewmembers brought down sustained and powerful ordnance onto the enemy ground positions in Vietnam and in Laos. In this case, the aircraft had become very much like the classic naval submarine: the pilot and co-pilot now functioned as the "captain" and "first mate" of the aircraft, while the "backenders" were now operating like the crew of a submarine. So, there were now military aviators on a "gunship" aircraft like the AC-130, whose specialized function was to monitor sensors, to compute the firing problem, and to operate the 105mm howitzers and Gatling guns against the enemy on the ground. There is no doubt that the definition of military aviator was changing as the "platform" was becoming a more significant aspect of the generic mission of "flying and fighting."

Astronauts initially were recruited from the ranks of the military and experimental pilots, especially those with extensive experience in flying high-altitude aircraft. These are the members of a community who were said to have the "right stuff." These astronauts (and cosmonauts) were assigned a variety of duties aboard the space vehicles (including the space shuttle and Soyuz). Later, these astronauts were joined by cadres of scientists and engineers who did not necessarily have experience as pilots. These were the "crewmembers" and "strap hangers" of the past, who now began to take more integral roles in space operations. The evolution of the "military aviator" has been manifested even more clearly in the continuing wars in the Gulf region of Arabia, including those in Iraq, Afghanistan...and into Pakistan. So, the UAVs (unmanned air vehicles) are now increasingly taking the place of the manned vehicles some of the core missions of the air forces. Thus, in the Afghanistan war, one of the most effective attack aircraft had no humans on board. Computer programmers and flight engineers now guide the UAVs towards a precise target and control the delivery of ordnance.

Like the rubric of "military aviator," the definition of "military geographer" has been rapidly expanding to cover a much wider area of scientific endeavors. Indeed, much of this work in this book represents an attempt to develop a more rigorous and relevant definition of the modern "military geographer" in the context of military space operations. Therefore, in this book, I will examine how the concept of "military geographer" has evolved during the as the expansion of our "geographic consciousness," and as a result of the ongoing evolution of military operations and in space exploration since the middle of the 20th century, and into the first decades of the 21st century.

My professional odyssey began with formal studies of geography in the early 1960s, at the University of Missouri at Kansas City (MO), under the tutelage of Professor Oberg. He taught me the intellectual value of using this unifying science as a tool for gaining an overall understanding of our world, by tapping into the other fields of science. Later, the faculty at the University of North Carolina graduate School of Geography further enhanced my knowledge of this unifying science. By 1972, when I returned from a tour of duty in Laos (a theater of the Vietnam War), I began to see three major changes in the science of geography – both in my capacity as an instructor at the Armed Forces Air Intelligence Center near Denver, CO – and as a member of the faculty at the U.S. Air Force Academy in Colorado Springs, CO. By this time, I had gotten into the intellectual habit of drawing from the other physical and social sciences to develop what I envisioned as Military Aerospace Geography.

This book is an attempt to apply these insights to the subject of military geography and military operations within the context of the entire solar system. It also is the product of the realization that the geography of our home planet, along with that of the other bodies in the Universe, is truly an integrated, dynamic system. Later, during the phase in my life as a computer scientist, I also came to appreciate that any analysis which begins from an anatomical or mechanistic point of origin must also recognize that all matter – from the atom to the Cosmos itself – interacts with the two basic states of being in the Universe, that is to say: energy and matter. It is this conceptual awareness that now enables geographers to conduct their traditional spatial-based scientific studies anywhere in the Universe. Starting from this basic premise, this book is an attempt to apply the methodologies and conceptual frameworks of geography to an analysis of military space operations.

THE EVOLVING SCIENCE OF GEOGRAPHY

The science of geography (including military geography) is mainly concerned about the location, distribution, and movement of phenomena within a given spatial construct. The concept of "place" is one starting point in the analysis of a geographic space. Another way to think about a place is to think of it as a "point." The geographic point, as we now understand it in the 21st century, can be as small as a micron, or as large as a galaxy within our Universe. Any interrelationship between points can be described by a "line," which is also a spatial construct. These are the "primitives" of space on which the science of geography is based. Added to this static model of geographic space, is the concept of movement between two points along a line or an orbital path. Further, repetition of lines and movement creates a field, which by propagation or dispersal, moves outward from a central point. There also is the type of movement that refers to some kind of linear change in position, from an original point to a destination point. In this sense, the concept of movement also encompasses the ideas of location and distribution. At a cosmic scale, everything, including the Universe itself, is moving. This implies that the notion of absolute location is only a temporary conceptual tool for taking a "snapshot" of the geography of a place, as it were.

The term "geography" (from the Greek "geo," which means "Earth") has traditionally referred to the scientific study of the Earth's surface (land and water) and atmosphere. It concentrates on the description and analysis of the spatial variations in phenomena that occur on the surface of our globe, and deals with their interrelationships and their significant regional patterns. In a sense, it can be said that geography is also concerned with the intrinsic properties and characteristics of the points and the lines that interact with the various spheres of a planet, including the atmosphere, lithosphere/hydrosphere-mantle-core system. Thus, geography can be said to be defined more as an analytical point of view, rather than the study of any set of natural or cultural phenomena. However, in attempting to understand planetary anatomy and physiology, the geographer is also interested in analyzing the processes that occur within the celestial body.

Although the term "geography" refers to the planet Earth in the strictest sense and by definition, in this book I will be using it to refer to all the celestial bodies within the solar system. This usage of the term is therefore a literary device that reflects the "sameness" of natural phenomena throughout the solar system. Also, geography has developed a "scientific language" which will prove to be applicable in our study of geography throughout the rest of the solar system. Again, it appears that the geographic concepts and models that have proven to be effective on Earth will also be useful in the study of the rest of the solar system.

On another conceptual vector, I can say that during the 50 years or so of my involvement with the science of geography, I have learned that various "geographers" practice these noble arts and sciences from many different perspectives, in many different ways, and under various rubrics and names. My own opinion is that the "geographer" is a direct descendent of the classic Greek philosophers who were, in a very real way, the first scientists, even though they relied more on intuition than the modern scientific method that is used today. The geographers of today share, for example, the same intellectual audacity that allows them to venture into the other physical and social sciences in their search for "knowledge" which can contribute to the overall understanding of our world, and now the other bodies of our solar system…and the Universe. I would argue that it is this sense of intellectual curiosity that transcends the artificial boundaries of scientific inquiry which will prove to be the norm in the 21st century, as the "Google Revolution" and the explosion of computerized data-bases make it easier to collate particular knowledge bases as well.

It also appears to be true that the many methodologies and techniques that are employed by the space explorers of today are very much like those that have been employed by the traditional explorers on Earth. Although these explorers may not have been called "geographers," they nevertheless have used the same set of perceptual and practical tools in their endeavors. Also, some of these "practical geographers" have had better publicists than others and therefore have been more effectively memorialized by the later generations of humans. Thus, we are most familiar with the odyssey of Homer, the travels of Marco Polo, and the exploration of Lewis & Clark expedition. In contrast, we are rather hard-pressed to identify the ancient "geographers" who carried out the first studies of the "region" of the earth that we now call the Nile River valley. One can only imagine what "geographic studies" were carried out by the first scouts and pathfinders who came into the Nile Valley from the other parts of the Sahara, which were undergoing rapid desiccation as a result of a climate changes.

One of the better-known, modern "geographers," who operated under the rubric of "explorer," was John Wesley Powel, who led a scientific study of the "Great American Desert" in the southwest United States. At the same time, there were the geographers who were affiliated with institutions like Royal Geographic Society of Great Britain. One thing that is common to these past geographers, however, is that they pretty much confined their activities to describing the portion of the Earth that was still unknown to their particular human subgroup. Another thing they all had in common was an approach to geography, which is often described as "simple description," and therefore lacking what might be called scientific "analysis" today. In any case, what all of these "geographers" have in common is the application of exact measurements, careful recording of observations, and the mapping of significant areal patterns of cultural and natural features they encountered in the course of their explorations, all of which can be considered to be significant levels of analyses. In a sense, these kinds of analysis have been the terrestrial equivalent of the initial round of explorations of the solar system that have been carried out by government agencies, such as the U.S. National Aeronautics and Space Administration and the European Space Agency.

During the middle of the 20th century, the evolving science of geography, not content with "mere descriptive" analysis, turned to "quantitative analysis" in order to reinforce their claim to being as "scientific" – as in the so-called "hard" sciences, such as physics. In reality, the utilization of quantitative analysis was really only an extension of the careful measurement and mapping that traditionally has been done by scientific geographers. One very important derivative of the drive to become a "hard" science was the necessary foray into the scientific studies that were being carried out at the same time in physics, chemistry, geology, biology, and botany, to name but a few. That is when geography moved beyond "mere description" and into "scientific" studies. However, more importantly, the science of geography has continued to reinforce its function as a unifying pursuit. In a sense, all aspects of matter and energy have become the laboratory for the professional geographer. And, at the mid-point of the 20th century, the field of astronomy would be another science that would become relevant to geographic studies.

From another conceptual standpoint, the science of geography offers the concept of the spatial "region" as a framework for studying and understanding human-environment interaction in a spatial manner, and within certain defined spatial areas. During my early years of study in the field of geography, I held an essentially two-dimensional predisposition with respect to space and time. Physical space, which is the common denominator of the science of geography, was held to be a flat, three-dimensional continuum, one that could be visualized as containing the arrangement of all possible point locations within that space. In this concept of space, Cartesian coordinates would be the best framework for analysis, and straight lines would be the most useful conceptualization of linkages between the points. This seemed to work well in the development of abstract models of location and linkage, but it fell short when it was applied to three-dimensional geographic regions.

The domain of geography (in every sense of the term) has grown tremendously since the 1960s. On one level, we now know more about our own planet, which traditionally has been the focus of our studies of "geography," than ever before. However, it was only in the 1960s that it was finally understood how the relative locations of the continents have been configured and reconfigured, as the crust of our planet has been created and recreated during its 4 billion years or so of existence. Indeed, the theory of plate tectonics has only been well-understood and actually studied empirically in recent decades, as new technologies for scientific inquiry have been developed. Earth-orbiting satellites now have provided a truly global perspective, as advances in optics and remote sensing have displayed the surface of the Earth with more detail and clarity than ever before in human history. Indeed, it has been the distance of the orbiting satellites from our planet that has provided a panoramic scale of observation that had not been possible prior to the beginning of the age of space exploration. Finally, the development of remote sensing in wavelengths beyond the visible light portion of the electromagnetic spectrum has given the geographer the ability to "see" the objects of analysis that were "invisible" before the 1960s.

Thus, at the most basic level, geography refers to all energy and matter in the Universe, in all its forms. More particularly, geography primarily deals with the spatial aspects of this energy and matter. When I first conceived of this broader approach to the study of "physical geography" my mind was still operating from a geocentric perspective. That is, I saw the conceptual outer boundary as the terrestrial atmosphere. However, with the advent of the age of space exploration, I now realize that "natural geography" includes all matter and energy, from the sub-atomic particles to the most massive geographic phenomena. The study of the universe has traditionally been the domain of the ancient philosophers and the modern astronomers and physicists. Now, the geographer also has entered this scientific arena.

According to physicists and other scientists, all matter and energy in the Universe is composed of a basic set of building-blocks: the atoms. To the geographer of the 21st century, this now is seen as the conceptual starting point for studying all the phenomena that is distributed spatially throughout the Universe, including our Milky Way galaxy and our solar system. So, the atom can be seen as being the essence of the "stuff" of geography – physical and cultural – at all levels of magnitude.

The geographer will recognize the atom as a spatial region. Its basic components include a proton and a neutron, which are contained within the nucleus of the atom. Continuing with the metaphor of the spatial region, we see that this "central node" is surrounded by various orbits of moving objects that are called electrons. The atom can also be viewed as an example of spatial regions which occur at the ordinary scale of human observation. Therefore, one also can "zoom in" on a particular node within a spatial system, in order to gain understanding of the internal anatomy and physiology of that geographic phenomenon. So, once again, it is the predisposition to see physical phenomena in a spatial manner that enables the geographer to contribute to the overall quest for knowledge and understanding of our physical environment at all levels.

Another conceptual tool of the geographer is that of abstraction, which I see as a way to place a singular phenomenon into some simpler framework for quantitative analysis and manipulation. A well-known example is the equation, $E=mc^2$ which simplified the relationship between all matter and energy in the Universe into a simple statement of abstract relationships. On the other hand, the plotting of phenomena as abstract symbols on a map can be a very effective and efficient way of communicating information about the "real world." Clearly, the abstraction of the phenomena into powerful symbols, which can transmit information that is both qualitative and quantitative, is now more necessary than ever in today's computerized environment of bytes, photons and pixels. Beyond that, the abstraction of the distribution of phenomena on a map also facilitates the application of mathematical and computer modeling of flowing phenomena, in which changes in the intensity of some property are being measured.

However, at the end of the day, geographic science is mostly about empirical observation, quantitative measurement, and analysis of spatial relationships. It also utilizes the power of the human brain to make great intuitive leaps, such as in the case of the theory of plate tectonics. As an example of the explanatory power of abstraction and analog thinking, I have found them to be useful to view the configuration of our planet as being like that of a golf ball. That is, consider a core center which is covered by layers of varying degrees of thickness and composition. Like the golf ball, this sphere, which we call Earth, apparently has put into motion by a secondary effect of an initial impulse ("Big Bang?") and it now spins and revolves among all the other heavenly "golf balls." Of course, abstractions are useful tools for simplifying phenomena and systems. I have found that the simpler and elegant the abstraction, the easier it is to model the real phenomenon; but, the "elegant" model may also prove to be more problematic when it comes time to apply the implications of the model to the real world situation.

One class of real-world phenomena that occurs on Earth involves the "flows" which can be observed in the form of molten rocks and minerals, or as volatiles (such as water), or in gaseous form. During my time as a geographer, I have noticed how the phenomenon of "flows" permeates the physical world which is my laboratory. In part, this realization has occurred as the result of decades of empirical observations; but it has also derived from my continuing study of the literature in the fields of physics and the other "physical" sciences. Thus, in the science of physics, for example, practitioners have long concerned themselves with response of matter to forces that are exerted upon them by heat and pressure. Another such form of matter that has been studied extensively in terms of flows is the movement of the glaciers that occur on the surface of the Earth.

One of the most familiar flows on Earth is that of water. Scientific studies of this natural phenomenon have been organized into the fields such as hydrostatics (water at rest) and hydrodynamics (water in motion). What has been learned from these basic studies of fluids has also been applied to other natural phenomena, such as the waves that are produced by the wake of ships in motion. The flow of air also has been found to have many analog properties to that of the flow of water, as have the plasmas of charged subatomic particles. At the level of everyday observation, geographers and other scientists have also found that other examples of flows in nature, such as the flow of lava from volcanoes have much in common with the flow of water. More generally, the science of physics has also provided many other insights of analogy that have benefitted the field of geography, as well as the other "Earth Sciences." For example, the notion of a "boundary layer" also is utilized in the study of the movement of air in the study of the atmosphere, as well as the layers of rock that are a major element in the methodology that is employed by geologists and archeologists in their work.

Natural flows of lava and ash which result from volcanic events on Earth have proved helpful to the studies of other planets in the solar system. Consider the greatest volcanic eruption in the modern history of North America, when Mount St. Helena erupted in 1980. Suddenly, we were presented with a replication event of what had happened at Krakatau in 1883, when a volcano on an island that lay near Java, Indonesia erupted. Interestingly, the profile of the eruption and the aftermath of Mount St. Helena event would have many similarities to what occurred at Krakatau a century earlier. In both cases, a large portion of the mountain collapsed and a landslide ensued. This was followed by an explosion of lava and gas that, together, killed 57 people in the immediate term, and expelled a gaseous cloud over much of the northwestern region of the United States. Out of this disaster much was learned about the effects of this phenomenon on the atmosphere and the surface of our planet. But, it also has sparked a lot of scientific studies related to the seismic and thermal phenomena that derive from such eruptions. Ultimately, the knowledge that has been garnered from these manifestations of volcanism is now being applied to volcanism in other parts of the solar system.

The Tools of the Geographer

It appears that some of the essential tools in the study of the geography of the solar system are the techniques of extrapolation and projection. On the one hand, the understanding of the geography of the solar system, as it has been unfolding since the start of the Apollo missions, is largely based on what has been learned by Earth-based telescopic observation, space-based observations, and in situ rovers. However, much of the conceptual and methodological foundation is based on what has been experienced here on planet Earth. This is now even more the case since man's landing on the Moon and the following series of missions to the other planets and their satellites, as well as to the Asteroid Belt and the Kuiper Belt of the comets. It seems that even the matter and energy which make up the Interplanetary Medium can be related to our experiences on Earth, and the knowledge of physics that has been developed on our planet. Thus, our understanding of gravity and electromagnetism in the solar system has grown out of our scientific tradition on Earth, as well as the laws that we have crafted here to explain the movement of the planets and the atom, which have been validated by our empirical observations in space. An example of such observations is the intensive study of the omnipresent and powerful presence of radiation in the Universe. One important way this is done is through the through the human brain's "trick" of extrapolation from the known to the yet unknown.

Consider two cases of the "known," in terms of the effects of radiation on humans and their machines, which occurred in 1945, with the dropping of the atomic bomb on the Japanese cities of Hiroshima and Nagasaki. One of the important aspects of this wartime disaster was the forensic data that was gathered regarding the effects of nuclear radiation on humans, animals, and structures. The initial lessons that were learned about the way nuclear radiation affects humans has resulted in universal guidelines, rules and regulations for nuclear radiation safety. One outcome of this was the set of strict instructions for the handling of all manner of radio-active materials during the succeeding "Cold War" period, as I would learn during my initial hitch as a medical corpsman in the U.S. Air Force. Now, the body of knowledge that has been accumulated from this catastrophic event is being utilized to study the effects of radiation on humans and their machines in space.

Another instance of extrapolation occurred during the worst nuclear power station disaster in history. It happened at Chernobyl, in the Ukraine, which was then part of the Soviet Union, in 1986. The initial meltdown of the nuclear core and the resultant dispersion of radio-active matter left 32 persons dead and dozens injured. Ultimately, the nuclear core was entombed in a sarcophagus of steel and concrete. A broader disaster occurred, subsequently, in the area of dispersion of radio-active material, which produced many mid-term and longer-term cases of radiation sickness and incidences of cancer. On the other hand, there also was produced a great amount of data that would be incorporated into the growing knowledge about the effects of radiation on humans and their machines. Now, much of what was learned from these and other nuclear events on Earth are being applied to the general problem of similar radiation in space. This knowledge also has proved extremely valuable in the effort to estimate the likely effects of nuclear and cosmic radiation on astronauts and space machines.

The most common type of extrapolation is the model, such as the map that is used in geographic analysis. The most common form of map is a generalized representation on a plane surface of any portion of a planet or other celestial body in the Universe. Usually it is presented at a reduced scale and is the principal means by which geographers organize the observed data and any relevant information that is derived from the analysis of the data. It also has proved to be the most convenient form of recording and cataloging geographic information, and of transmitting information to ultimate users, such as aviators and astronauts, for example. Maps also provide a very effective means of visually comparing phenomena between regions. With the advent of the orbiting platforms throughout the solar system, it is now possible to comprehend the area of interest in real-time, and to input the data into a computer system for processing, updating, and outputting as usable information to the user.

In another sense, maps are the equivalent of mathematical models, which also strive to reduce empirical data into a conceptual whole. In my experience, the use of the mathematical model can be extremely helpful in the understanding and communication of geographic phenomena in the real world. Its corollary, quantification, is also necessary for developing computer-based models of reality. Inherent in the quest for quantification is the assignment of numerical values to the various characteristics and properties of the real-world data that are being observed. These represent the data that is inputted into computer models. The objective is to manipulate and convey the tremendous amount of geographic data that is now available, due to the growing ability of sensors and scanners to gather and input data from geographic "spaces" into an analytical system. Moreover, not only is the amount of raw data increasing, but also the complexity of the data in terms of space and time, and intensity of occurrence. This form of analysis is now essential because many aspects of the natural world can be usefully considered as being continuous, or what might be called "time-lapse" phenomena. Then there is the issue of probability in dynamic interrelationships and the ability to define an adequate degree of approximation in predictive models. All these dynamics raise the complexity of the equations to the level of reality. Thus, the efficacy of a formula (or model) that expresses relationships within a total reality depends on the accuracy of the elements of the formula, and the reality of the relationships that is asserted.

One commonly-used technique for time-based analysis, which deals with "variable factors" along a time vector, is the flow chart. This is usually a graphical display, composed of a series of boxes that represent some kind of stepwise progression of interaction, among individual components or subsystems of a system. These boxes, and the arrows that connect the boxes, show how the subsystems interact within the whole system. Another approach is the construction of a mathematical model, which also incorporates a set of equations that are based on approximations and calculations of probability of the variables. These describe the interactions within a system in quantitative terms. Sometimes these variations can approach the level of chaos. Modeling is one way to deal with this variability that is present in any natural or cultural phenomenon. In short, it is a way to reduce the complexity of the overall "space" of data that is being contemplated.

Computer technology has been especially useful in the more complex three-dimensional modeling process. It is essentially a computer-assisted variant of the traditional manual modeling process of preparing geometric data for producing graphics, as well as stand-alone models. Similarly, mathematical models, which involve an analysis of a system through the use of mathematical concepts and language, are essential to dealing with the much greater number of data points that are needed to create a three-dimensional model in the solar system environment. Along with the advances in computer hardware, there have been great strides in the science of computer programming, which involves a set of instructions that are given to the computer system.

The technique of simulation also has been used to predict various conditions in space exploration, through the comparison with the various sites on Earth which manifest the same characteristics and properties that have been discovered elsewhere in the solar system. Some of these natural simulation sites occur on the dry surface of our home planet; others are found in our oceans. These natural simulators have been located in certain extreme Earth environments such as the oxygen-starved Himalayas of Asia, the arid and extreme diurnal environments that are found in such places as the Atacama Desert of South America and Death Valley in California, and in the frozen continent of Antarctica. All of these have served to prepare space explorers from Earth for what they might encounter on the terrestrial planets and the Moon. These natural simulators have been especially valuable in preparing the space explorers for the conditions of extreme temperatures and low oxygen that are encountered in space and on other planets. Consequently, the polar regions of the Earth have been utilized to provide simulation training to the astronauts and cosmonauts, as well as the crews of the space stations, such as the "Skylab" project. Artificial simulations of expected environmental features in space include the atmospheric chambers and low-gravity devices that have been manufactured by humans on Earth. Perhaps the best example of a man-made replication of natural forces in the Universe are the event-cases of atomic bomb detonations and the meltdowns of nuclear-reactors that have occurred since the middle of the 20th century. These have provided some of the best data regarding the hazards of the radiation that derives from such events.

———

The term "analysis" is used in many ways, by many of the different sciences. I have encountered these in my professional pursuits and studies, as well as in discourse with my colleagues in the military, academia, and in business. This term has been used, in my experience, to refer to the separation of a whole into its component parts; the identification or separation of ingredients of a substance; and a statement of the constituents of a mixture. All of these describe a process or action which gives unity to the various parts or components of the whole "system." Another inference which I draw from these various definitions is that an analysis almost always takes place for a specific purpose. An example of this would be the careful study or analysis that takes place in order to identify a list of targets that are to be attacked by military aircraft or missiles. Another would be an analysis of the planetary core-mantle-magnetic field systems throughout the solar system.

At a more abstract level, the systems approach has proven useful in the analysis of a complex whole, as well as its elements or components, and their interrelationships. It is also valuable in conducting a careful study of a state of complexity by reducing the number of variables of a problem. The desire for clarification in the transmission of information also is manifested by the notion that the objective of analysis is often to make the complex more simple and, hopefully, more understandable. So, a comprehensive analysis of the relevant systems may serve to enhance the efficacy of communication, especially in the context of "real-time" communications in military operations, where the communicating is now increasingly being done by computers and telecommunications systems at nearly the speed of light. In such a technologically-advanced environment, it is desirable to have analyzed all the components of the activity before the system is put into operation.

———

The systems approach also can be appreciated as one of the most powerful conceptual tools for understanding the inter-relationships and inter-connections of the parts of the whole of any phenomenon. It is arguably at the core of the modern science of military geography and of military aerospace operations. Ever since the advent of the so-called "operations research," which was conducted by the British and Americans in World War II, this perspective for analysis and description of phenomena has become pervasive in all manner of science and technology. It refers to the application of scientific methods to the management and administration of organized military, governmental, commercial, and industrial processes. It is not a science in itself, but rather the application of a scientific methodology to the solution of a myriad of problems. It focuses on the performance of organized systems taken as a whole, rather than on the individual parts taken separately. Thus, operations research originally was concerned with improving the operations of existing systems, rather than development of new ones. The systems approach also involves the utilization of logic, mathematics, statistics, as well as more recent developments in communications theory, decision theory, cybernetics, organizational theory, and general systems theory.

Another feature of the methodology which is called "systems analysis," within the context of military space operations, is that it is oriented toward certain operational functions, such as the planning the trajectory of a space mission, or designing a rocket launch event. It may also involve the development of an appropriate rocket-fuel/payload ratio for a given space mission. Simultaneously, there are analyses which are applied to problems regarding the development of necessary life-support systems for space operations. Then, there is the whole domain of engineering analysis that is done for the development of materials and energies, guidance and control subsystems, and avionics for space vehicles. All these types of analyses are done with specific objectives that are related to the developing optimum hardware, software, and human systems that might be needed for the successful achievement of mission objectives.

The technique of systems analysis is a valuable tool for making sense of what geographers and the military operations people perceive as the "real world." Thus, it is an effective tool for organizing the seeming chaos of phenomena which is perceived in reality. In both of these situations, it also provides a very effective approach to the process of ordering raw data into abstractions wherein the phenomena which one observes in reality is converted into a spatial network of points and connecting lines. Thus, a general definition of a geographic system is that it consists of interconnected and interacting nodes, which provide the dynamic structure for processing the inputted data, which is then outputted as information that contributes to a defined objective or outcome. All of this has led me to believe that this systems approach to analyzing military aerospace operations and the physical environment in which they take place, is the most efficacious one for the military geographer. It works on many levels of abstraction and reality; and it works in the many dimensions of the physical world in which we live.

It may be reasonably argued that the systems analysis approach is one of the most important conceptual methodologies that have helped to advance the study of geography as a science. It begins with a point of view; an abstract rationalization of geographic phenomena as an abstract set of nodes and linkages, all operating as an interdependent and interrelated spatial system. Geographic studies of natural and cultural systems can be initiated at any level of subsystem within the overall system and projected outwards toward any other point in the system or subsystem. The systems approach to geographic studies also can be utilized to analyze a space-time slice of a system or a stepwise series of such events, thus adding a fourth-dimension to the analysis. Thus, systems analyses lend themselves to the dynamic tracking of flows and fluxes that might occur in within a region. On another trajectory, one of the great fruits of this systems approach to geography is the ability to facilitate the "digitization" of geographic data, thus providing the ability to utilize all manner of sensors and the processing power of computers to solve geographic problems. The conceptual power that is inherent in the systems approach to geographic analysis already has been used in many of the specialties within the field of geography, including political geography, economic geography, and military geography.

One effective approach for analyzing this increasing level of complexity in geography is to study each of the various geographic subsystems from an anatomical or structural perspective. Thus, for example, some geographers have, for purposes of analysis, organized the Earth into various "natural" spheres which may be seen to radiate from the core of the planet. These are generally categorized as the atmosphere, the lithosphere, and the hydrosphere. Some geographers also recognize a fourth natural sphere: the biosphere, although at the present time, this only applies to Earth. Here, the domain of the biosphere is seen as encompassing a domain, extending from the atmosphere and downwards to the deep-sea vents of the ocean. This domain of organic life-forms on Earth is now studied as an interconnected and interdependent system, that is, an ecosystem. It also can be seen as a possible paradigm for detecting and understanding the nature of life-forms on other bodies of the solar system.

———

Some geographers have focused on the past 10,000 years of interaction between human activities and the natural layer of geography, resulting in "cultural geography." It includes the present man-made system of human settlements of all sizes and types, from the smallest village to the largest megalopolis that exists throughout the Earth at the present time. It also includes the cultural systems that are interwoven into the overall cultural geography that humans have developed on Earth. Among these are the transportation-communication subsystems, which include such cultural geographic features as harbors and airports; railways, highways, and canals; as well as the places (nodes) which they connect. These geographic objects also are seen as being manifestations of human history, especially with respect to the production and exchange of goods.

One notable example of such a cultural geography system is the ancient Silk Road that facilitated trade and cultural exchange between Europe and the Far East, via the Eurasian mainland that lay between them. A more recent example of such a communication system is the World Wide Web which now connects nodes located around the globe and even areas of the solar system outside our planet Earth. Cultural (man-made) geographic systems are also among the most interesting subjects of study by the military geographer. For example, these make up virtually all of the typical "system of targets" that is presented to the air reconnaissance aviator and the bombardier or weapons aviator in military aerospace operations. But this rich mine of targets has it dark side for the military aviator: the air defense systems that can prove quite hazardous to the military aviator who is attempting to strike the passive targets which they protect. This subsystem is essentially made up of anti-aircraft weapons (Air Interceptors, AAA and SAMs) and the detection-tracking-fire control subsystems that enhance the precision and, therefore, the lethality of the weapons.

––––––––

Scientific geography has traditionally relied heavily on the plotting of all manner of data on media which are called maps. It is a specialized form of communication which emphasizes the spatial information about the phenomenon that is the subject matter. Once communicated, a major value of the map is to provide special insights which are drawn from spatial patterns of location and distribution throughout an area. Pattern recognition itself is a form of information-reduction and the assignment of visual or logical patterns to classes, based on the features of these patterns and their relationships. Such information has proven especially useful in many human endeavors, including military operations. Consequently, even in the early years as a student of geography in the early 1960s, I remember learning the "nuts and bolts" of plotting phenomena of all types onto maps or other cartographic devices.

It was a labor-intensive process that involved a lot of transposition from empirical observation of data to the plotting of abstractions of information onto a map. As an example, I remember learning how to use dots of varying diameter and hue to portray geographic phenomena. The art of contour-line drawing was also highly valued because of its power to show densities, three-dimensional perceptions, and gradations of physical phenomena such as elevations. It also could be used to portray relative degrees of power, potency, or intensity. Then there was the plotting of distribution symbols onto plastic transparencies, which could then be projected onto a screen for display. Ultimately, these techniques would facilitate the ability to utilize the power of the computer to develop mathematical models for more sophisticated spatial analyses.

It is well-understood that geographers, like all other scientists, gather data from empirical observations in order to begin the process of analysis. These are then posted onto maps of one kind or another in order to convey geographic "information." Then the geographer proceeds to mull over what he or she sees on the maps in much the same way that economists ponder economic curves. In doing so, many tools of logic and reasoning are also employed to derive "information" from the data that have been gathered. This often leads to the formulation of "spatial hypotheses" which provide guidance for the design of future studies. Most importantly, geographers use the knowledge and understanding that they have gained from their maps in order to try to predict what is to be encountered for the first time. This is what was done by Marco Polo, the Spanish Conquistadors, and the Captains Lewis & Clark. Indeed, one of the most powerful of these tools in this application of logic to geographic data is that of inferential thinking. This refers to the intellectual process of inferring the nature and operation of the unknown from what is known already.

It is apparent from archeological and anthropological findings that humans began to develop a "mental map" of locations, both in absolute and relative terms, almost from the beginnings of the human history. So, even in the most abstract sense, the concept of location has been a signatory aspect of human consciousness. Also, it appears that the understanding of "place" has become a conceptual vessel for understanding the properties, attributes and characteristics of a site, which could then be communicated by the observer to others. The idea of "place" also has gone from the qualitative to the symbolic and quantitative dimension; which has enhanced the ability of humans to communicate perceived inherent and intrinsic qualities of a place to other individuals, at a later time, and over long distances. However, if the observer were to describe the geographic context of the point location, it would be useful to assign some measure of reference to the places within the area in question. And, in doing so the observer would then begin to think carefully about the linkages of between points, so as to enrich the "mental map."

The use of abstraction in spatial analysis can be seen as being part of the trend toward the employment of "quantitative methods" in the 1970s. This is when a tangible "place" on the surface of the Earth could now be reduced to an abstract "node" and given a numerical value. This would become a powerful as a tool for conducting all sorts of "quantitative analyses" of geographic phenomena, particularly the utilization of sampling techniques, determination of probabilities and multi-variable analyses. In a sense, the science of geography is about going from the particular to the general; from the qualitative to the quantitative. It was also a movement toward the development of theories and hypotheses which would place geographic studies within the mainstream of the scientific studies. Such conceptual thinking will make it easier to transfer the knowledge that is gained from the study of Earth to all other globes in the solar system, and beyond.

The interweaving of geography within the matrix of sciences, both physical and cultural, has also provided powerful tools for those geographers who have chosen to work in the field of military geography. For me, it represented the nexus of scientific geography, air intelligence, and computer science that has provided me with the basis for studying the various aspects of military aerospace geography in the 21st century. The knowledge and skill-sets that I have acquired as a geographer have also provided me with a somewhat unique perspective when analyzing military operations. One aspect of this is the focus that geography puts on the inter-relationship between human cultural activities and the "physical" environment within which they occur. Another is the persistent recognition that these activities always occur on an abstract plane of one sort or another; at a level of abstraction that enables the geographer to deal with the primitives of point and line in a quantitative manner, and to develop hypotheses that can then be evaluated against the real world variables that introduce entropy in the model.

In the past, it was acceptable to think about distance in terms of static and absolute units in the conduct of spatial analysis. These have included the standard units of distance, such as miles or kilometers, depending on the protocol that was being used and on the context of application. However, as methods of transportation progressed, the reference to distance has changed as a result of the utilization of "enhancers," such as the horse and, later the automobile and the aircraft. The notion of relative location and relative distance began to take on more dynamic meanings. Thus, for example, aviators have become used to thinking in terms of "fuel consumption" distance, as opposed to an absolute measure of distance based on latitude and longitude, even in the age of the GPS. This trend towards "subjective distance" in spatial analysis is still in progress, and is probably the most fundamental change in the history of geography, especially in the age of space exploration.

During my tour of duty in Laos, I realized that the military aviator faced a multitude of positive and negative variables that determined the "effective distance" between a takeoff point and the target destination, within any given mission. Positive variables included a favorable atmospheric envelope at takeoff. Conversely, higher temperatures occurring after sunrise were a hindrance to the takeoff phase of the mission, for example. In the case of our "Cricket" command and control missions over Laos, the thrust that was generated by the four engines of the C-130 was a positive factor. But, at the same time, the combination of the weight of an extra 8,000 lbs. of fuel, the capsule load, and a backender crew of seven or more, was a negative factor. The presence of clouds and smoke, as well as the threat of air defenses, produced another negative factor. The same positive/negative ratio is now seen in the space missions of today.

———

The geographic concept of "place" also is important in all military aerospace operations. Just as with the concept of distance, there are also inherent absolutes within places and those that are relative compared to other places. The former refers to such characteristics as the spatial dimensions of a place. One application of this concept is seen in the mapping of ground targets for military strikes. Typically these are plotted on a chart in the form of circles of varying size, which may be relevant to the decision of the type and amount of ordnance to place on the target. Now, in space, the concept of place is being utilized to analyze such geographic phenomena as the craters on the Moon and Mars, as well as on other bodies. Thus, in the era of space exploration, we can see the utility of central place theory to the analysis of "places" in the solar system wherein gravitational force is the operative variable for the spatial distribution of the planets, moons, asteroids, and comets, as well as the "belts" of natural and artificial satellites.

The mapping of places within separate geographic spaces appears to distort spatial relationships because we know intuitively that absolute space (which is within our normal experience) is "normal," and other spaces are somehow deviant from the norm. Conceptualization of such contexts takes us outside our normal idea of spaces. Indeed psychological studies have shown that the spaces in which people "live" are much more apparent (to the particular observer) than "real." Places have a number of relevant contexts, each of which is a different space. This can be seen in the

Mercator map of the world, which is actually a distorted image of the Earth's surface, caused by the attempt to project from a globe, onto a flat surface. This distortion results in the size of Greenland, for example, being portrayed on the Mercator map as much larger than the continent of South America. To be sure, the goal of identifying and mapping different "non-absolute" spaces is an increasingly important part of contemporary geography – especially in the era of the exploration of the solar system and beyond. For one thing, we are being informed by cosmologists and astrophysicists that the spaces that we, as geographers, have constructed on portions of the surface of our terrestrial globe can no longer be taken for granted as corresponding to "space" in the other parts of the Universe.

Another aspect of spatial analysis can be the study of the intrinsic properties and characteristics of a particular place; that is, the natural and cultural phenomena that are occurring at a particular place, at a given time. In the years following the beginning of the Space Age, this geographic introspection also has to include all the various planetary bodies within the solar system. At a very basic level, the analysis of location can refer to the study of the absolute location, relative location, or patterns of points – with reference to a given grid scheme that is superimposed on an area. However, the interest of the geographer usually goes beyond the mere cataloging of points (places). A secondary step in the analysis might then involve the plotting of the "absolute locations" of the member of the population on a matrix or grid, in order to give each member an "address" of sorts. Then, at some point in the analysis, the nature of the relations between various points, or the linkage characteristics of the phenomena will be plotted onto a map to provide some useful information regarding the "ground truth." Ground truth also can occur in layered or gradiated fashion, which adds a third dimension to the spatial distribution of phenomena on the actual surface of a planet or other celestial body, for instance. Moreover, the distribution of phenomena can occur within any geographic configuration of space – in both two and three dimensions. In some cases, geographers analyze the correlation or other significant relationship that might exist among the members of the statistical population. That is to say, to determine the controlling variable in the relationship.

With respect to military aerospace operations, the mission objective is almost always the controlling factor in the analysis of an optimum site for surface bases. Consider the decision to locate several SAC air bases just south of the Canadian border during the Cold War. Absent the mission imperative to minimize American response-time in case of a nuclear first-strike by the Soviet Union, it is doubtful that such a huge military installation would have been constructed in such a forbidding natural environment. This latitudinal zone happens to be one of the coldest places in the continental United States, with winter conditions of about nine months duration. It was not uncommon for temperatures to fall to -40 (F), and the fierce winds made it seem even colder. Also, blizzard conditions at ground level, which occurred several times a year on average, reduced visibility to zero and caused white-out conditions. It was a constant struggle and involved heroic inputs of resources to enable humans and machines to maintain optimum operational effectiveness. So, why was this base located in such an inhospitable environment? The answer is: "mission imperative."

Spatial analysis is becoming increasingly dependent on the domain of the electromagnetic spectrum and the technologies that have derived from it. One can reasonably argue that electromagnetic radiation is now the most important aspect of the geographic environment with respect to military aerospace operations, particularly as they are being expanded into the rest of the solar system – and beyond. This region marks the confluence between energy and matter that make up the geography of the Universe. As indicated by the term "spectrum," such radiation occurs within a series of frequency/wavelength packets of charged particles called photons; these move throughout the Universe in a wave-like motion, and at virtually the speed of light. These photons often are contained within "fields" or "fluxes" which interact with the magnetic fields of many of the planets and, thereby create many of the geographic features that are found throughout the solar system (such as the Aurora Borealis on Earth).

Humans have learned how to utilize the properties and attributes of these photon flows to create new tools for studying the Universe. More relevant to military operations in space, we have also learned how to utilize the nature of these photon waves for communicating and navigating, and also for detecting, tracking, and guiding ordnance onto specific targets. At a greater level of precision, we have harnessed the frequency/wavelength attributes of the wave itself in order dynamically measure differences in intensity, in order to automate many aspects of military space operations. The most commonly utilized frequency/wavelength spectra are those we call visible light, infrared radiation, ultraviolet, and radio phases of the broader electromagnetic spectrum. Thus, the electromagnetic spectrum can be said to be the new realm of the physical world that has become part of the inventory of subject matter for the geographer. To deal with this subject-matter, the modern geographer now has many tools that have been developed in the areas of optics and other sensor systems which can function in all the phases of the electromagnetic spectrum. There also have been continuing advances in the understanding of subatomic particles and energy that have given the modern geographer a means for studying the spatial nature of geographic objects of all sizes.

Moreover, it appears to be the case that, since the beginning of the exploration and study of the whole heliosphere, the same concepts and methodologies that the geographer has utilized in the study of the planet Earth, are now proving to be valid everywhere else in the solar system. In fact, what has been occurring in space studies indicates that what has been learned on Earth can be successfully transferred to the studies of the other parts of the solar system – and vice versa. For example, the advances in the science of optics and in remote sensing have improved the ability to detect and analyze phenomena throughout the solar system. By the same token, the science and technologies that are related to spectrometry can be applied equally effectively everywhere that the laws of physics apply. Geologists, geographers, and those who specialize in many of other terrestrial disciplines have discovered that this principle of commutative application applies to all matter and energy sets in the Universe. The reality of the matter is that both types of geographers – terrestrial and astro-geographers – will continue to learn from each other as each carries out their scientific studies of the common subject-matter.

According to current scientific thinking, in the first instant which followed the "original" event known as the "Big Bang," the Universe consisted only of hydrogen, helium, and a trace of lithium gases. Then, after a rather short "cosmological" time, dust particles were added to the mix. Since then, for the past 14 billion years or so, the proto-geographic space – the Universe – has been constantly growing and changing. Gases and dust particles have been congealing to form celestial bodies of varying sizes, some of which eventually migrated to a corner of a galaxy of stars we call the Milky Way; and within that gravitational domain, a particular star, our Sun was born. The next phase of development of our solar system then began; this was the solar system that has spawned the Earth and the other planets, as well as all the other celestial bodies within the region of the heliosphere.

Since the middle of the 20th century, human scientists have acquired many new technologies for reconstructing the history and the nature of the development of this solar system. The most important of these now include the remote sensors which we call telescopes; as well as the in situ instruments, such as the microscope. The telescope enables astronomers and astrophysicists to view the past development of the solar system by the remote observance of the light emitted by the most distant celestial bodies in the Universe. The working assumption being that we can calculate the age of the celestial body by measuring the time it takes for the light they emit to reach the observer on Earth. Similarly, we have now proceeded beyond the ability to observe and measure from afar to being able to utilize close-up laboratory techniques to decipher the temperature, pressure, and chemical composition of the subject matter from space. One of the most powerful of these tools is the mass spectrometer, a device that illuminates the "spectra" of the molecules of just about everything in the Universe, including those atoms and molecules that are characteristic of "life," as we know it, or can imagine it. Today, in the 21st century, many unmanned spacecraft and deep-space probes are equipped with mass spectrometers; and they are revealing much new information about the stars of our Milky Way galaxy, and the bodies of the our solar system.

Many of the advances in the understanding of our own planet, Earth, are now proving to be useful in the geographic analyses of other celestial bodies in our solar system. This is part of the overall growing appreciation of the similarities between Earth and the other planets of the solar system; thus resulting in the broadening of the domain of science of geography in the 21st century. One result of this recognition of the commonality of planetary geography can be seen in the recent geographic studies of the Earth, which utilize empirical data that has been acquired from space missions. So, with the advent of human exploration of space and the development of the tools that are necessary to carry it out, there has been the development of an "astro-geography" wherein the geographer now can utilize the concepts and methodologies that have been developed on Earth to study the other planets and bodies throughout the solar system. One of the interesting developments that have derived from this broader domain of geography, as it turns out, is that the concepts and methodologies that have served the geographer well on Earth can also be applied effectively throughout the solar system.

In this book, I make use of several basic geographic concepts to study the military geography of space operations. The first has to do with the idea that all human activities (including military space operations) are affected by the physical environment (both natural and man-made) in which they are carried out. However, the traditional "man-land" interrelationship which has been studied by geographers has tended to be seen at the ordinary world scale of analysis on Earth. Thus, for example, the earliest such analyses tended to deal with the human activities which economists refer to as "primary" economic activities, such as agriculture or mining pursuits. Later, the human activities which made up the "man" part of the relationship also included manufacturing and resource acquisition, especially during the course of the Industrial Revolution of the 18th and 19th centuries. There was a further expansion of the domain of geographic studies when "man" began to fly in the atmosphere and, once again when humans began to operate in space. So, a pattern has developed in which continued in which, as the scope of human activities expand in scale, so does the concept of "land" in this geographic equation. It also appears to be the case that the expanding "land" part of the equation is no longer so easily conceptually separated from the "man" part. That is, certain human activities may occur within subsets of the overall domain of human activity, but each of these subsets continue to be affected, to a greater or lesser degree, by the whole Universe. What we are left with, therefore, is the reality of the geographic concept of "land" which is expanding toward the cosmic end of the scale.

Therefore, I argue that the traditional scope of the field of military geography also is no longer adequate in the Space Age. One reason for this is that the domain of potential military areas of operation, at least during the 21st century, is realistically likely to include the inner portion of the solar system, including the region that would be encompassed by Earth, the Moon, and Mars – as well as the artificial satellite belt and the near-Earth asteroids. Later, it is quite possible that the scope of military geography will extend throughout other areas of the heliosphere. One early manifestation of this expansion in what might be called the practical area of military aerospace operations, is the now thirty-year existence of the U.S. Air Force Space Command, which is carrying out the explicit mission of developing practical systems and procedures to "organize, train, equip, maintain, and provide space and cyberspace operations forces." This mission statement appears in the USAF Almanac (Air Force Magazine /May 2011). The Air Force Space Command was established in 1982, and it has continued to be involved in the management of many space operations. Its most important domain of operations at this time is the artificial orbiting satellite system.

This leads to one of the main theses in this book: that geography – both natural and cultural – affects the operations of military aviators (and now astronauts), in many ways and to varying degrees. Prior to 1971, I would have limited my focus on this relationship to the "macro" aspects of geography on Earth, such as the larger phenomena in atmosphere, lithosphere, hydrosphere, and the biosphere. This also was an era in which military aerospace geography dealt almost exclusively with the troposphere and the terrain parts of the environment. It was a time when the geographer usually made only observations within the visible-light portion of light emissions, utilizing the visual acuity of the human eye – either unaided or aided by the optical devices such as eyeglasses, telescopes and microscopes. Visible-light photography was also used to memorialize what was seen, and "databases" of such imagery would be stored in archives of "hardcopy" photographs or films for future reference.

Now, however, after four decades of studying the geography of military operations, and now looking back from the vantage point of the 21st century, I have come to the conclusion that the "geography" that interacts with military aerospace operations also includes the "micro-geography" that occurs within the relevant environment. This level of geography can be comprehended only at a microscopic scale. So, the current view of the subject-matter of "geography" now also focuses on phenomena at the atomic and even sub-atomic scale. These encompass geographic phenomena such as subatomic particles and energy photons; of organic and non-organic cells; and of electronic bits and pixels. Thus, it can be said that the operational scale of geographic science has evolved from an analysis of the visible geography to that which includes the common denominator of energy and matter.

This heightened conceptual vision of geography, I believe, enables the military geographer to apply the traditional scientific methodologies that have been developed in the field of geography to deal effectively with any aspect of military operations in outer space, as well as on Earth. But to do so requires the erasing of artificial boundaries between other sciences – especially physics – but also astronomy and geology, among others. So, now the science of geography can delve into the domain of quantum mechanics, the science that deals with the behavior of "geographic" phenomena at the subatomic scale. This area of physics seeks to describe and account for the properties of the atom, and its constituent electrons, protons, and neutrons, as well as the particles that operate within the atom itself. Quantum physics also deals with the interaction of these sub-atomic particles with one another and with the spectra of intertwined electrical and magnetic fields that interact with the particles of matter.

On the other hand, there are some elements of "geography" that can be said to be "universal," if you will, in their occurrence. In one sense, some of current manifestation of the phenomena that we call "geographic" can be attributed to billions of years of actions by natural forces that have occurred since the "Big Bang" some 14.5 billion years ago. In the interim between that original event and today, it can be said that there have been a great number of other applications of physical and cultural forces that have resulted in the present geography that interacts with military operations within the atmosphere and in space. That is to say, the physical and cultural geography that military geographers are confronted with at the present time, has been the result of approximately 14 billion years of natural and cultural geographic events that have been occurring throughout the solar system.

The Geography of the Atom

The atom is the basic unit of matter, according to physicists. But the perspective of the geographer, the atom can be seen as being similar in many ways to the geographic region. That is to say, we can comprehend the atom as being a functional region with nodes and linkages. Like the region, the atom has a structure which is animated by several interacting processes. And, within the atom there is a population of nodes that reside in certain zones and interact internally with each other, both within the atom and the neighboring atoms. So, it would appear that the conduct of geographic inquiry also should take place at the scale of the atom, because this is the building block for all geographic phenomena. The atom is the realm of the particles, protons, neutrons, and electrons that are held together by chemical binding forces, while also being pulled apart by the forces of decay and the destabilization caused by finicky electrons. The geography of the atom includes such things as particles, photons, and chemical elements that also are in constant motion and reformation; and are only in a quiet state for short periods of time. As is the case with all geographic phenomena, the atoms and the subatomic particles also can be plotted on virtual maps, such as the electromagnetic spectrum and the periodic table of elements.

Isaac Asimov, in his classic work: Atom: Journey Across The Subatomic Cosmos (1991), describes the evolution of the atom. He postulated that all of the positive charge and nearly all the mass of an atom are concentrated in an extremely tiny nucleus. The negatively charged electrons that surround the nucleus have almost no mass, yet they occupy nearly all the volume of an atom. Others have suggested that the smallest positive-ray particle is the unit of positive charge in the nucleus. This particle, called the proton, has a charge equal in magnitude to that of the electron. James Chadwick (1932) discovered a particle with about the same mass as a proton, but with no electrical charge. This particle was called a neutron, and inasmuch as the neutron is electrically uncharged, it can penetrate the nucleus of an atom without being deflected (which has become especially relevant to military space operations). With the discovery of the neutron, the list of "building blocks" we need for "constructing" atoms is complete. It seems to me that the atom is the quintessential component of the geography of the Cosmos. Its origin can be traced back to the Big Bang event, which scientists calculate as having occurred some 14.5 billion years ago. According to conventional scientific theory, just moments after the Universe had begun it essentially consisted of a small, hot, and dense gas of charged particles. Once the proto-plasma had expanded and cooled enough, the atoms were able to form. These first atoms already contained the nuclei (protons and neutrons) and the electrons with which we are familiar.

———

The traditional understanding of the anatomy of the atom is that it is comprised of several parts: the nucleus and the electrons that revolve about it, very much in the way that the planets and other bodies revolve around the Sun in our solar system. Within the nucleus there reside protons and neutrons. Atoms which have a low atomic number contain approximately the same number of neutrons and protons. However, as the atomic number increases, the number of neutrons inside the stable nucleus becomes greater than the number of protons. At that point, the nucleus reaches a point where it is no longer stable. The interactions between the nodes, as well as the processes that drive them in an atom also have become better understood. For one thing, it seems that the electrons move around the nucleus in "fuzzy" orbital paths, not unlike the way the planets tend to meander within broad orbital paths that describe an elliptic rather than a circle. There also appears to be a spatial relationship between the nucleus and the electrons that might be defined as "dynamic tension," which is determined by gravity and a countervailing repulsion force. The upshot of this kind of relationship between the nucleus and the electron is that, under certain conditions and from time to time, an atom loses an electron or gains one from another atom. Also, the nucleus tends to move one state of energy to another and, in the process emits charged particles into its "environment."

There are other dynamics that are occurring within the atom. To begin with, electrons have an electrical charge, and the charged particles are constantly moving and accelerating. It also has been determined that electricity may be disseminated through space, with the same properties as those of light. At the turn of the 20th century there were several advances that led to the further understanding of an atomic structure as an electrical phenomenon. It also is now known that the positively charged portion within the atom is a relatively tiny, but extremely massive atomic nucleus. This core is surrounded by circling electrons, held within the atom by electromagnetic attraction. The resultant stationary charge will produce only an electric field in the surrounding space, but an electric charge in motion also produces an electromagnetic field. This is because the flow of energy through free space or through a material medium occurs in the form of the electric and magnetic fields. These produce electromagnetic waves such as radio waves, visible light, and gamma rays. In quantum theory, the resultant electromagnetic radiation refers to the flow of photons through space. These photons are packets of energy that always move at the speed of light.

Atoms also interact dynamically with neighboring atoms. They often will form a network and thereby produce a molecule which takes on certain properties and a distinctive identity. A molecule is defined as a group of atoms that are chemically bonded together in various configurations. The process of reconfiguration occurs when atoms form bonds by sharing electrons and by gaining or losing electrons to other atoms. All atoms of a given element are alike, but atoms of different elements combine in fixed proportions. An atom is the smallest substance that makes up the element and it has the same characteristics of that element. The molecule also is the mechanism by which chemical elements are formed. If a molecule contains more than one element, a compound can be formed. Compounds are formed when atoms of different elements are combined in fixed proportions. One outcome of such rearrangement of atoms is called a chemical reaction. Chemical reactions occur when different atoms and molecules combine together and split apart. During these chemical reactions, new product atoms are not created, and old reactant atoms are not destroyed. As part of this reconfiguration of atoms, no atoms are created or destroyed or broken apart in a chemical reaction.

Once scientists acquired the technology to actually view an atom, it soon became apparent that it is a highly dynamic system, with inputs and outputs of nuclear and electromagnetic energy. Also, one can observe that the number of electrons in an atom is variable and transitory. The number of protons in a nucleus varies as well, and it has become the basis for the cataloging of the elements;

that is to say, the atomic number determines the kind of atom it is. We now say that it is not the mass, but the number of protons, that determines the "signature" property of an element. Furthermore, we now know that atoms are mostly comprised of empty space. The atoms that comprise matter never actually touch each other; rather, the closer they get to one another, the greater the repulsion there is between the electrical charges on their component parts. The electromagnetic force is vastly stronger than the force of gravity. Indeed, the electromagnetic force from a tiny magnet would overwhelm the gravitational attraction of the whole Earth. The nucleus that makes up the vast bulk of the matter in an atom is relatively tiny. But the integrity of the atom is maintained by the force of strong interaction. It is the attractive force of strong interaction and repulsive force between protons that is responsible for the stability of the nucleus.

Therefore, it would seem that the atomic region experiences the dynamic tension between stabilizing and destabilizing forces, just as all other functional regions. We have seen that electromagnetism and gravity are stabilizing forces. However, unstable radioactive isotopes can be produced in many ways: fusion, neutron bombardment, radioactive decay of neighboring elements, and bombardment of neighboring elements with charged particles. These represent a destabilizing dynamic that is related to the nucleus of the atom. As is the case with all atoms, if the number of neutrons within the nucleus exceeds a certain amount, they will reach a state where they are higher than what is required for maintaining the stability of the nucleus. At this point, the nucleus will become unstable, and in that circumstance, there would be an attempt to convert the excess of neutrons into sub-particles, which then would be expelled from the atom. This process is one form of radioactive decay.

The various processes that occur within and between atoms are rationalized by the periodic table of elements. An element is defined as a substance in which all the atoms have the same atomic number; that is, all the atoms of a given element have the same number of protons. For neutral atoms (those without an electrical charge), the atomic number also gives the number of electrons. Apparently, atoms interact with one another, whenever possible, and rearrange themselves into various configurations. So, by categorizing the various electron arrangements in atoms, chemists have found that they can understand why the periodic table is arranged as it is. That is, it is arranged on the basis of chemical reactions that, in turn, depend on the electron arrangements in the outermost shell. It also has been found that if particular elements are heated, they do not produce a continuous spectrum, as the Sun does. Instead, they radiate light in separate wavelengths, so that the spectrum consisted of number of bright lines, separated by stretches of darkness. When sunlight is sent through the relatively cool vapors of a particular element, the vapors absorb only those wavelengths that they emit when radiating. The upshot, in terms of military operations, is that metrics like the atomic number of an atom can be used to detect a particular target type.

———

The atom interacts closely with the force of electromagnetism, which is one of four fundamental forces in nature (the other three being the strong interaction, the weak interaction and gravitation). Thus, the electrons that orbit the nucleus of the atom are bound by electromagnetic wave mechanics into orbits around atomic nuclei to form atoms, which are also the building blocks of molecules. Electromagnetism also is concerned with the forces that occur between electrically-charged particles, including the phenomenon of electromagnetic fields. Wherever there is a flow, a current of energy on the move, there is always a magnetic field generated around the flow. When there is a magnetic field, there is always current, and where there is current there is always a surrounding magnetic field. The function of the magnetic field is to even out the distribution of the energy until it becomes completely spread within any closed system.

Electromagnetic radiation also exhibits a multitude of phenomena as it interacts with other charged particles of atoms. One of the attributes of the free-flowing photons is that they radiate outwards within a given medium. Because these flows exhibit many of the properties of ocean waves, they are therefore often described as "waves" of energy whose frequency past a given point can also be measured, in the same way that frequency of ocean waves passing a given point is done. The length of these waves is usually measured from crest to crest as they flow past a given point; the unit of measure for each length is the meter. A magnetic field also can be described as a region of changing and intertwined electric and magnetic fields. In nature, these fields flow like untamed rivers. Now, however, humans have learned how to direct the flow of these electro-magnetic currents, through the amplification and stimulation to produce very cohesive and a more powerful flow of electrons. Thus, "LASER" (Light Amplified and Stimulated Energy Rays), has become the universal metaphor for this human manipulation of natural energy flows.

Of major importance to military operations is that the flow of radiated electromagnetic energy can be amplified, stimulated, and made more coherent by artificial means, in much the same manner as flowing rivers can be constricted. The flow can also be "deflected." One manner of deflection, "electron scattering," can also refer to the deflection of the path of electrons as they pass through a solid (such as a metal, semiconductor, or an insulator). Deflections are really just manifestations of the effects of collisions of electrons which are caused by electrostatic attractions or repulsions of electric charges. Overall, the structure and behavior of the flowing electrons that occur in the state of electromagnetic radiation has become increasingly important in the development of what is called electronic warfare. Another element of natural "geographic" phenomena that is presented to the military geographer is the so-called "magnetic field" that occurs in nature when electrons interact with such things as iron. The outcome of this interaction is usually a "magnet" of some sort. One thing that these magnets do is to provide some degree of cohesion and discipline to the energy flows that are produced by this interaction. Thus, for example, within the medium that is our own planet, Earth, magnetic fields give direction to compass needles in a generally predictable way.

Another source of radiation is the result of what is called the radioactive decay process and, it too impacts military operations. This involves a situation in which an unstable nucleus or the atom itself, splits or otherwise transforms into some other nucleus or atom by emitting ionized particles in an attempt to regain stability. In the most common form of decay, known as gamma emission, gamma rays are radiated. There also is alpha decay, which is a type of radioactive disintegration in which some unstable atomic nuclei dissipate excess energy by spontaneously ejecting an alpha particle. Then there is beta decay which refers to any of three processes of radioactive disintegration by which some unstable atomic nuclei spontaneously dissipate excess energy and undergo a change of one unit of positive charge without any change in mass number. The three processes are electron emission, positron (positive electron) emission, and electron capture.

Similarly, ionization refers to any process in which electrically-neutral atoms or molecules are converted into electrically-charged atoms or molecules. Ionization is one of the principal ways that radiation, such as charged particles, transfer energy to matter. In ionic bonding, electrons are completely transferred from one atom to another. When atoms lose an electron, another type of bonding occurs. Covalent bonding occurs when the chemical activity of an atom is determined by the number of electrons it has in its valence shell. When the valence shell is complete, the atom is stable and shows little tendency to combine with other atoms to form solids. Only atoms that possess eight valence electrons have a complete outer shell.

––––––

Another way to express the fundamental nature of atoms is to say that they are also basic to the construction of the chemical elements which form all matter in the Universe. The connection between atoms and the elements of the Universe is that the latter are made up of atoms. At the present time there are 112 known chemical elements. A chemical element is any substance that can't be decomposed into similar substances by ordinary chemical processes. On the other hand, most naturally-occurring substances are not elements; that is, they can be broken down into the various elements that make them up. Combinations of elements are known as compounds. The elements that make up the compounds are always present in definite proportions within an atom. The most abundant atoms are carbon, nitrogen, oxygen, and hydrogen. Hydrogen, the most common element in the Universe, and the major feature in living organisms, was produced by the Big Bang event. Heavier atoms, such as carbon and oxygen, developed in the stars, between 12 billion and 7 billion years ago, and were subsequently propagated forcefully throughout space when stars exploded. Ultimately, these stellar explosions were powerful enough to facilitate the processes that created the elements that are heavier than iron.

Another facet of the atom is its association with electromagnetic radiation phenomena. In a sense, electromagnetic radiation marks the boundary between the visible "macro" geography and the "micro" geography" of the microscope and the spectrometer. As indicated by the term, "spectrum," these occur within a series of sequential energy segments, which are distinguished and measured in terms of a ratio of wave length and the frequency that passes a certain point of observation. The most commonly utilized wavelength/frequency phase is the one we call visible light, infrared radiation, and the ultraviolet phases of the broader electromagnetic spectrum. The modern geographer now has many tools that have been developed in the areas of optics and other sensor systems that can peer into all the phases of the electromagnetic spectrum. There have also been continuing advances in the understanding of subatomic particles and energy that have given the modern geographer a means for studying the spatial nature of matter and energy – of all sizes.

In summary, it would seem that the domain of the subject matter of geography has broadened in terms of the kinds of phenomena that are seen as relevant for geographic analysis. However, the subject matter for geographic analysis has also expanded along the spectrum of scale; that is, on the one hand, the Cosmos is at the upper end of the scale, and the sub-atomic particle is at the lower end. In my earliest days as a geographer, the subject-matter of our studies was generally confined to the visible macro-matter which could be seen by the human eye and its artificial enhancers, such as binoculars and visual-light cameras. Mountains and sand dunes, forests and prairies, or urban centers and other "physical" manifestations of topography invariably came to mind when thinking about the traditional domain of geography. A greater understanding of the ideas and concepts which has been developed in the science of physics has also given me a greater awareness of a "new" subject- matter, including gravity, electromagnetism, and the strong and weak forces.

THE GEOGRAPHIC REGION

———

I see the mental construct of the "region" as another step in the geographer's quest to create spatial order out of seeming chaos. Without this insight, the phenomena which appear on the surface of the Earth (or other body in outer space) would appear only as random instances of natural and cultural phenomena to the observer. Thus, even an obvious pattern of the stuff of geology, biology, and botany that might be encountered on the surface of the Earth might not be recognized as much more than collections of random phenomena, in which the intrinsic properties would be the focus of study. One conceptual tool of the geographer, which has proven to be universally efficacious for rationalizing seeming randomness on any given area, has been the spatial "region." But within the context of the overall region, there is another type of region that has arisen since the beginning of the virtual revolution that is related to the development of evermore powerful computers – the "virtual region." One example of these virtual regions that are now relevant to the study of geography is the electromagnetic radiation spectrum. This type of region is manifested in the fields of electromagnetic radiation that has become a primary battlefield in modern warfare.

On the other hand, some of the most useful techniques that I have encountered in doing regional geography studies have been borrowed from other sciences. One of these has been the scientific approach to the geography of historical cultures, which is being utilized by the archeologist. Thus, the "archeological site" of that discipline has many commonalities with the "region" of the geographer. According to Philip Barker in his book, Techniques of Archeological Excavation (B.T. Batsford Ltd., London 1995), he describes how the term "context" is used by archeologists in the modern era. Like the "region" of the geographer, "context" is used as the preferred way to study a particular area of the surface of the Earth. On another level, Batsford says that "the skeleton of any recoding system must be the site grid."

———

Regardless of the conceptual origin of a region, it is true, by definition, that all regions have boundaries. That is to say, when constructing a region in the physical world or in one's mind, it is imperative that a boundary line of some sort be defined by some limiting factor. This has been one of the most important areas of study by geographers. Since the earliest time in human development, the groups intuitively have sought to draw a boundary around the perimeter of the group's hunting ground, settlement, or agricultural bread basket. The practical limits of this type of territorial imperative depended on the amount of game or other resources that lay within an area that could be practically traversed and defended. Resources might also have included bodies of water or a salt deposit, for example. Later, when humans began to practice agriculture, the boundary might have been defined by the "uniform" area of a crop that could be cultivated there. As human technology advanced, the "practicable" size of the economic-political region widened, but it was still constrained by similar factors related to the presence of some perceived natural resource.

So, it can be said generally, that a uniform region is defined as a contiguous area within which certain phenomena are measurably different from phenomena without the region. Further, there is a line or a zone of connected points along a periphery, where there is a transition towards an adjacent region. With time, the boundary may become more permanent, depending on the inherent group's ability to establish a valid claim on a given territory – either by military prowess, generally-accepted law or custom, or some other social mechanism. As civilizations became more connected to a region, certain regional "nodes," such as religious sites, became part of one's region so-called "sacred places" which have to be protected or fought over, as the case might be. At this point, the region is said to be a cultural region. Indeed, since the 8th century, three major religions have come to define the "Holy Land" of Palestine as their own "sacred ground." The nerve center for all these religions is now Jerusalem, which is now undergoing a great conflict in terms of where to draw boundaries. In the 1920s, another "ideological region" was constructed in Russia. This region would expand to include the so-called "Warsaw Pact" nations that comprehended the Communist World in Europe following World War II. After the end of the Second World War, the Communist Region expanded to include China, Vietnam, and even the island of Cuba.

One thing to take note of here is that boundaries are usually ambiguously defined; the criteria for their location can be as subjective as the nature of the regions that they are supposed to define. In some cases, a boundary is as plain as one's nose. Thus, a region on the surface of the Earth may be bounded by a river, a coastline, a mountain range, or some other objective phenomenon in our range of empirical observation. At other times, the "region" and its "boundary" is only in the eye of the beholder. The vast plains of Central Asia or the Great Plains of the United States, for example, have had have boundaries drawn around them by other, less direct, means. These might be determined by the distance a horse can travel in a given amount of time. Or, it can be defined by measuring an angle subtended by an external fixed point, such as a star, or some other external object. The latter methodology has been utilized by "surveyors" since time immemorial in human history. However, regardless of how the boundary is defined, it is only as real as its acceptance by those who reside on either side of the line.

In some cases, the boundary-drawer chooses to emphasize the "reality" of the boundary by constructing walls of one sort of another to mark its existence. One famous example of such boundary definition by man is the Hadrian's Wall. This was a wall that was initially constructed by the Roman Emperor Hadrian in 122 AD, to emphatically define the boundary between the region of Roman-Britain, and the region of the "Barbarians" to the north. It is likely that its course was set by local Roman military commanders and their engineers and, therefore, might have been determined more by tactical considerations of military defense operations. By the same token, there is no "natural" DMZ between North and South Korea; its location along the 38th Parallel is simply a reflection of the political and military realities of the time. Indeed, analyses of boundaries by political geographers have shown that they are not always a narrow "line" on the maps and on the surface of the Earth. Thus, some boundaries are considered to be "zones" of transition with respect to some accepted criteria.

If the observer is standing on a point on the surface of the Earth, one empirical boundary might be the nearest horizon. This is where my geographer's eye focuses first; it is the most basic of all boundaries on Earth (and, as it turns out, on every other planet in the solar system). In any case, it traditionally has referred to the apparent junction of earth and sky, as seen by the Earth-bound eye. Interestingly, this empirical horizon can be extended simply by moving to higher ground. This higher vantage point can be a mountain or a hill. On the other hand, the high ground can also be constructed by humans, as in the case of the pyramids that occur throughout the world; this is an example of "cultural geography." An even greater comprehension of a horizon can be achieved by the utilization of the concept of the "great circle." This is defined as the great circle of any celestial sphere that is formed by the intersection of the celestial sphere with a plane, tangent to the sphere's surface at the point of observation. This has been the practical boundary between the notion of a "flat earth" and a more realistic view of the earth as a globe.

———

Back in the early 1960s, when I first began my studies in the science of geography, I was introduced to the concept of regionalism as a way to comprehend, analyze, and make scientific conclusions about natural and cultural phenomena that occurred within a given area on Earth. I learned, as an initial step, that a "region" referred to a cohesive area on the surface of the Earth. This concept of cohesion implied that there could be defined phenomena that were more alike within a given area than outside it. One type of region, the uniform region, could be discerned according to the occurrence of a set of natural or cultural phenomena which seemed to predominate mostly within a given area. Examples of these would be "Corn Belt" (natural) and the "Bible Belt" (cultural) regions of the United States.

Other regions might not comprehend a given set of phenomena within a single, coherent area. That is, a population of phenomena that makes up the "region" might consist of similar, but discrete units or nodes that are scattered throughout a larger area. In this case, what provides the unity that makes them a region is a commonly-shared characteristic, or a common purpose. An example of this type of region would be the U.S. Air Force Strategic Aerospace Command of the Cold War era, which consisted of many discrete air and missile bases that were deployed throughout the United States. More recently, the system of facilities that are utilized by the U.S. space agency (NASA) are scattered throughout many places on the surface of the Earth and in space as well. Both of these are said to be "functional regions," because they are perceived as a spatial unit by virtue of a given purpose and function.

Another characteristic of the geographic region concept is that the size or circumference of a region is variable. That is to say, the term "region" can refer to a small area of economic activity, such as an agricultural or mining site, or it can refer to a population center. Similarly, a political region can be as small as a village or as large as a nation-state, or an empire. Thus, I came to understand that, ultimately, a "region" really exists only in the mind of the beholder. Now, in the age of space exploration and other operations, it also appears that both physical and cultural regions can be found throughout the solar system. Throughout this book, I will be using the conceptual vehicle of the region to deal with the basic question of how the natural and cultural environment affects military aerospace operations throughout the solar system.

Most regions on the surface of the Earth are comprised of some mixture of physical and cultural elements which are interacting to produce a geographic system. Thus, for some 10,000 years or so, humans have been working and combining with the physical environment in order to make a cultural environment for themselves. Some see this process as placing an "overlay" on the physical geography of a place. Others see this as an "alchemy" that produces an entirely new environment and region for the region-builder. One type of resultant geographic region is the political or economic region. Another type of cultural region is the military region or domain which exists in the mind of the military geographers. As far as they have been concerned, the surface of the Earth is a palette on which military operations can occur. Thus, rather than focusing on the agricultural activities that may be going on in a portion of the surface of the Earth, they concentrate on the military activities that are occurring within it. Regardless of the size or scope of the region, the military geographer has traditionally sought out the best ways to rationalize the military activities that are occurring within any defined area. This has resulted in many studies which have been designed to develop effective criteria for the definition a particular operational region. These can be generalized under the rubric of, deciding "what is inside and what is outside" one's region. In the case of the military aerospace region, this can refer to a particular order of battle or an enemy air defense system within a given area, as examples.

––––––

One such military aerospace region, which I personally experienced, occurred over an area of the country of Laos, the so-called Plainne des Jarres (PDJ) in northern Laos. It was during my tour of duty in Laos I that I was introduced to this region of military air operations. It was a particular area that encompassed space on (and above) the surface of that portion of the Earth. In this "area of operations," U.S. military aircraft, as well as air resources controlled by the Central Intelligence Agency, carried out their operations in support of allied indigenous ground forces. It was also an area which was defined by the continuous orbit of our C-130 command and control aircraft, which was given the call sign of "Cricket." Its elliptical orbit covered an area that was bounded by the Laotian-Chinese border on the north, Vietnam and Cambodia on the east and southeast, and Thailand on the west and south. Within this overall military aerospace region, there were several smaller operational subregions, such as the system of fortified posts that was centered on "Lima Site 20A" (The operational command center of the CIA for the PDJ).

Laos was a major theater of the Vietnam War for military air operations by the U.S. Air Force and the U.S. Navy in Indochina. It generally was situated west of the Annam Mountains, which divide the Vietnam peninsula along a north-south axis. Within the overall Laotian theater "region" there were geographic subregions that further defined it. Thus, there was the China-Laos border subregion to the north, the so-called Ho Chi Minh Trail to the east, and the Mekong River valley which delimited it to the south and west. And within the overall region there were political-military regions such as the "no-fly zones" and the zones of restricted air operations that were as "real" as the international borders that actually existed only on a map. Another such political-military subregion was a corridor along western Laos. It was the site of a highway that was being constructed by the Chinese to connect China to Myanmar (then Burma) and Thailand. This was a highway that was under active construction by the Chinese throughout my tour of duty in Laos; U.S. air forces were forbidden to even fly within 5000 meters of this construction project. This experience reinforced in me what I had learned in graduate school: that is, regions are merely "logical constructs" that are defined by the human mind, even though they sometimes overlaid areas of the Earth's surface that were bounded by such physical features as rivers or mountain ranges. But more to the point, I learned that military aviators have an innate ability to comprehend military air operations "regions" on a very pragmatic level. Whether they are air reconnaissance or attack aviators, they develop the ability to measure their particular spatial "workshop."

At the turn of the 20th century, humans became able to actually fly through the atmosphere and even to conduct sustained operations there. This also provided humans with the ability to move about in three dimensions of space, not just the limited two-dimensions of surface movement. Then, during the latter half of the 20th century, humans began to operate in the relative void of outer space, where the particular gravity regime made it possible to hover practically forever, or until the space vehicle should collide with another body so that it falls out of an orbit. These technological advances in locomotion also have been of great significance the science geography; it meant that the classic two-dimensional model of the spatial region would have to be expanded in terms of our understanding and analysis of this new spatial reality.

All of these changes in the meaning of distance, speed, and time in a three-dimensional space lead me to believe that the concept of the geographic "region" itself must be reformed and brought into congruence with the realities of military aerospace operations in the 21st century. Up until turn of the 20th century, the idea of the military region, in particular, was essentially a two-dimensional one, except in the case of mountainous terrain. Most often, the military region of the pre-20th century could still be defined in flat spatial terms which were measured by lines along x and y axes. Thus, one would tend to measure the area of a particular region of study in terms of square inches (centimeters), feet and yards (meters), acres (hectares), miles (kilometers), and so forth. Only when the analysis the region included mountains and extreme local topography was a third dimension, altitude and elevation, taken into consideration. This was the region of traditional military operations and the "geography" that they comprehended. Maybe this is the game of chess was used as an analog model for the conduct of warfare. Since 1914, however, the military region has evolved into the Military Aerospace Region (MAR), which now has a third dimensional aspect.

Our understanding of the three-dimensional region also has continued to grow with the development of so-called over-the-horizon (OTH) radar systems. These radars emit electromagnetic energy in the radio wavelengths. When these are directed over-the-horizon, toward the sky, they are reflected or "bent" by the ionosphere, so that the waves are returned back to the surface of the Earth. Thus, by bouncing radar and radio waves off the ionosphere which surrounds the Earth, one can actually leverage the distance capability of the OTH radar system. The important point in this discussion is that all this cumulative technological advancement has been developing and enhancing the notion of a three-dimensional military air region.

Imagine a region that is more like a cylinder rather than a two-dimensional, horizontal map; or even a three-dimensional topographic one. Now consider that this vertical region that can be portrayed like a skyscraper building with many levels, and that each of these levels is more homogenous within (with respect to certain variables) than the levels below and above it. This is how I perceive a three-dimension military air region. Each of the subregions of the overall cylinder can be described and analyzed in terms of such geographic phenomena as atmospheric pressure, wind patterns, cloud cover, and precipitation, electromagnetic radiation, solar and other extraterrestrial radiation, and a host of other such variables. To this physical geography, add the mix of AAA (Anti-air Artillery), missiles of all types (surface-to-air and air-to-air), other flying vehicles of all kinds, and a host of other denizens that will interact with the military air operation within that subregion – and vertical neighboring subregions. All together, these elements make up the cultural region that I call the military aerospace region. Given this paradigm, the geography of military air operations provides an analytical tool which can be used to consider the multivariate reality that interacts with military air operations in the 21st century.

Like all regions, the military aerospace region exists only in the mind of the military geographer or analyst. Thus, it might be said that the surface of the Earth with all its "physical" complexity has no effective existence until it is "perceived" by an intelligent entity, such as the humans mind. These are the only perceivers, as of now, who have the ability to process and communicate what they are seeing to other humans. The innate power of perception that is present in the modern human has been constantly enhanced by tools and machines which humans have created throughout the history of mankind. The natural power of perception of the human has also been enhanced by artificial means throughout human history. It began with the telescope and other optical technology which amplified the power of the human eyes. The telescope itself has gone through much iteration, as technology has advanced in the 18th through the 21st centuries. All this kind of observational technology has enabled astronomers and other scientists to study the heavenly bodies that resided outside the domain of Earth, but there was not developed a comparable technology to study the surface of the Earth (including its atmosphere), until the 20th century and the advent of space exploration.

In the era following the end of the Second World War, it was the United States and the Soviet Union that generated the many of the advances in observational technologies that have enabled scientists to study atmosphere and surface of our planet from the perspective of the astronaut. Both of these countries contributed to this extraterrestrial perspective of our globe. This was the era of the so-called "Cold War" between the United States and the Soviet Union. And it was during this period in history, that the U.S. Air Force Strategic Aerospace Command's B-52 Stratofortress and the Minuteman ICBM (Intercontinental Ballistic Missiles) systems both contributed to the further expansion of the three-dimensional military air region, both horizontally and vertically. At the same time, the high-flying U-2 air reconnaissance aircraft of the U.S. Central Air Intelligence Agency was instrumental in pushing the limits of this even farther into the upper limits of our atmosphere. But it was the satellites that were placed into orbit around the Earth that created the giant leap that put military aviation into outer space. In time, these orbiting satellites and the manned and unmanned space vehicles have helped push the upper limits of the military aerospace region into the realm of the furthest reaches of the solar system. This represents the ultimate three-dimensional, cylindrical military aerospace region. The height and breadth of the region now extends to the farthest points of the solar system, and beyond. It is a region that is measured in not only miles, but light-years as well.

Perhaps the most useful measure of a military aerospace region is its size. However, in my role as an air intelligence officer and a geographer, I came to realize that the definition of the "size" of a region was as subjective and relative as the concept of "distance." Thus, the aviators to whom I gave pre-mission briefings at the squadron intelligence shack in Udorn, Thailand during the Vietnam War were most interested in the "functional" size of the Plainne des Jarres (PDJ) area of operations.

That is to say: what is the flying time with respect to my particular aircraft, and within the context of my particular mission? The answer was not quite as straightforward as one might think, at first blush. Sure, the aircraft has a known speed capacity over a given time, in the abstract. But in the real world of the military aerospace region, there are "friction-factors" that tend to elongate the size of the military aerospace region. In Laos, there were the storms and the smoke of Hmong farmers who practiced slash-and-burn cultivation methods within the PDJ. These were practical obstacles in the flight path that had the ultimate effect of increasing the size of the PDJ. Then there were the Communist air defenses that made it necessary for the aviators to set a less than direct flight path within the PDJ. It was these "friction-factors" that the military aviators also wanted to hear about during the pre-mission briefings.

In Laos, the military aviators took a number of pragmatic approaches to the measurement of the size of the PDJ, in order to calculate the true flying distance from point to point within this military aerospace region. Thus, when I presented aircrew pre-mission briefings for the 7th Airborne Command and Control Squadron aircrews I would typically begin the briefing by displaying a small-scale (detailed) map of the entire area of operations. On this briefing map, I tried to depict all the natural and man-made phenomena that might have tended to elongate the distance from base to the target, for instance. Among these would be dangerous weather conditions and any known enemy ground-based air defenses sites. Each of these occurrences could have extended the "effective distance to the target area. To the extent that they would have required the military aviator to engage in avoidance maneuvers, they would have created a virtual "friction-factor" that would have extended the actual distance from base to target by the friendly aircraft.

In my studies of ground-based air defense systems during the Vietnam War, I came to conclusion that the military aerospace region construct could not be adequately portrayed by the traditional two-dimensional maps. A third dimension included such phenomena as the "flak region" that was created by the North Vietnamese Army, both in North Vietnam and along the Ho Chi Minh Trail whose course largely straddled Laos and the Vietnams. This three-dimensional region of anti-aircraft ordnance (including both artillery shells and surface-to-air missiles) was defined primarily by the vertical reach of the extensive air defenses that the NVA had installed to defend this region against American and Allied air attacks. Many of these weapons were guided and controlled by radar systems whose job it was to provide early warning of enemy air attacks and, as the attacking aircraft approached the Communist ground targets, to provide guidance and control to the anti-aircraft artillery and surface-to-air missiles.

The NVA air defense systems can also be seen as examples of both functional and uniform regions. In the latter case, it was a uniform region that was populated by bullets, projectiles, and missiles. There was also a form of vertical zonation in the flak regions that reminded me of the terraced crop-lands that I had seen along the sides of mountain in Guatemala. In such mountainous topography in the lower latitudes, there form horizontal micro-climates in which zones of agricultural products occur. Because temperatures fall as one ascends a mountain-side, the zones of relatively uniform micro-climates are somewhat analogous to the zones of climate that form, latitudnally from the Equator toward the higher latitudes. Within each of these vertical zones there is a micro-climate that presents agricultural conditions that are nearly analogous to the latitudinal zones on Earth. Thus, at the 6,000-foot level of altitude, Guatemalan farmers are able to cultivate crops that are similar to those grown at higher latitudes.

In the same vein, the NVA developed vertical zones of air defense ordnance, both in North Vietnam and along the Ho Chi Minh Trail. At 2,000 feet of altitude, or below, there was a vertical zone which contained a "population" of relatively small caliber ordnance, such as that from the 12.7mm heavy machine-gun and the 23mm Anti-Aircraft Artillery (AAA). At higher altitudes, there were vertical zones of heavier 37mm and 57mm AAA ordnance. Mixed in with the vertical zones of AAA, there was a similar vertical zonation of surface-to-air missiles (SAMs) which ranged from the smaller SA-7 shoulder-mounted, heat-seeking missiles to the heavier SA-2s which had effective range of as much as 60,000 feet in altitude. So, in effect, the NVA had created a "cultural region" that was analogous to the zones of micro-climates that form along the sides of mountains at lower latitudes.

Another aspect of the vertical NVA air defense zonation was the "cloud" of electromagnetic radiation that formed over the battlefield. These included the emissions by the radar and infrared transmitters that were associated with the air defense flux of ordnance. Some of these emissions were designed to detect and track attacking aircraft, and to guide the delivery of much of the anti-aircraft ordnance that was aimed against attacking aircraft. Adding to the formation of this electromagnetic cloud were emissions of electromagnetic energy that were sent out by the attacking platforms to detect and counter the electromagnetic radiations that the ground-based air defenses were emitting. This measure-countermeasure dynamic is the essence of "electronic warfare" (EW) and it often was ongoing throughout the course of an offensive air attack against ground targets. One low-tech manifestation of this EW was the flares and simple camp-fires that the enemy forces used to confuse and distract the high-tech tools of infrared radiation that were employed by the attacking aircraft. Another low-tech EW countermeasure (although usually unintended) was in the form of the dense smoke that was produced by slash-and-burn cultivation on the ground.

As noted before, one of the most important vehicles for organizing the phenomena within a military aerospace region is the map or navigation chart. It appears as though, for at least 10,000 years of our human history, we have been drawing some kind of pictorial "model" of a spatial truth that they want to communicate to others, or to simply catalog all that they know about a place. The formal name that has been given to this practice is cartography. It is an ancient device for storing data about a place, and an efficient way of communicating this geographic information to others, or to simply memorialize it for political, economic, or social purposes. Some maps are drawn for purposes other than to simply represent the location of ground truth. There are, for example, maps that are drawn to navigate from one place to another; these are generally referred to as charts.

The concept of "scale" has also been essential to the study of regions. In a counter-intuitive way, the maps and navigational charts that are compiled to describe regions are said to be "large-scale" when they focus in on a smaller subset of an area. Conversely, when map covers a larger area on the surface of the earth, with less detail, they are referred to as "small-scale" projections. Thus, a map that depicts greater detail over a smaller area is said to be a "large-scale" map. This shows how maps of all kinds are defined in the eye of the beholder. It also shows how maps can "telescope" to show extensive areas or, conversely, to show smaller areas in greater detail. In short, maps are simply expressions of mental constructs with reference to the actual surface of the earth.

Often, one can show relative differences in "intensity" of occurrence of any given phenomenon over throughout a two or three-dimensional space. An example of this would be temperatures across a given portion of the surface of the Earth. When these absolute locations of a given phenomenon or value are connected with a line, the line is called an isoline. Thus, a topographic map reveals the "relief" of an area by means of these contour lines that represent equal elevation values. Each line passes through the points of the same elevation. By the same token, an isobar is a line that connects points of equal barometric pressure. Both of these types of contour lines which are drawn on an aeronautical chart can help the aviator to plot a course which avoids dangerous relief phenomena, such as a mountain peak, for instance. Also, some fusing systems on bombs and other ordnance are timed for activation by certain changes in the atmospheric pressure over the target.

This method of representing topographical values can contain biases due to the subjective decisions regarding the placing of the contour base points, especially when the measurement criteria are not well defined, as in the case of a variable thermal reading at each point on a surface. Thus, a contour map that is created by actual "boots on the ground" techniques is really only an approximation of the true relief regime that is being mapped. There is also inherent interpolation with the traditional methods of creating contour maps. There also are "contour maps" that refer to phenomena such as meteorological, economic, and other non-topographical data. It is generally an effective way to map the distribution and relative intensity of any phenomena that occurs on the surface of the earth. Geographic distribution maps are one of the most effective and efficient ways to describe the incidence of data on the surface of the Earth (or other planet).

For the modern military geographer, contours are a matter of direct influence on military aerospace operations. This aspect of the surface of the Earth, for example, is an important influence on the takeoff and landing phases of military aerospace operations. My experiences with the takeoff problem calculus at Long Tieng – the headquarters for CIA-led guerrilla operations against the North Vietnamese in northern Laos – taught me a vivid lesson of the effects of topography, and its portrayal on a map, when I took off from this airfield. I had some reinforcing experiences as a passenger onboard commercial aircraft that took off in close-quarter mountainous terrain at Tegucigalpa, Honduras and at Merida, in the Andes mountains of Venezuela. All of these produced some "hairy" takeoffs and landings on landing strips that were snugly ensconced in mountainous terrain, with only a relatively small, tilted landing strip – somewhat like a naval carrier.

Sometimes the military aviator is forced to deal with contour maps from the perspective of the ground forces. Unfortunately, these forays outside the normal aviator's element are not always planned and, sometimes, are related to a search and rescue of a downed airman. These are unwanted, but not unplanned situations in combat air missions. These "downed airman" emergency scenarios are planned for in great detail and are the subject of much training by military aviators and the search and rescue people. I recall the intensive training at the Pacific Jungle Survival School on Huk Mountain, near Clark Air Base, the Philippines. All military aviators that were destined to serve in combat went through a course of training that included lessons in "escape and evasion" procedures, as well as procedures for Search and Rescue (SAR). One segment of training was in the use of contour maps and compasses in order to successfully traverse the terrain; a kind of "orienteering" exercise, if you will. The equipment that a combat aviator carried for this kind of situation essentially consisted of a weatherproof soft rubberized contour map, a compass, a handgun (mostly for signaling or, in an extreme need, for self-protection), and an electronic beeper which automatically sent out signals on a designed SAR frequency. Now, in the 21st century, these have been augmented by devices that receive pinpoint data about one's position from a system of orbiting satellites above the Earth. Indeed, some of the cutting-edge electronic devices, which can fit in the palm of one's hand, now enable a downed airman to engage in secure communication with a SAR team in the air, or with friendly forces on the ground.

In Laos, I became intimately aware of the importance of good contour maps in the weapons delivery phase of military aerospace operations. For one thing, air intelligence and operations depended on contour maps in planning the appropriate ordnance for the task at hand. As an example, 500 pound "dumb" bombs and other such ordnance was equipped with special extenders so that they would explode at a given height above the ground or vegetation. To properly "fuse" this kind of ordnance, a precise mapping of the target terrain contouring has to inputted or set. For the most part, this kind of intelligence was derived from existing topographic maps; some of these were accurate, but many were not, because they had been drawn up from ground surveys that were not always accurate reflections of reality. Since the end of the Vietnam War era, topographic maps now are drawn through the use of satellite reconnaissance and computer-assisted cartography. Indeed, the contour data is now plotted in "real time" and outputted to visual media, or transmitted directly into analog micro-processors, which can then control actuators.

The importance of having accurate contour maps was made acutely clear to me during my tour of duty over Laos, in the Vietnam War. In the early months of 1972, the cutting-edge F-111 fighter-bomber was introduced into the Laotian theater of the war. This was presented as a platform that could fly and fight practically "on the deck" and thereby operate under the enemy air defense radar systems. And it could do this at very high speeds. What made it possible for the F-111 to do this was its "TAR" (Terrain Avoidance Radar) system. This was an onboard system which relied on real-time data which was fed to an integral flight-control computer, which then gave commands to the flight-control system of the aircraft. Over time, I began to discern a pattern with respect to these cutting-edge platforms. Soon after their introduction into northern Laos, these aircraft were beginning to be lost at an alarming rate, and under mysterious circumstances. Indeed, following the loss of the third F-111 in Laos, I became aware that the F-111 missions over Laos were terminated and the aircraft were grounded. At the time, there was no information that was made available to me as to the reasons for these actions. It was only after I had returned to the United States in April of 1972 and begun my new duties as an instructor at the Armed Forces Air Intelligence Center near Denver, Colorado that I learned what had happened. While doing research for my class preparations, I learned (from now unclassified information) that the F-111 Terrain Following (or avoidance) Radar which enabled the aircraft to operate at less than 100 feet over the terrain (along the peaks and valleys) did so because of an advanced radar system that was designed to sense the surface of the earth and to upload this data almost instantaneously to avionic systems which controlled the altitude of the aircraft. Unfortunately, the designers of the TAR system had not taken into account the nature of the limestone subsurface of the land in northern Laos.

From my study of the physical geography of that part of Laos, I learned that its geology, particularly in the central provinces of Bolikhamxay and Khammoun, is that of karst topography. This limestone geologic regime can also be found places like the Yucatan Peninsula in Mexico, and the cave region of Tennessee in the United States. This kind of topography is often characterized by the presence of sinkholes, or "cenotes" as they are called in Spanish. These are created as leaching water carves out "holes" in the limestone just below the surface of the earth. Sometimes these sinkholes are covered by a relatively thin cover of thatch and soil that does not appear as surface to the Terrain Following Radar; instead, the system reacts to the returns from the bottom of the sinkholes as though they were the operational "surface." As a result, the F-111s were being "pulled down" (and crashing) into the actual surface of the earth, which was below the thin cover of thatch and mud that sometimes forms over the sinkhole. In other words, the "contour map" that was being transmitted to the TAR control system was giving it spurious information regarding the terrain. Since that period, technology has advanced and the accuracy of contour depictions has been made increasingly more accurate.

———

In addition to the idea of the "region," one of the most powerful approaches to geographic analysis is the concept of the "man-land" inter-relationship within a region. The basic idea of this conceptual approach is that every human activity must be done within a natural and a cultural environment. However, the relationship is not one of total passivity on the part of the human species. Man's proactive utilization of the natural environment may be said to have begun during the early centuries following the end of the last Ice Age – some 10,000 years ago. However, it is only in the last two or three centuries, that man has been able to do significant proactive "relating" to the Earth, as science and technology has evolved to the point where humans now have the capacity to significantly affect the natural environment.

In the early stages of the post-Ice Age era, the man-land relationship has been described within the context of particular kinds of human activity, such as hunting-gathering and agriculture. The idea has been that the environment (land) offers to humans and their technologies a widening range of options – but within certain limits of physics and nature. Thus, for a long time in human history, agriculture could only be practiced where nature provided adequate water supplies. The Nile, Indus, the Tigris-Euphrates, and the Yellow (Huang Ho) river valleys were the earliest examples of where this geographic contract was entered into between man and land for the development of human civilizations. The so-called Industrial Revolution of the 19th century in Europe and North America represents another phase in this sort of contract between man and the land. It requires a much more extensive mix of natural resources and human resources to function efficaciously. In this part of the book, we will examine another sort of human activity – military aerospace operations – and its relationship to the natural environment, throughout the solar system.

However, in the age of aerospace operations, there is the growing awareness that the concept of land needs to be expanded to include the geographic environment exists above the Earth's surface.

To be sure, geographers have always factored in the effect of the Sun's solar energy on man's activities on Earth. And of course, there has been the recognition of the effects of the Moon's gravitational pull on large bodies of water on Earth. Now, however, with the empirical experience of the solar system in the age of space exploration, there is now a whole new "man-land" equation to be studied by geographers. The first iteration of this awareness of a whole new dimension of the "man-land" interrelationship was the "macro-geography" aspect of the solar system. That is to say, it was discovered that an environment exists beyond the spatial limits of our planet. At the first series of empirical glances, this was seen to be a "land" that consisted of planets, asteroids, and other space bodies. These were geographic features that the terrestrial geographer could easily relate to, given our experience with the "physical geography" of our home planet.

———

Two other features of physical geography that have become more vivid in space are the electric and magnetic fields, which often interact to form electromagnetic fields. I also have found the concept of the "field" to be extremely powerful as an analog to geographic region. That is, I see the field as a model that is useful for analyzing spatial regions throughout the solar system, from many perspectives and levels of scale. The "field" is a concept that originally was developed in the science of physics. However, during the past decades, many other disciplines, including the science of geography have adopted the conceptual vehicle of the "field" as a tool for explaining many of the phenomena which they observe. So, even though the basic concept of the field came from physics, geographers have developed independent theories, data models, and analytical methods with respect these phenomena.

As a geographer, I see a field as being like a football field. It has a bounded "space" within which the resident "population" acts and moves according to some known rationale. One classic model in geography – the Central Place Theory – defines the bounds and the topography of a field (or region). It then attempts to explain, and predict, how a population of marketplaces and consumers will behave in spatial terms. In the case of the CPT model, the controlling variable is the motivating force of profit maximization. Another kind of "central place model" that is studied by astronomers and physicists, among others, is the solar system, in which the population of planets and other celestial objects behave spatially under the discipline of the gravitational force.

In physics, the term "field" may refer to a discharge of electrons that are emitted from the surface of a material to another point in space. The emissions are usually depicted as a series of parallel lines that travel from one pole to another along lines that form continuous, forming closed loops. Or, the term "field" may refer to a gravitational field which provides the force that causes particles to clump together and thereby form the larger objects in the solar system. The process of aggregation begins with discrete particles of material and then, because gravity and the weak and strong forces cause every material particle in the Universe to form an attraction to other particles, with a force whose strength of attraction is directly proportional to the product of their masses; and inversely proportional to the square of the distance between them. Geographic fields also can be seen as aggregates of certain particles which form uniform fields (regions) that are homogeneous with respect to certain phenomena or properties of phenomena. Other populations may include features of artificial or aggregate fields, as well as statistically constructed properties of aggregate groups of individuals, like population density, or potential (such as the probability that a person at a given location will prefer to use a particular facility like a shopping mall). So, it seems that geographic fields can be seen as being synonymous with the spatial region. In this book, I will use the term field and spatial region interchangeably.

Within the context of military geography, the concept of the field can be used to analyze the orders of battle, such as the likely spatial deployment of enemy ground-based air defenses. Indeed, one of the most useful attributes of the models that have been developed from the concept of the field in geography is the ability to not only describe the given spatial distribution of a population of phenomena, but to actually make predictions about it with some reasonable level of statistical confidence. That is because the "field space" can be used synonymously with the idea of the spatially dependent variable that forms the foundation of geostatistics. Both scalar and vector fields are formed are found in geographic applications, although the former is more common. The simplest formal model for a field is a function, which yields a single value given a point in space. That is, $t=f(x, y, z)$. Or, to put it another way, a model for a field enables one to discern the controlling variable that can explain why a member of a certain statistical population can be located at certain locations within a spatial area, and at a given place and time. As we study the military geography of the solar system we will discover several forces that can operate as controlling variables for several kinds of fields that occur throughout the region. Two of the motivating forces are gravity and electromagnetism.

Viewed from the perspective of the geographic field, some of the most important to the study of military geography in the solar system can be said to be the electric and magnetic fields, and their common manifestation as electromagnetic fields. Perhaps the best-known of these is the planetary magnetic fields that envelop and protect many of the planets and some moons in the solar system. On Earth, the geomagnetic field serves to protect human civilization from most of the harmful effects of most of the cosmic and solar radiation which is emitted by our Sun, and the stars that lie beyond the solar system. From the perspective of the geographer, the electromagnetic fields may be treated as another instance of a geographic feature in the Universe. We also can analyze the electromagnetic field as a geographic phenomenon which interacts with human activities, including military operations.

One of the characteristics of charged particles is that they sometimes remain static. At other times, they will move through space in what is described as a wave-like motion, consisting of electromagnetic radiation; a movement that is sometimes also referred to as a "flux" of charged particles. It is a byproduct of the dynamic tension and friction that occurs during the movement of electrons through a medium. The electricity half of this wave can be perhaps understood through the use of the analogy of flowing water. Thus, "electric current" refers to the flow of electrically-charged particles, including subatomic electrons (which have a negative charge), protons (which have positive charges), and ions (which are atoms that have lost or gained one or more electrons). Magnetism, for its part, is a phenomenon that is most associated with the motion of electric charges. Overall, the wave can occur in many forms: it can be the flow of an electric current in a conductor or in a circuit of wire; or it may take the form of the movement of charged particles moving through space; and it can be seen in the motion of an electron within atomic orbit.

According to current scientific theories, electricity and magnetism are simply two aspects of one common phenomenon in the Universe. However, the electrical and magnetic forces behave differently. Electrical forces are produced by electric charges, whereas magnetic forces are produced only by moving charges and act solely on charges in motion. Electric and magnetic forces can be detected in regions known as electromagnetic fields. These fields can exist in space at distances far from the charge or current that generated them. Also, electric and magnetic fields travel together through space as waves of electromagnetic radiation, with the changing fields mutually sustaining each other. Electromagnetic waves themselves travel through space independent of matter, and these waves occur along a continuum known as the electromagnetic spectrum, in which they vary as to wavelength. These have been given unique names to distinguish them and for purposes of analysis and application. These are: radio, television, microwaves, infrared rays, visible light, ultraviolet, gamma and x-rays, in diminishing length of wave and increasing quantity of frequency.

The fields that are generated by the transmitters or other emitters of electromagnetic radiation can be analyzed in the same manner as any geographic region. Thus, the deployment and other spatial characteristics of the transmitters and receivers within the region may be analyzed. The following are examples of some of these electromagnetic regions that would be of interest in the context of military geography. Consider that when I taught air intelligence courses at the Armed Forces Air Intelligence Center at Lowery AFB, Colorado in the early 1970s, I had only a general awareness of the "aerospace" military geography that was becoming ever more influential in the conduct of military air operations. In 1972, I was constructing my first lesson plans for teaching air operational intelligence. As I did so, it became increasingly clear to me that the military geography of a portion of the surface of the Earth was transcending the traditional geography of rocks, vegetation and weather. Indeed, as military air operations began to rely more and more on portions of the electro-magnetic spectrum for carrying out strategic and tactical air operations, the "geography" of a military area of operations, or environment was becoming much more complex and complicated. What this meant for military aerospace geography, in particular, was that the scope of the discipline was widening to include phenomena and processes beyond the domain of ordinary experience. Particles and energy waves now became part of the study of geography, in addition to the traditional physical phenomena of topography, vegetation, and climate. One type of phenomena that was becoming relevant was the field of matter and energy that is encompassed by the model of the electromagnetic spectrum.

———

The spatial dimensions (or geography) of the radio portion of the electromagnetic spectrum can be plotted on a map of any particular region of a planet or in three-dimensional space. The region can take many forms, but here we are focused on the military operations region. On this map, we can plot the location of various relevant phenomena which happen to lie on the surface of a planet, the atmosphere, or in space. The most-frequently occurring of such nodes on our map of a communications system will likely be the transmitting/receiving stations. After we plot all these nodes on our communications map, we will perceive that there is a pattern which seems to have some positional logic. That is, there seems to be a set of variables which determine the relative location of each node a network. After some careful study, we will realize that the relative distance between two communications nodes has much to do with the power of a given transmitter to send radio signals, that is to say, their area of influence. This power, of course, is a result of the amount of electrical power that is available to a transmitter to propel the electromagnetic waves through a medium, in the form coherent waves, to any receiver in its field of propagation. Part of the relative distance between two such communications nodes would also be a function of the sensitivity of the receiver to the wavelength of the signal that is being transmitted.

The observer will also note that not all the nodes will be of the same size or intensity as they are portrayed on the map. Some of the nodes will appear as relatively larger dots on the map, or with a darker hue of color. This is because the cartographer has decided to depict the relative transmitting power of each of the nodes by depicting them as a hierarchy of black dots of varying size and hue. This is one way to denote that the wattage power is greater for the larger dots and proportionately smaller as the wattage power is reduced. The velocity and direction of radio signals can also be depicted on a map through the use of vectors. These could be in the form of vectors whose length would indicate the range, as well as the direction of the beams or the overall field. It could also indicate the location of a particular transmitting antenna. At this point, a map becomes a chart which can be used by aviators to garner information for the purpose of air navigation, such as the location of electronic beacons which will available within a given flight path, for example. Many of these beacons can also actively transmit constant information that can be received by any receiver that is on the appropriate frequency channel. Ultimately, what is being depicted is map of a communications system (or field) which accurately portrays the spatial profile of a certain kind of electromagnetic network or system.

Historically, the first application of the electromagnetic spectrum to military aerospace operations was at the "lower" (longer wave; higher frequency) portion of the spectrum: Radio and RADAR. This is where most of the concepts, methodologies, and terminologies related to this new technology originated. Radio refers to the transmission and reception of signals which consist of electromagnetic waves of certain lengths. These waves travel through space in a straight line, which meant that early radio communications were limited by the obstructing terrain or, ultimately, the horizon. However, advances in the technology ultimately made it possible to bounce the transmitted radio signals off the ionosphere and, later, off artificial communication satellites that are in orbit above the surface of the Earth. Obstructions aside, one of the major advantages of the longer-wave portions of the EM spectrum is that they are able to travel extremely long distances – even beyond the solar system. Thus, both radio and radar technology is being used today to communicate with space explorers and space vehicles. In passive mode, it is even being used to listen for possible intelligent life beyond our own galaxy.

By this time in history, the radio spectrum was also being utilized for air navigation, as radio transmitter stations began to be extensively deployed as electronic light-houses to guide aviators. In the United States and other parts of the world, radio-phased stations were emitting beams of EM energy which guided the flight of aircraft, just as the historical light-houses had provided guidance for ships at sea in previous centuries. During my experience in military aviation I have been aware of such electronic "light-houses" as the TACAN and MSQ sites of the second half of the 20th century. These have been used not only to direct aircraft to their desired destination, but also to provide directions for the application of ordnance on targets. One of the best known MSQ site during the Vietnam War was located on a mountain top in northeast Laos. Its main function was to provide radar direction to the B-52s were sent to attack targets in North Vietnam's Hanoi-Haiphong sector. Evidence of the effectiveness of this radar site was the high priority which the North Vietnamese placed on its capture, at a time when they decided to launch a major offensive against South Vietnam

One can argue that the most important element of the radio and radar technology is the transmitting and receiving antenna. This is the device that projects the radio-frequency energy that is beamed into the surrounding space, or a portion of it. The more powerful (watts) the generator, the greater the distance the signal can be transmitted, assuming it does not encounter an obstacle, like topography or atmospheric phenomena. Conversely, the antenna can also be used to receive signals from other transmission sources. But there is another factor that determines the ultimate distance that a signal can travel; that is, the cohesiveness of the radiation. This is, ultimately, a function of wattage power, which determines the transmitter's ability to concentrate the radio energy into a cohesive beam; the more cohesive the beam, the greater the "reach" of the transmission. On Earth, radio waves usually travel in straight lines, but they can be bounced back from a field of charged (ionized) particles lie above the atmosphere. This is the so-called ionosphere surrounds the Earth at altitudes between 30 and 50 miles. They have a variety of effects on the radio transmissions in the atmosphere, but their most significant is that it reflects the waves back to Earth, thus extending the range of the original transmission far beyond lines of sight. And, as space explorers were to discover, there are other ionospheres around some of the other spatial bodies that can either hinder or be utilized in similar fashion.

Radio waves can also be adapted to a specialized purpose. Thus, in the form of RADAR (Radio Detection and Ranging) these wavelengths were used during WWII to develop a network of receiver antennas which were especially sensitized to signals that were bounced off incoming enemy aircraft. This was a military application of the discovery that a radio-wave signal could be "bounced" off an object that happened to be within the field of transmission. Then, by calculating the time it takes for the return signal to reach its receiver component, the operator could determine not only the distance but also the direction of the incoming signal. Moreover, a catalog of aircraft radar "signatures" made it possible to determine many other characteristics of the target. Also, a map of the spatial distribution of the radar stations enabled planners to work out a grid of coordinates to enhance the ability to allocate interceptor resources and to direct their reaction vectors. In the case of the Battle of Britain, this "map" was actually a scale model of the battle region, which depicted all the relevant nodes and their dynamically occurring locations.

Generally speaking, the term "RADAR" refers to any device or system consisting of a synchronized radio transmitter and receiver nodes, which emits radio waves and their reflections for display on a map-like medium. This single pithy definition describes the classic radar system. It can be used even today to define radar systems that are used to detect and locate such objects as aircraft, as well as surface features on Earth and the other planets. On another dimension, the basic model of the radar dynamic can be abstracted further to an algorithm involving and input-processing-output operation, which is common to all operational systems. Beyond that, this same algorithm can be applied to other phases of electromagnetic radiation, including the infrared and ultraviolet phases of the spectrum. Thus, radar, which lies along the radio phase of the EM spectrum, was only the precursor to the future battles that would be waged in other phases of the spectrum, such as the infrared, visible light, and ultra light portions. Radar utilizes the radio portion of the spectrum, and its usage would become the prototype for the conduct of all electromagnetic warfare during World War II, and beyond.

To the geographer, perhaps the most significant aspect of the Battle of Britain would be the spatial parameters of the scenario. One of these is that the spatial area of the radar stations and the Royal Air Force interceptor bases make up a geographic region. In this case, it was what is called a functional region (as opposed to a uniform region of internal phenomena). That is, the air defense region comprehended a system of interacting and interdependent nodes whose relative locations were designed to further a specific objective. With this spatial system, the British were able to send their air interceptor aircraft to a precise "rendezvous" point in the three-dimensional space-time point of interception. The radar and radio wavelengths were utilized in a coherent fashion to generate information to all of the nodes of the system. The intelligence also included information with regards to all aircraft, both enemy and friendly. Therefore, through a technology that was based on the electromagnetic spectrum, the British air defense system was able to process inputs from the radar sites (as well as visual sites) to track enemy aircraft, in order to direct RAF interceptors against them. This system of radar detection and tracking proved highly effective and the RAF was able to thwart the German air offensive against their homeland.

As a result of these preparations, the RAF had in essence constructed a model of an air battle area that was a fairly true representation of the actual air battlefield. What the British had done was to create a dynamic scale-model of the air war that was actually occurring over southern England during the Battle of Britain. The main parts of this model included a two-dimensional representation of the various aircraft that participated in the serial air battles over England and the English Channel. There were also representations of the network of radar stations that were distributed along the southern shores of England. These elements of this "game board" were fashioned so that they could be moved according to instructions that were derived from the data inputs that were provided by the radar sites and the human observers as well.

Utilizing radio and radar, the command and control function was able to monitor the air battle in almost real-time fashion. That is to say, the data that the command center received from the early-warning radar sites via radio was quickly converted to usable depictions of the location of the enemy attack aircraft. These depictions were then constantly updated on the board; the data was also posted on a wall chart which kept track of not only the attacking aircraft, but also the available RAF interceptors. Thus, the British has learned how to utilize the electromagnetic spectrum attributes to detect and track objects in the air, and to communicate their findings to the elements of their own air defense assets. That is, they were able to do this by using the radio/radar aspect of the electromagnetic spectrum to not only utilize single instances of observation by radar sites, but also utilized the capabilities of two-way radio communications to develop the first truly integrated air defense system. It would be the basic model for future such systems; but the future systems would become much better at observing, reporting and reacting to air threats.

Thus, the initial manifestation of the electromagnetic battlefield had to do with the problem of coordination in an aerospace environment. Communications has been integral to the successful conduct of air operations since the early years of aviation in the first decades of the 20th century. In the early days of military aviation communications between aircraft and the ground were dependent on visible "eyeball" signals: this included such devices as painted signs on buildings and other such devices. But, at some point during the early years of aviation it became obvious that there would have to be effective communications between aircraft and between the aircraft and the ground operations. Fortunately, the era of the telegraph and "telegraphy" had shown that EM energy could work effectively as a medium for transferring not only Morse code but also actual sounds. The telephone and other inventions had shown how sounds could be transmitted via electromagnetic media. All of this can be seen as a precursor to the telemetry systems of the space age.

Another aspect of the Battle of Britain experience was its example of an input-output regional model. Thus, the radar returns from the enemy aircraft served as input data to the system of RADAR early warning stations. The output data from the radar stations was then relayed to the command and control center, where it was processed. That is, the captured data was incorporated into a model wherein the data was converted into useable information. That output was then directed to a tracking and ranging subsystem whose output resulted in a desired action – that is, RAF fighter aircraft being positioned in the optimum place in order to destroy the enemy bombers. This sequence of events would later be found in every "guided aircraft/missile/probe" situation, in which a radiated energy stimulus would elicit a response, and a feedback radiation from the original receiver would be used to shape any follow-on stimulus. In any case, I have come to the realization that military air battles – on Earth, in space, or on another space body – will henceforth, be waged along the various phases of the electromagnetic spectrum. The stimulus-response loop will also continue to be expedited by the computer (and its microchips), which operates at the speed of light.

The concept of frequency with respect to electromagnetic radiation was also seen in the utilization of the infrared portion of that spectrum. In this case, frequency became known as the "heat signature" of particular matter on the ground. Examples would be the frequency signature that was generated by vehicle engine blocks or by air defense guns and missile sites. Not just matter, but also certain cultural activities, such ammunition sites or POL (Petroleum, Oil, Lubrication) tank complexes. The general point I am making here is that we learned to use analogue "observations," that is, the radiation waves related to objects, rather than relying on less reliable "eyeball" observations, especially at night or in bad weather. Thermal detection of infrared energy is based on the conversion of a temperature change, resulting from such radiation falling on a suitable material into a measurable signal. Two such devices that are used are the "Golay" detector that employs reflection of light from a thermally distorted reflecting film onto a photoelectric cell, while a bolometer exhibits a change in electrical resistance with a change in temperature. In both cases, the devices must respond to very small and very rapid changes.

The analysis of the military region can also be extended to other portions of the electromagnetic spectrum of radiated energy, such as the infrared set of wavelengths. In this case, the map of the region would probably focus on the location of individual military-related objects on the surface or in the air, as opposed to the transmitting stations. These would be seen as a geographic distribution of heat-emitting objects which would represent targets of either the camera or the missile. The depiction of this reality would also require the development of a model which would be able to deal with a three-dimensional and dynamic time-space situation. On the surface of a planet, the two-dimensional map would probably be adequate for plotting the locations of moving targets (e.g. vehicles) whose infrared signature is known and stored in a computer database which could be accessed in real-time fashion. Once again, modern computer-based modeling and projection technology could be used to instantly project a digital map of the actual situation on a scope as targeting information to the weapons operator of the attack platform in a space over the surface of a planet or other body in the solar system. The weapon itself could then be equipped with an input-process-output feed-back system which would use the infrared signature of the target to "home-in" on it and destroy it.

One thing I have learned about the use of the EM spectrum in air warfare is that it is a double-edged sword. That is to say, the spectrum can be used by air and ground defense systems as well as by air attacker systems. This so-called "electronic warfare" really came of age during the Vietnam War, in particular and the Cold War, in general. An example of this phenomenon was the electronic warfare duels that took place the Vietnam War. There, radar and infrared technology eventually came to be utilized in many different air combat situations. In Southeast Asia, it was most used in tactical operations. Fighter-bombers like the F-4 Phantom used radar systems for both navigation and weapon-control functions. Interceptor aircraft used radar to locate, track, and destroy enemy aircraft. Specialized attack aircraft like the F-105 were also used to detect, lock-n, and guide anti-radar munitions against radar-controlled enemy anti-aircraft artillery (AAA) and surface-to-air missiles (SAMs). The ordnance of the F-105s or another attack aircraft would then "ride" the beam back to the originating system.

Thus, in this back-and-forth battle of electronic systems, the North Vietnamese radar made their anti-aircraft weapons more accurate and lethal. On the other hand, the radar fire-control beam emission also provided an excellent "highway" for air-to-ground missiles which were fired from specially-equipped attack aircraft. So, the scenario that took place over and over was that of electro-magnetic measure and a corresponding counter-measure. It was like a grim and lethal game of chess. Thus, in the particular case of the radar portion of the EM spectrum, throughout Laos and Vietnam, NVA air defense weapons which were controlled by radar would typically paint an Allied aircraft with a radar beam – then pass data as to the location and direction of the aircraft onto the fire-control subsystem of the AAA or SAM. This information would then be used to direct the fire of the air defense weapon system. In a perfect world, this process would ideally make the air defense system virtually and prohibitively dangerous for the air attack aircraft. However, in the real world, radar hunter-killer systems were now being placed on-board specialized aircraft like the F-105 Wild Weasel and others.

The electromagnetic "bulls-eye" eventually went beyond the radar portion of the spectrum of radiation. There were EW duels that utilized the infrared portion of the spectrum. It was in this realm of electromagnetic energy that I began to be aware of the concept of "signatures," which refers to the distinctive pattern of energy radiation that can be attributed to known types of targets. It was this aspect of infrared energy that enabled attack aircraft to home in on a very discrete sort of target, based on its infrared or "heat signature." Thus, tanks and trucks within the Ho Chi Minh Trail could be detected, even at night. On the other side of the coin, however, attacking aircraft also presented a distinctive heat signature to such air defense weapons as the shoulder-mounted, infrared-guided surface-to-air missiles that proved so lethal in the final years of the Vietnam War, and in the USSR-Afghan War of the 1980s.

The visible-light portion of the EM spectrum also has been used in military aerospace operations by following the era of World War I. Sometimes referred to as "electro-optical" applications, it began with the iconic ground-based searchlights of World War II. But, soon after the end of that war, scientists were able to develop the technology to "excite" the light molecules and to focus their radiation. This technology is now referred to as "LASER" (Light Amplification by Stimulated Emission of Radiation). My first acquaintance with the application of stimulated, amplified, and coherent electromagnetic energy actually occurred during my tour in Laos as an airborne intelligence officer. It happened when U.S. Air Force attack aircraft began using laser energy to guide (provide a path) specialized receiver units that were attached to bombs and other ordnance. Later, I was to discover that the technique of stimulating, amplifying, and forming energy into a highly concentrated beam would be used with respect to electromagnetic energy along other phases of the spectrum, not just visible-light. Apparently, the limiting factor in the application of this technology is the power source for producing the "aser" effect. That is to say, in a package that is small enough to be carried aboard an aircraft (or a space vehicle). During my tour of duty in Laos, I encountered the application of laser technology to military aerospace operations. Its main use there was to take advantage of the powerful and coherent beam of light energy to designate a target, and to then provide a "path" for the smart ordnance which was equipped with a suitable sensor-actuator system. Later, this extremely long-range and coherent beam of light would be used to carry electronic messages throughout the solar system. What has made this applicable to such military aerospace operations has been the development of sensors that are able to lock-on and follow the path of energy onto the target (or receiver) that is being "painted" by the beam.

During the course of the Vietnam War the emitter was primarily carried onboard aircraft. Typically, one aircraft would carry the laser generator, and would be designated as a pathfinder. A second aircraft would be armed with the laser-guided munitions which would be fired along the laser path onto the enemy target. Now, it is small enough to be carried on satellites and other platforms, even in space. This was the beginning of the era of the "pathfinder" laser guidance of airborne ordnance which is still being utilized in the 21st century. One of the major manifestations of this kind of battle was the "Wild Weasel" program in Vietnam and in Laos during the Vietnam War. This was a program that was designed to degrade or destroy the enemy radar sites that controlled SAMs in these areas of operation. The icon of this program was the F-105 fighter-bomber; it was especially equipped with radar-detection equipment so that it could detect any SAM-control radar that became active. It was also equipped with radar-homing bombs which could ride the path of the laser beam which painted the target.

———

Another manifestation of the electromagnetic spectrum is found in the plasmas. Plasma is usually defined as an electrically-charged conducting gaseous medium, in which there are roughly equal numbers of positively and negatively charged particles. These charged particles are produced when the atoms in a gas become ionized. On Earth, we are mostly familiar with three states of matter: the solid, liquid, and gas phases. The transition from one to another is basically a function of increasing temperature; that is, as the constituent atoms and molecules achieve ever greater freedom of movement. Moreover, when the atoms in a gas are heated even further and accumulate enough energy, they decompose into ions and electrons, and the gas becomes partially or fully ionized. At that point, the plasma state is reached; it is now in the fourth state of matter.

Plasmas carry electrical currents and they are more influenced by electromagnetic forces than are they are by gravitational forces. Their flow can be accelerated and steered by electric and magnetic fields, thus allowing it to be controlled and applied. These same high-energy particles also exist in the magnetosphere which, is a region in the atmosphere conductivity caused by ionization of incoming solar energy particles. Plasmas appear to be important in determining the behavior of charged particles. In a sense, virtually all of the visible matter in the Universe exists in the "plasma state." In physics and chemistry, plasma is a state of matter, similar to gas, in which a certain portion of the resident particles are ionized. Ions are formed by the addition or removal of electrons from originally neutral ions by other ions. Thus, it could simply be a case of electron swapping among positive, negative, and neutral atoms. Plasma is so energetic that in space it consists solely of ions and electrons. It is only when it is cooled that the atoms or molecules which predominate in the forming gases, liquids, and solids that we are accustomed to on Earth is possible. When the charged plasma becomes electrically conductive, it responds strongly to electromagnetic fields and becomes a plasma field itself. These fields can be thought of as being geographic phenomena, just like the topographical features of the atmosphere, lithosphere, or hydrosphere. They are found in the Sun, as well as in the solar wind, planetary magnetospheres and ionospheres, auroras, cosmic rays, and the electromagnetic radiation that is emitted when charged particles accelerated.

Therefore, it can be said that plasma has properties quite unlike those of solids, liquids, or gases. Indeed, it is considered to be a distinct state of matter in its own right. For one, it is much "hotter" than gas and, like gas it does not have a definite shape or volume unless it happens to be enclosed in a container. On the other hand, it differs from gas in that, under the influence of a magnetic field, it may form definite structures. Plasma is a complex object, manifesting variations in the way electrons in excited state relax to lower energy states after being recombined with ions. These processes emit light in a spectrum that is characteristic of gas being excited. Plasmas and dust are both ubiquitous ingredients throughout the Universe. They contain dust grains as well as charged particles. Voyager observations in the early 1980s showed that phenomena in the rings of Saturn which could not be explained solely in terms of gravitation might be composed, at least partially of "dusty plasmas." These are also thought to occur in the rings of all the other Jovian planets, and in the interstellar dust clouds, among others. These are also thought to occur in the asteroid belt, in cometary comae and tails. The grains range in size from micron-size to larger particulates; in a continuous range from macromolecules to rock fragments. The grain charges are determined by the plasma potentials and can fluctuate.

Outside the Earth's atmosphere, the dominant form of matter is plasma. Nearly all the visible matter in the Universe exists in the plasma state, occurring predominantly in the Sun and stars, and in interplanetary and interstellar space. Indeed, the Earth, itself, is immersed in a tenuous plasma called the solar wind and is surrounded by a dense plasma called the ionosphere. The conductivity of hot ionized plasma is extremely high, and the coronal temperature decreases only slightly with its distance from the Sun. One particular form of plasma is the solar wind, which emanates from the outermost region of the Sun's atmosphere (the corona), consists of a hot, ionized gas. This plasma is extremely hot, but has extremely low density. The corona has no definite boundaries; rather, it is a continually pulsating object whose size and shape, at any given time, and which is affected by the Sun's magnetic field. Where magnetic fields are strong, the coronal material's outward flow is stymied, and it becomes trapped. The resulting high density and temperature that occurs, is due partly to this trapping-effect and partly to heating pressures of mostly the solar flares. Where-ever a fissure in the magnetic field occurs, the hot material escapes through the so-called coronal "hole." Analysis of solar wind data shows that coronal holes near the equator are associated with high-velocity streams in the solar wind, and in the recurrent geomagnetic storms that are associated with the return of the holes.

One particular electromagnetic field is the solar wind which is emitted by the Sun. As it flows and propagates from its origins, magnetic field lines are drawn outward from the surface. Meanwhile, traveling at a speed of about 300 miles per second, the ionized particles that exist within the wind will eventually reach the outer areas of the heliosphere, as far out as the orbit of Saturn. It takes four days for the solar wind to arrive at the planet Earth. The magnetic field lines that are emanating from the Sun describe a spiral; a flattened, roughly circular system of particles, which are concentrated along the spiral arms. These arms are thought to be produced by traveling density waves, which compress and expand during the spiral action. This solar plasma field can also be found flowing around and interacting with the magnetic fields that surround some of the planets and other objects in the solar system. The magnetic field lines which do not break up continue to maintain the overall path, and surrounding gases move along with it. On some planets, other magnetic fields form, like eddies along a river and can exert a continuing effect on the atmosphere of the planet. On the whole, the total mass and angular momentum carried away by the solar wind itself, is relatively insignificant, even over the lifetime of the Sun, at least at present levels. A higher level of activity in the past, however, might have played a role in the Sun's evolution, and stars larger than the Sun are known to lose considerable mass through such processes.

———

Geographers and other earth scientists have been developing other devices – called sensors – for "seeing" electromagnetic energy radiation of varying wavelengths that lie beyond the visible-light portion of the spectrum. Those sensors that can detect and measure radiation remotely are called "remote sensors." Probably the most significant utilization of EM radiation has been, and continues to be, the radio part of that spectrum of radiation. The early applications in terms of radio and radar are still immensely useful in the present age of space study and exploration. Since then, in the latter half of the 20th century and into the 21st century, there have been developed many other military applications for some of the other phases of the electromagnetic spectrum of radiated energy. One of these, in the infrared portion, is now commonly used for night-time reconnaissance, target detection, and a host of other tasks. Other applications of the visible-light portions include electro-optical and laser technologies. All of these applications to military operations create what amounts to an electromagnetic battlefield in any area of military operations.

In the Laotian theater of the Vietnam War, for example, the military aerospace operations included the high-flying B-52 bombers which conducting their "Arc Light" missions over northern and eastern Laos, as well as over Vietnam itself. These were the high-fliers which operated at altitudes well-above 30,000 feet, and which dropped their thousands of conventional bombs via radar guidance. This tactic usually placed them above the range of most of the enemy ground-based air defenses, except for the SA-2 surface-to-air missiles. This was a case in which both sides utilized remote sensors to detect and guide their munitions. The B-52s were guided by remotely-located radar "beacon" sites that emitted energy beams to provide navigation aid to the target, and then indicate electronically the optimum point at which to drop its bomb load. The North Vietnamese SA-2 surface-to-air missiles, on the other hand, were guided to the bombers via radar control.

Then there were other the "specialized aircraft," such as those that were designed to seek out and destroy ground-based radar sites with radar-guided air-to-surface missiles. Still other electronic surveillance aircraft trolled the battlefield, in search of any radar or other electromagnetic emissions that could be provided as electronic intelligence to other friendly aircraft or ground stations. Then there was the ground data that was captured by ground-based sensors which were deployed all along the Ho Chi Minh Trail and which subsequently emitted relevant "ground truth" data to orbiting electronic intelligence aircraft, for subsequent relay to the Igloo White computers at Ubon, Thailand. Ultimately, this intelligence would be used by "truck-killer" aircraft, such as the A-26 (and its follow on, the A-10) to target trucks, tanks, and personnel clusters along the trail.

Stacked within these layers of aircraft, like cards in a shuffled deck, were the "top guns" of the PDJ. These were the F-4 Phantoms that carried out tactical attack missions, air reconnaissance missions, and force protection ("Mig-Cap") missions. In the first instance, these were F-4s that were configured with the right mix of ordnance to conduct attack missions against targets on the ground. The Phantoms also provided unarmed aircraft, like our command and control EC-130 with protection against sudden attacks by marauding Mig-19s emanating from bases in North Korea, and invariably under radar control by their home bases. These F-4s were configured with air-to-air missiles that relied on radar or infrared-based devices for guidance and control. This, in summary, describes the importance of the electromagnetic spectrum to military aerospace operations within

Earth's atmosphere. It almost certainly will be an indispensable component of any military space operations throughout the solar system.

One of the most important characteristics of electromagnetic radiation is that it can be manipulated to form coherent and powerful beams. Thus, a beam of visible-light, whose electrons have been stimulated and made extremely coherent can be utilized for a wide range of applications in military space operations, including communications, navigation, detection and tracking of objects, and even as offensive weapon systems. Rays can also be beamed on certain screens to project imagery from great distances in space and planetary objects. Another way in which the electromagnetic spectrum of radiated energy has been useful in military aerospace operations has been in the development of "maps" of the battlefield environment which are based, not only on what is seen through the prism of the visible-light portion of the EM spectrum, but also the other prisms, such as those that are in the infrared or x-ray portions of the spectrum. These prisms make possible the development of specialized maps of a battlefield region which can only be "seen" with the aid of sensors that are tuned into the particular wavelengths of electromagnetic energy that is being radiated by objects in the area of interest. One example of this is the infrared map that is produced through the use of infrared-sensitive film in remote aerial reconnaissance operations.

In summary, we have seen how the flow or propagation of electromagnetic energy through a medium or space can form recognizable "regions" in the geographic sense. One very useful method for rationalizing this broadened view of physical geography is via the mental construct that we refer to as the electromagnetic (EM) radiation spectrum. Each range of spectra within the overall spectrum may be visualized as a virtual subregion of the whole region. On the other hand, just as the Latitude-Longitude Matrix is used to give rational structure to the surfaces of spheres, including the planet Earth, the EM spectrum provides a rational structure for understanding the various manifestations of electromagnetic energy in the Universe. It also provides a structure for understanding that physical reality; and it provides a "common language" that facilitates the efficient communication of its realities in the accomplishment of tasks. In the same way that geographers and military aviators refer to a particular latitude-longitude number, especially when it refined down to minutes and seconds, everyone understands the particular point on the surface of the Earth (or planet) that is being referred to. By the same token, when geographers and military aviators refer to a particular wavelength and frequency number, everyone understands which portion of the electromagnetic spectrum one is being referred to. There are many combinations of waves and frequencies along the overall spectrum of radiation. Nevertheless, it has proven useful to consider the continuous spectrum as consisting of discrete phases, which can be plotted along a continuum that is referred to as the electromagnetic spectrum. Applications of these ascribed characteristics of the electromagnetic spectrum have been indispensible in the development of military aviation.

––––––––

The solar system also can be seen as a geographic field that is populated by both natural and cultural objects. In this field, the planets and other celestial bodies make up a natural population whose locations and movements are basically controlled by the variable that is called gravitational force. This natural force is not particularly strong, compared to electromagnetism, but it is ubiquitous and relentless. Many physicists, including Albert Einstein, have likened the gravitational field of the Universe to a space-time "mesh," within which bodies of varying mass cause distortions, or warps, if you will, to the theoretical ideal plane of the field. Like the geographic region, the gravitational field at any point in space-time also has subregions. These are the tributary gravitational fields of influence that are formed within all objects, according to their absolute and relative mass. That is to say, relative to the mass of the other objects around them. Actually, all the other massive bodies in the near the "hinterland" or neighborhood of an object appear to create turbulence or perturbations in the spatial nature of any gravitational field. I see this kind of disorder in a uniform gravitational field as being analogous to that which occurs if one throws, not one stone into placid pond, but multiple stones at the same time, in close proximity to each other.

This hierarchy of gravitational influence can be depicted as a series of "spheres of influence," which emanate from various points of origin throughout the Universe. As a geographer, I am struck by how much such a map of the influences of gravity from various sources is similar to map drawn by political geographers to depict the spheres of political influence, which emanate from nation-states on Earth. Indeed, if one were to follow this analogy, the stars, planets, and other bodies within our solar system could be seen as projecting an "influence" over all other matter within this region. And, just as military-economic power determines the geography of the relative influence of political units on Earth, what physicists call "mass," determines the size and degree of the gravitational sphere of influence of each node (planet and celestial body) in our solar system – and throughout the Universe. Again, mass is seen to be multiplied by the degree of distance between any two bodies. In other words, the closer to bodies are to each other, the greater the degree of influence. One universal fact about gravity is that it occurs, permeates, and controls the overall spatial operation of the Universe.

According to this disordered and dynamic model of the cosmic gravitational field, the attractive force of these indentations is what is experienced as gravitational force on all objects, including humans. So, we can say that the gravitational field is ubiquitous and its effect is felt throughout the Universe, in varying degrees of magnitude and intensity, depending on the portion of the space-time geography in which it happens to occur. In the language of physics, gravity is a "force" which acts on a given body in very well-defined and disciplined ways. More generally, the nature of the gravitational force is that it is like a mesh-fabric material, exhibiting elevations and depressions according to the mass of the nearest body. From the standpoint of any object that lies within this mesh of gravity, the attractive forces that are experienced are really due to the tendency of the fabric to come together around the object, much like objects of varying mass and weight on a trampoline net.

In any case, gravity is still considered to be the universal force of attraction that operates on and between all matter, no matter how big or small. It has been the great "controlling variable" in the Universe almost since the very beginning – at the instant of the Big Bang event. It also has continued to perform this function during the subsequent development of the Universe and the spatial relationships of all the other spatial systems that have derived. Given this set of realities, a geographer would, almost instinctively, attempt to draw a three-dimensional map of this force to describe the spatial distribution of all objects throughout the largest region (the Universe), and its subregions, such as the solar system. This view of gravity can be visualized as a "mesh" fabric, which extends equidistantly and in all directions; what he calls a "space-time event." I understand it to be a type of region; one with both elements of space and time interweaved within it.

However, the geographer is also likely to approach the study of the magnetic field as a spatial region that is populated by charged particles; charged particles that behave spatially as a function of magnetic forces. Magnetic fields also often interact with electrical fields to form an electromagnetic field. Both of these can be found operating within and around the planets and other celestial bodies. They may be represented by continuous lines of force or magnetic flux that emerge from one pole and move toward an opposite pole of the field. The density of the lines indicates the magnitude of the magnetic field. One insight that might be contributed by a geographer is that this is a region which is continuously changing – depending on the position of the observer within the mesh. On another conceptual dimension, this "gravity-mesh" region can be seen as a incorporating nested and inter-related "spheres of influence," where the influence is the result primarily of gravity. In the classic geographic region, the various nodes in a given region are seen to have a greater intensity of some characteristic (such as gravitational influence) than other nodes.

Thus, we see that the influence of the gravitational force is a result of the mutual attraction that exists between all masses of matter in the Universe, including the smallest subatomic particle. Gravity is also like oxygen in its pervasiveness within the atmosphere of the Earth. However, unlike oxygen, gravity occurs throughout the Universe and not just in the atmosphere of the Earth (as far as we know at this time). This universal phenomenon is probably one of the best-understood by scientists, but in terms of it practical relationship to military aerospace operations, it is still a subject that is being learned. It is something that we continue to learn more about as we venture out into the solar system and beyond. The most obvious role that gravitational force plays is to maintain a certain "relative position" among the particles on the spectrum of mass. On one end, there are the stars, planets, and other heavenly bodies, and on the other the quarks and other subatomic particles, all maintaining a kind of dynamic tension, if you will, that is enforced by gravity and the magnetic forces of repulsion. To the military aviator and the astronaut, however, whether on Earth or in the solar system, gravity is most important because of the ways in which it affects military aerospace operations at all levels. Within the atmosphere of the Earth, gravity manifests itself as a downward "pull" or a horizontal "drag" on the artificial flying machines which the aviator pilots. In outer space, away from larger bodies, the main effect of gravity on military space operations – after escaping the Earth's gravitational pull – will derive from the lack of gravity – until the space vehicle enters the gravitational sphere of another celestial body. That is, the planets and other large bodies in the solar system will also exert a gravitational pull, depending on their individual mass, dimension, and relative distance from other bodies. One of the creative ways in which this gravitational force "pinball machine" is being utilized by space operations folks today is the called the "sling shot" effect to enhance integral propulsion systems. It appears that the local gravitational regime on the various bodies of the solar system will also influence target-area navigation and weapons utilization on various venues there.

In the case of planetary operations throughout the solar system, the effects of gravitational forces on military space operations will again manifest themselves in many of the same ways as they do on

Earth. One difference will be in the degree of force and variations in the ways gravity will interact with other forces in the atmosphere or on the surface. Generally, the problem facing the astronauts is how to keep platforms in a controlled state of dynamic tension with the planet's gravitational attractive force, in order to maintain a kind of controlled state of gravitational dynamic tension. Within a planetary atmosphere, the solution is found in a form of "lift and thrust," which is produced by some form of propulsion technology. The laws of gravity are not nullified; they are simply accommodated.

Within the atmosphere of the planets, this homeostatic situation is disturbed when military variables are added to the equation. One such variable is the addition of weight on the air platform, such as additions in the number of crew members, and ordnance loads; that is, war materiel, bombs, missiles, and guns. These were eventually brought under the general rubric of "payload." And, the basic state of homeostasis, or dynamic equilibrium, also began to breakdown as aviators sought to expand their operations in both vertical and horizontal vectors, that is to say, faster and higher. Now the accommodation that had been reached with gravity would become more complex; and it would require more technology and human ingenuity to sustain these more ambitious aviation operations. One outcome of the increasing demands of military aviation has been the practical realization that gravity operates on a body (human and machine) from all directions, and the greater the desire to move in one direction, the greater is the reactive effect of gravity on the body.

Whether within the gravitational field of attraction of a planet, or in the near zero gravitational environment of interplanetary space, humans in space readily discover a general crucial factor with respect to the exertion of increased demands of the gravitational field at any point in the solar system. This becomes immediately and powerfully clear when humans venture into the interplanetary medium of the heliosphere. Generally speaking, the force of gravity imposes greater stresses on humans and their operations in space because it is so different from the gravitational environment of our home planet. Indeed, the need to respond to the forces of multiple and multidirectional gravity forces on the human astronauts has prompted the development of a whole new medical specialty: that of the aerospace flight surgeon.

Even though we may not understand completely all the nuances of the gravitational forces in the solar system, we do know some of the limitations and opportunities that it provides to space operations, including military space operations. One area of these environmental factors has to do with life support in the zero gravity milieu of space. Another such development is the portable life-support environment; specialized flight suits, which are designed to maintain an optimum level of oxygen and pressure in a portable "atmospheric environment" and to protect the wearer from any hazardous environment elements. This portable life support system is also designed to keep the aviator's blood flowing as it would on the surface of the Earth, and to keep the blood from pooling in or leaving vital organs (brain and heart) without oxygen-bearing blood. The main idea is to keep the aviator/astronaut conscious, alert, and able to move the arms and legs normally.

Another technological solution to the problem of extreme effects of gravity has been to equip modern aircraft/spacecraft with a variety of systems to counter these forces. All of these responses to the challenges of gravity, with respect to military aviation, throughout the solar system, are just a few examples of movement along the man-machine continuum that has come to be called "bionics." It would be moved further along in space flight when the aircraft itself would become the manifestation of the man-machine amalgamation. Thus, in the 21st century, the control of the aircraft has been automated and computerized to the point where "intelligent" computerized systems can now takeover when the human might not be able to control flight manually. So, in the end, it may be that the training that is provided to the military aviator/astronaut that will prove to be the most effective way to deal with the challenges of extreme gravity throughout the solar system.

On Earth, many military aviators have pushed the edges of the gravity envelope of our terrestrial atmosphere. One of these was the quintessential "test pilot," Chuck Yeager. He was an icon of this era in which airplanes flew ever higher and faster, in what would turn out to be the precursor to the era of manned space exploration. Brigadier General Yeager was the first man to exceed the speed of sound in flight. He was one of a group of pioneer pilots that was selected to test-fly the secret experimental X-1 aircraft that was built by the Bell Aircraft Company to test the capabilities of the human aviator, in a fixed-wing platform, against the acute aerodynamic stresses of sonic flight. On October 14, 1947, he rode the X-1, which was attached to a B-29, to an altitude of 25,000 feet. The X-1 then separated and utilized rocket-power to rise up to 40,000 feet in altitude. He also became the first man to break the sound barrier, which was approximately 662 miles per hour at that altitude. On December 12, 1953, he established a world speed record of 1,650 miles per hour in an X-1A rocket plane. Then, in 1962, the first U.S. astronaut went into orbit around the Earth. This was also the first time that a human spaceship had achieved a "free fall" condition in which a body would move freely in any manner in the presence of gravity.

Newton's laws show that a body in free-fall follows an "orbit," such that the sum of the gravitational and inertial forces equal zero. All objects associated with the spacecraft, including any crew and other contents, are accelerating, that is, falling freely, at the same rate in Earth's gravitational field. As a result, these objects do not "feel" the presence of Earth's gravity but, instead, experience a state of weightlessness, or zero gravity. So, it seems intuitively obvious that the force that we call gravity has always been a major factor in aviation, beginning with the first attempts to overcome its downward tug on balloons and the drag on the forward movement of the heavier-than-air platforms during the 19th and 20th centuries. We also are familiar with the phenomenon that occurs when we whirl a stone in a sling around us. First, we feel a tug of war between ourselves and the end of the sling. What this means is that a countervailing effect against gravity is occurring from an opposite direction. This is called a "centrifugal force" effect which counteracts the effects of gravity in any situation. The variables involved include the length of the string and the mass of the stone on the end of the circular plane ("distance"). Also, the faster the stone is whirled around the person, the stronger is the force ("velocity") that is needed to keep the stone within the plane of the motion ("orbit").

MILITARY SPACE OPERATIONS

U.S. military commanders and the civilian leadership have expressed their recognition of the importance for developing military capabilities in space (Joint Publication 3-14, 16 January 2009) in the 21st century. There is a realization that military, civil, and commercial sectors of the United States are becoming increasingly dependent on space capabilities to deal with potential vulnerabilities in the space age. The threat is often couched in terms of the potential for "purposeful interference" with our nation's artificial satellites which are orbiting the Earth, as well as other U.S. assets in space and on the surface of the planet. The reasons for this concern include the proliferation and increasing sophisticated nature of space capabilities and products that have military utility in the world today. Any of these could be used by an adversary – whether a nation-state or some private entity, such as a terrorist organization.

Recent experience with respect to the space launching and rendezvous operations by private corporations in near-Earth space has shown that a large infrastructure is no longer required to project destructive power into the solar system. In other terms, the minimum cost of access to the military space battleground has been lowered considerably. This emerging situation has made the challenge of ensuring U.S. military dominance in space through the maintenance of the ideal military space level of superiority is becoming more immediate and critical. The state of space dominance is defined as having: (1) the freedom of action which ensures maximum maneuverability and responsiveness to a threat event; (2) force security against any threats from an adversary; and (3) the ability to employ the element of surprise and continuing level of action against an adversary.

Some of assumptions that will be relevant to the conduct of military operations in space are similar to those that hold in naval warfare on Earth. Thus, it is expected that any spatial expanse within the solar system is not a self-contained battlefield. Rather, the military space domains are all part of a comprehensive medium (like the atmosphere, or ocean) in which warfare is conducted. In the same way in which the oceans of the world constitutes one of the media through which nations project power, the interplanetary medium will perform the same function throughout the solar system. Thus, the space fleet will likely be deployed in a pattern that is best for protecting the trajectories which become the lines of communication in the solar system. This will require the maintenance of a warfighting readiness in every domain of the overall military space domain that will develop. Furthermore, there will be a need to maintain a robust space-lift capability and a forward deployed presence in order to have a quick-response capability at all points within the solar system.

Where ever humans have gone on Earth, they have created an "overlay" on the natural geography of the planet that is called the cultural geography. Going back as far as some 10,000 years ago, when the Earth was experiencing dramatic global climate changes, we saw the refugees from the Sahara regions, which were becoming arid, migrating to the greatest oasis in the region – the Nile River valley. Within about 5,000 years or so, the refugees from the erstwhile paradise that had been a northern region of Africa, had developed the valley of the Nile in much the same way that modern developers turn farmland into urban suburbs. So, the Egyptians, as they came to be called, set about reorganizing the natural "chaos" into an ordered civilization. They built houses and plotted the land into units that were deemed best for farming the land on either side of the Nile River, and especially throughout the fertile delta region. They then set about converting a natural drainage river system into a major transportation and communication system. In time, they were able to produce more food than they needed, so they constructed storage facilities to conserve the surplus to get them through the lean times. The surplus grain became wealth that could be used to fight wars and to construct pyramids.

What the Egyptians had done was to overlay the natural order that was the Nile River region with a political, military, economic, and social entity. A similar process can be seen to be occurring in the other parts of the solar system. In the Nile-Red Sea region it likely began with the dispatching of "scouts" and "pathfinders" from the shrinking oases of the Sahara. Over a period of many years, successive waves of these refugees from the desert continued to settle the Nile Valley and contributed to the development of the wealth of the nation. Subsequently, through the "man-land" relation that developed along the Nile Valley, an empire that was formed around 3,000 BC, following unification by conquest. The cumulative wealth then was used to build monuments, and to build infrastructures in order to provide for the general welfare of the population – especially the priesthood and the pharaohs who maintained order in the empire. The downside to all this wealth, however, is that it had to be defended against rival empires and other outsiders. So, at some point, this necessitated the development and maintenance of a military institution to do the job. This is the pattern that has repeated itself throughout the history of humans, since the beginnings of recorded time.

There is a considerable body of literature which has dealt with the reasons why humans engage in warfare. These range from biological arguments which ascribe a "warrior gene" to individuals, to social arguments that point to a long process of acculturation in societies that seem to impel them towards warfare. So, we have those who see a linkage between animal and human behavior as an explanatory factor. This often leads to the "territorial imperative" notion which then provokes aggression and defenses based on territory or resource-base. At some point, humans seem to have transcended such "primitive" motivations for warfare and began to ascribe their naked pursuits of territory, resources, and plain-old booty, to higher motivations – that is, those related to religion and ideology; honor and duty; and so on.

Following the European discoveries of "new worlds" in the Americas, Asia, and Sub-Saharan Africa, as well as the Australian lands, the voyages of discovery began to gear up for a more organized approach to acquisition of territory and resources. Using whatever theological or economic argument at their disposal, they essentially carved out other-folks' land and declared their territories to be part of an "empire." Once the real-estate conflicts were more or less settled, these European imperials sought to extract as much of the natural resources, both mineral and animal, as well as agricultural, in order to increase the Mother Countries' wealth. Some referred to it as enhancing the "wealth of nations." It is this period in the history of warfare that may turn out to be the most relevant to the space age in the 21st century. If this model of human behavior were to continue into the other parts of the solar system, it is likely that the individual nation-states and non-governmental entities will engage in a highly-competitive rush to claim the resources of the asteroids, moons, and planets of the solar system. This will represent the implementation of the 17th century model of conflict by the European nation-states in the Americas, Asia, and Africa. On the other hand, if the paradigm of the Antarctica Treaty (1959), the North Atlantic Treaty Organization (NATO) and the United Nations were to be followed, there is a chance that military space operations would resemble that of a multi-national peace-keeping force.

In point of fact, I believe that military operations are in a period of transition – from the conduct of air operations exclusively within the atmosphere of the Earth – to the operations that are taking place in a transitional "geospace." This is the frontier zone that begins on the surface of the Earth, and extends out to the portion of outer space that lies just beyond the zone of the Earth-orbiting artificial satellites. Most of the earliest military space operations are taking place within the parameters of this geospace. This first stage in the development of military space operations actually began with the systematic placing of artificial satellites in near-Earth orbit in 1958 by the Soviet Union, which was followed by the deployment of these artificial satellites by the United States in the 1960s. This expansion of military presence into space seems to have occurred almost as a collateral effect that derived from the race to develop ever more powerful and accurate intercontinental ballistic missiles by the principle antagonists in the Cold War.

In any case, the deployment of Sputnik and the U.S. satellites in low-orbit around the Earth soon took on a life of its own. Since then, thousands of these orbiting satellites have been deployed by a variety of nation-states and private organizations. Generally, these systems in space perform several different functions, which can be summarized as being: (1) communications systems; (2) navigation-aids; (3) reconnaissance platforms; (4) telescopic platforms; and (5) remote sensor platforms. So, it is now possible for any nation-state or private agency to have "eyes and ears," as well navigational beacons, and communications relay stations, as it were, in orbit about 200 miles above the surface of the Earth. This obviously has added a new dimension to military aerospace operations as well.

The next phase in space operations by the United States, Russia, and more recently, China, has essentially been what can only be characterized as a "clandestine" program of military space capabilities, under the umbrella of "civilian" agencies, such as NASA. Thus, we see that the next step in this transition toward a greater level of military space operations has been the gradual honing of the process of the actual planning for military space missions. This, I suggest, is when the civilian NASA program began to develop models for the development of a formal military space program by the United States. The first analog "military" system was the space reconnaissance infrastructure that was developed by NASA to capture data and generate intelligence about the inner region of the solar system – in preparation for actual space exploration operations. With this information, NASA and other space agencies were then able to actually organize the resources and plan the missions that would be needed to achieve the initial assigned missions (such as landing an astronaut on the Moon, for example).

During the early years of its existence and acceptance of its assigned overall charge in the 1960s NASA set about to develop a physical infrastructure on the surface of the Earth. Its first objective was to build the capability to launch a rocket and payload (capsule) from point A (Earth), and to then land on point B (the Moon), and finally, to return to Earth. Consequently, the launch facilities and the mission control complex and the attendant support functions that would be needed to sustain this ongoing effort were constructed at a site in Florida – and staffed by all sorts of human resources. Thus, at Cape Canaveral, Florida the space agency built the launch apparatus and the support elements to launch the rockets and their payloads toward the low-Earth orbit, and then the Moon, Mars, and the outer solar system planets and their moons – and beyond. Indeed, on September 9, 2012 it was reported in the news that Voyager 1, after 35 years mission activity throughout the solar system, was going out beyond the heliopause.

Although the NASA space exploration program ostensibly has been done for "peaceful" purposes, there has always been an unstated military corollary to its development of systems that are necessary for manned space flight. Take, for example, the "human survival and physical comfort" systems that have been developed by NASA, as a paradigm for the development of potential for the adaptation of civilian resources toward military purposes. The elements of the space environment which affect the health and safety of the human astronaut include the danger of the space vehicle being hit by hurtling asteroids, meteorites, or even fast-moving space rocks. Minute sand-like or silt particles also can pose dangers to the human and the space vehicle, especially for longer missions where the cumulative exposure to these objects can foul life-support and operational systems. Also, the longer a human remains in space, the greater are the increases the cumulative effects of zero gravity on the bones and muscles, making them become very much like those in older humans on Earth. Then there is the constant bombardment of cosmic radiation in space, where humans no longer enjoy the protection of the both the terrestrial atmosphere and the geomagnetic sphere. Just as nuclear radiation causes the destruction of human cells, so does the cosmic radiation in space. Already, some of these effects have been manifested in retired astronauts who present brain-damage and other health problems, according to NASA reports.

Much of NASA research and design activity has been focused on the ways and means for providing for the necessities of human life in space. Studies have been done on the feasibility of storing, on board the space vehicle, or in pre-placed space caches; or on systems that can be engineered to recycle resources, or even manufacture them out of the chemical "whole-cloth" of space itself, through nanotechnology. The problems of long-term boredom on longer space missions, as well as physical wasting in a zero-gravity environment, however, will require more subtle and sophisticated responses. Even in the early years of human space exploration during the last decades of the 20th century, astronauts and cosmonauts have been undergoing extensive psychological testing and behavior analysis, in order to determine their ability to deal with the challenges of long-term space flight. This imperative to find ways to maximize the human element in space operations has spawned much scientific and engineering research and testing. It has even produced a new science by the name of "bioastronautics," which bears a strong resemblance to the familiar "flight surgeon" activities of the 20th century on Earth.

There is also already a wealth of documentation with respect to the effects of boredom on humans on Earth. Many films, for example, have dealt at least superficially with this matter. These have depicted the experiences of prisoners of war who have been held in isolation for many years at a time. Similarly, there have movies and books that have told the story of convicts who have faced a similar state of long-term isolation. Survivors of both kinds of experiences in the "hole" have reported that the worst aspect of their ordeal was the "boredom," and the loss of time and spatial awareness. On the other side of the coin, there is the paradoxical pain of long-term isolation and forced interaction – as is the case that has been reported by crewmembers of nuclear submarines. In the latter situations, there has been the paradigm of relative isolation, forced personal interaction, and unforeseen short periods of stress, danger, and terror. Personally, I believe that the experience of the nuclear submarine mariners has many analogs to the problem-matrix that faces long-period astronauts in space.

In all of these types of situations, humans are put under prolonged periods of low-level psychological stress, often brought about by boredom and attention tunneling. The latter has been shown to create critical mistakes during an emergency situation. One example of this phenomenon of attention tunneling, and its effect on the ability of humans to make correct decisions in emergency situations, was the Three Mile Island episode. To deal with these problems, as they relate to long-term space exploration, the U.S. National Aviation and Space Agency (NASA) has developed a comprehensive selection and training program to recruit potential astronauts with the appropriate personality makeup and skill set needed for long-term space exploration. The selection protocol resembles that of modern marriage-broker internet websites that endeavors to enable on-line marriage matches. Prospective space explorers are subjected to an extensive battery of practical psychological tests to ascertain their personality profile. The most significant of these have taken place in such analog environments as the Atacama Desert of South America and the cold tundra of northern Canada.

These personality tests are designed to determine the applicant's ability to deal with long periods of boredom that are punctuated by intense periods of emergency stress situations. Also, the prospective long-range space explorer is evaluated for his or her ability to deal with the threat of death, unexpectedly, over long periods of time. (Probably like the situation of a person on "death row" whose appeals have run out). The probability of suicidal tendencies in high-stress situations like that of a condemned prisoner has to be considered. Again, the possibility of murder in spacecraft caused by such a high-stress situation has to be considered. Thus, a new breed of astronaut or cosmonaut is required. Personality types such as those described as "drivers," or analytical types who tend to think outside the box, as well as people who are motivators, and team builders are to be prized over the "adrenaline junkies and cowboys" of the early years of space exploration. To further ensure the efficacy of long-term space exploration, NASA has provided team-building training for prospective space explorers by sending them through a simulation program in the wilderness, where the trainees are given a series of outdoor survival problems to solve. The objective is to maximize team bonding and problem solving. This is very much like the basic training programs conducted by the various branches of the U.S. Armed Forces in which the recruit is stripped of his individualism and is taught to work with his or her mates as a team. The space explorer recruits, however, were being groomed to deal with problems related to exotic, hostile environments, rather than military opponents.

One practical outcome of the NASA space program to date has been the development of a proto-type analog model for military operations in space. This is especially so in the case of the cultural (or man-made) geography which can readily be converted to military purposes. Consider that the physical infrastructure of the NASA space program, including the launch facility and the mission control facility, which has been constructed to carry out the mission of space exploration, can also be utilized in the development of military space operations in the future. We see a historical example of this conversion-adaptation model in military air operations in the case of the famous Doolittle Raid on Japan during WWII. Here is an example of how, for starters, civilian air operations technology was "militarized" when aircraft designs were first being converted to military purposes. The U.S. Army Air Corps B-25 bomber, which was the platform that was chosen to carry out the attack, had originally been designed to take off from land bases. Now the aircraft was modified for launching from a U.S. Navy carrier ship, and the bomb bays were filled with fuel tanks. The aircrew also had to be trained to take off from an extremely short runway in an ocean environment. All of these actions are examples of how platform systems that are designed for one purpose may be converted to another. It is therefore reasonable to expect that the same pattern of adaptation and modification will occur with respect to the "civilian" NASA operations and any future military aerospace operations in space.

One of the most important strategic decisions in the early planning of the NASA space program was the one concerning the architecture of the main spacecraft platform. That is, it was decided to send one large platform (spacecraft), consisting of several modules, as an integral unit, which would be launched by one powerful rocket propulsion system. The other alternative that was considered was to launch the lighter modules independently in Earth orbit, where they would be reassembled for the main trip to the Moon. This original strategic decision would provide the blueprint for the rocket systems that would follow, at least during the initial stages of the space exploration enterprise. It would also help determine the parameters for the construction of the "cultural geography" of the NASA space program, that is, the man-made physical infrastructure of the system.

Thus, that decision to make one heavy launch, as opposed to many lighter launches, has determining the configuration of the basic launch and auxiliary support geographic region that has been developed by NASA. There was to be essentially one launch complex, around which the resources would be situated, rather than a multiplex of launch complexes, with each one operating in concurrent fashion, and launching subsets of the rocket-payload system to a rendezvous point in orbit, where the whole system would be assembled. In other words, the regional system would have only one or two major nodes, to which secondary nodes would be linked, rather than a multi-nodal spatial configuration. The U.S. military has essentially adopted this model in its military space program. This is an example of how a military decision can affect the "cultural" military geography which interacts with the relevant military aerospace operations.

Another nodal subsystem of the whole space operations system would include a network of orbiting satellites, including the GPS constellation of 31 orbiting satellites that provides time, location, and velocity data; these are critical real-time data, which provides communications and navigation support to military space operations. GPS signals are especially critical to the guidance of precision space weapon systems. In other words, the cultural geography of the NASA space operations system had grown, by accretion. This accretion process had been completed through a series of rocket launches in which the payload was the assembled and ready-to-operate satellite. The linkages between the ground-based nodes and the orbiting nodes would be primarily electronic, though the rockets provided a physical linkage. This was an era in which there was much attention being paid to the development of more efficient electronic communications among the growing number of nodules in the space internet. This, in turn, stimulated the creation of even more nodules in the system, whose purpose was to do research, development, and actualization of the hardware and software which would make the communications subsystem more efficacious. Still other nodules were added, to develop improved fuels, materials, and propulsion subsystems. A study of the FYI 2013 Budget proposal by NASA (nasa.gov) shows that the development support systems for space operations is alive and well, and that it is moving into a new phase where it can support deep-space operations, as well as those in the inner-solar system.

————

During the past three decades, the U.S. Air Force also has been developing a space operations system which somewhat parallels that of NASA (albeit for different purposes), but differing mostly in philosophy and strategic objectives. In a sense, the U.S. Air Force is taking the final steps along a spectrum from an atmospheric-based air force to a true aerospace force which is operating on Earth and in space. Like NASA, the U.S. Air Force space program began with policy directives from the President of the United States and the Congress. Thus, a program to achieve the policy objectives has been put into place, in much the same way that NASA did at the start of the 1960s. Some of the military space system parts have already been put into place, as the U.S. Air Force Space Command which has been assigned overall responsibility for U.S. military space operations. This book is a look-forward into what the next phase of the U.S. Air Force operations in space might be like.

Consider the first iteration of assumptions regarding future military operations in space that appear in the early U.S. Air Force operational planning literature: (1) some operations might be staged directly from the surface of the Earth; (2) operations may persist for weeks or even months; and they may be carried out at any time of the day or night; (3) the specific site of the tactical area of operations is unknown beforehand; (4) weapons must be highly-accurate in order to minimize any collateral damage; (5) the platforms that deliver the weapons must also be able to defend themselves against space interceptors and surface-to-air missiles; (6) the adversary may be either in the form of organized national forces, or it may be rogue "guerrilla" or terrorist groups; (7) targets may be fixed or mobile, and may be well-concealed and; the enemy order of battle may consist of a variety of weapon systems, with multiple capabilities. As one reads the literature, therefore, it appears that future military space conflicts – at least during the 21st century – will bear many resemblances to the small, but locally-intense "asymmetrical warfare" that has been occurring in the early decades of the 20th century, especially in the Gulf War and Afghanistan War context.

The Air Force Space Command actually has already implemented the first phase of its first strategic assignment – which is to develop a satellite-based communications, navigation aids, and reconnaissance, and early warning systems. The last of these systems has been designed and deployed to guard against ICBM launch threats and to monitor nuclear explosions on Earth. It has done this, in part, by working with NASA and its established resources to deploy several satellite platforms in orbit around the Earth. I believe that this "joint venture" model will probably continue to function well into the 21st century, even as the military community takes on more missions to secure the systems that are developed in space. In a sense, it is developing into a paradigm that is similar to the "base and area security" model that has been developed by such military thinkers as Colonel Karl Woelz, and others.

———

If we use the NASA model that has been developed to carry out its assigned missions of space exploration and scientific analysis, which has been developing for the past half-century or so, the outlines of a similar military space system can be better understood. Thus, the Earth-based infrastructure of bases which NASA has developed includes: a space launch facility (now named the 21st Century Space Launch Complex); a space communication and navigation facility; a human space flight operations node and associated launch-support services; and a rocket propulsion test facility. Also, deployed throughout the surface of the Earth, are tracking and communications relay stations, which monitor the space missions which are operating in the solar system and feed a continuous stream of data and imagery to the mission control centers of the system. Together, these comprise the main ground components of the overall NASA system.

These support systems are utilized by various dedicated mission programs, including the Space Shuttle (now the STS) program and the International Space Station, the Hubble Space Telescope, and others, which are provided launch services and space and flight support. There is also an extensive network of terrestrial and orbiting communications nodes and the associated hardware and software subsystems that are needed to drive the input-process-output cycle with the terabytes of data that are generated by the fleet of manned and unmanned spacecraft in space. These include the International Space Station, the Hubble Space Telescope, as well as the growing fleet of orbiters, probes, and other space vehicles that are operating in the solar system.

Before any military space operations can be maintained in the solar system outside the Earth, there must be an appropriate operations support system in place to support the mission. Once again, NASA has provided a model which can be utilized in the development of a military space system of operations in any areas of operation outside the Earth's atmosphere. To begin with, it is axiomatic that every aerospace system, regardless of its area of operation, must have an efficacious navigation and communications system for situational awareness, trajectory planning, command and control, and guidance and control of spacecraft. NASA now has in place such a comprehensive operations support system. Its foundation is the Deep Space Communications & Navigation System, whose basic mission and function is to track vehicles in deep space – that is, at planetary and lunar distances. It does its job through the use of a variety of combinations of radio and optical techniques. The particular mix of these two techniques depends largely on where the spacecraft is located along its flight path. Thus, during the cruise phase of a mission – the time between insertion into the interplanetary transfer orbit and the final approach to the target body – Earth-based radio-metric tracking techniques are generally used.

Radio-tracking involves many of the aspects of the familiar air traffic control systems that operate on Earth. In the space environment, radio-tracking systems provide highly-accurate orbit information, both in terms of the trajectory of the spacecraft and the orbital of the target and any bodies within the mission environment. Just as occurs at thousands of airports throughout the Earth, the radio (radar) tracking systems in the space environment provide accurate orbit information to support any midcourse trajectory corrections of the space vehicle. Then, at the time of approach phase of the mission, the Earth-based tracking systems can be used in conjunction with on-board optical images of the target or one of its neighboring bodies juxtaposed against a known star background, to provide complete navigational awareness of the target environment. Thus, the combination of radio and optical techniques can function in a way that is similar to the familiar LORAN (Long-Range Navigation) and MSQ (targeting-beacon) systems that have been used in military aerospace operations on Earth.

In the case of space operations, both NASA (as well as other space agencies) and the U.S. Air Force have deployed global tracking and communications stations to maintain command and control over the various space assets that are now in orbit or are on a trajectory towards an operational orbit. The system of navigation and communications that has been developed for space operations by NASA includes an extensive network of Earth-based and orbiting communications nodes, along with the associated hardware and software needed to capture the trillions of bytes of data that are generated by the fleet of space vehicles, both manned and unmanned. The Space Communications and Navigation (SCAN) network of Earth-based systems is deployed world-wide to provide tracking data and the vital linkages without which satellites could not transmit their data to Earth in any useful manner, or be effectively commanded and controlled by the mission control facilities. These nodes provide day-to-day, real time global coverage and are deployed in various remote locations, such as on Antarctica. Another system, the Tracking and Data Relay Services (TDRSS) nodes provide tracking and data-relay support of the communications satellites that are in geosynchronous orbit; these are located at such locations as White Sands (New Mexico) and Guam. These ground terminals operate in conjunction with Earth-orbiting TDRSS to provide signals between all spacecraft in the area of operations. These systems also provide navigation services to all space missions.

The Russians also have constructed a simulation facility in Moscow, to provide realistic training for prospective human explorers, in planning for a future mission to Mars. Named the Mars 500 project, it is housed in an erstwhile warehouse on the campus of the Russian Institute for Biomedical Problems. Several thousand applicants to the program have been culled down to only seven men. They will endure a long-term period of isolation in a structure that has been configured like the space vehicle that will be used in the mission. In this restricted space – or complex of modules – the crew members will go through an intensive training and testing regime which will last 520 days. This is the approximate time period of a return trip to Mars, including a planned 30 day layover on that planet. The objective is less about working out the technical aspects of the voyage than it is about ensuring that the crew members will be able to deal effectively with the physiological and psychological challenges of such a mission. In any case, fortunately – or unfortunately – depending on one's viewpoint, the exploration of the solar system and beyond does not really depend on our ability to travel "en persona" in that hostile environment. There are a couple of alternatives to manned space exploration. One has been utilized for almost 500 years – the telescope. Another that has evolved as part of the space exploration experience of the past 50 years, or so, is the unmanned space exploration option.

In the same way as NASA had done when it was given the responsibility for developing a system that would be able to ultimately land a human astronaut on the surface of the Moon, the U.S. Air Force Space Command also set about to construct the appropriate physical infrastructure to enable the accomplishment of its own assigned missions. The following is a descriptive analysis of the elements of this cultural geographic region and system. To begin with, the U.S. Air Force Space Command is already operating a sophisticated command and control system that is aimed toward military space operations. Some of the assets are deployed on the surface of the Earth, while others are in orbit. Together, these components of what amounts to a situational awareness and response system, enable the United States to initiate active operations in space in response to a hostile event. The elements of this system include: the Air Force Satellite Control Network; the Maui Optical Tracking Identification Facility; the Ground-Based Electro-Optical Deep Space Surveillance System; the Passive Space Surveillance System; and the Rapid Attack Identification Detection Reporting System.

Those who are familiar with the systems that were developed by the United States, during the Cold War, to deal with the threat posed by the Soviet Union, will recognize these various functional subsystems of an overall system that maintains readiness to deal with a similar (but still hypothetical) threat situation from space. In the case of the Maui Optical Tracking Facility we see a descendent of the Distant Early Warning system of radars which was designed to detect and track all objects (natural or manmade) within its radar "aperture." Then there was the North American Aerospace Defense Command (NORAD) system whose function was to keep track of our friendly assets, as well as the enemy "objects." There was also a similar system to catalog and track the enemy's strategic naval forces, especially the submarines with nuclear payloads. And, there was the subsystem of tactical interceptors which operated under the control of NORAD, and which were ready to intercept any threats to the North American continent. One major advance in the present U.S. Air Force system is the advanced global positioning system that provides the ultimate "situational awareness" on any battlefield.

Nuclear detection satellites had their origins amidst the international agreements of the early 1960s whose aim was to ban nuclear tests in the atmosphere, under water, underground, and in space. To verify adherence to these agreements, the United States and the Soviet Union developed a satellite-based system to detect explosions in the upper atmosphere, in near-Earth space, behind the Moon, and in deep space. This task was made even more difficult by the natural background of solar radiation, cosmic rays, and the Van Allen radiation belts. The ultimate solution to the set of problems that are related to the main objective of detection of nuclear explosions in the trans-atmospheric region was a constellation of orbiting satellites which could detect an explosion as small as 10 kilotons, at distances of as much as 100 million miles from Earth. The satellites carried sensors to detect the X-rays, neutrons, and gamma rays that are emitted by the fireball associated with an explosion. During its period of operation, the satellite-based systems also have developed a considerable body of knowledge with respect to the detection of the signatures of the various types of radiation that are associated with nuclear explosions that is useful in military space operations.

The Vela series of satellites that were sent into orbit during the 1960s and 1970s is an example of how NASA and the U.S. military establishment find mutual benefit from each other's operations and development. Each Vela spacecraft is equipped with radiation detectors which were sensitive to X-ray and gamma-ray emissions. To complete the architecture of this specialized reconnaissance system, there were also a network of ground stations that were deployed to relay the satellite data to command and control functions on Earth. Although the primary function of these satellites, which were launched in pairs, was a military reconnaissance one, the Vela program made possible significant astronomical discoveries, including the detection of a powerful X-ray source outside the solar system, which could, in turn, affect military space operations in the solar systems in the future.

The Maui tracking system is routinely involved in numerous observation programs and, it has the capability for projecting lasers into the atmosphere as well. During its operational history, it has discovered several asteroids. It combines large-aperture tracking optics with other visible-light and infra-red sensors that enable the system to collect data on near-Earth and deep space objects in the solar system. The 3.67-meter telescope and advanced electro-optical system is the largest optical telescope that is designed for tracking satellites. It detects and tracks large numbers of objects at a time very accurately, but it is nevertheless agile enough to also track low-Earth orbiting satellites, as well as ballistic missiles. It employs sophisticated sensors that include an adaptive optics system, radiometer, spectrograph, and a long-range infrared imager, with which the telescope tracks manmade objects in deep space and performs space object identification data collection. One particular feature that makes the telescope so effective is its ability to remove the Earth's atmospheric distorting effects.

As can be seen from the above, one line of development has been the implicit buildup of military space "reconnaissance" capabilities by the United States, in tandem with the NASA systems. These originally consisted of a network of telescopes and communications receiver stations which were deployed around the surface of the Earth, and in space as well. These were a network of telescopes that could operate in many frequencies of the electromagnetic spectrum to detect, track and range electromagnetic emissions throughout the solar system. These "optical" sensors were either aimed toward the Earth, or were oriented outward to the outer solar system, depending on the origin of the potential threat radiation. Since the turn of the 21st century, the U.S. military establishment and the "spy agencies" also began to place their telescopes and other sensors. At the present time, these are aimed either at the surface of the Earth or the belt of artificial satellites itself. Among the most important of the space-based telescopes which has effectively become an "early warning" system is the group of infrared telescope platforms that were placed above the obscuring effects of the Earth's atmosphere by the National Reconnaissance Office, and those of the U.S. military which are designed to detect the plume that is associated with a launching of an ICBM.

On the other hand, the on-going developments of the space age has also brought changes in the perspective of the optical and other sensors that had previously been developed to study the skies. Now, the full array of sensors that utilized the electromagnetic spectrum was beginning to be carried aloft on air vehicles of all types. This trend has culminated in the array of the orbiting satellites that would turn their attention to the surface of the Earth. Meanwhile, other space vehicles, such as the space probes and space stations, would again turn their sights 180 degrees; now oriented towards the realm of space beyond the gravitational domain of the planet Earth.

The advent of the artificial satellites which were placed into orbit around the Earth also changed the perception of cameras and other sensors in a new and revolutionary way. For one thing, these orbiting reconnaissance platforms were now able to take their "pictures" at a much higher distance above the surface of the earth than had any previous high-flying spy planes. And, because they are controlled by computers at ground bases, they could adjust their focus in much the same way as the microscope. Their distance from their subjects on the ground also allowed them to take their photographs and other images in an extremely panoramic manner, which enabled their cameras to take in a much larger swath of the surface of the Earth than any other high-altitude imagery systems. Finally, the satellite-based photographic systems have been able to take in much broader fields (in terms of the electromagnetic spectrum) than the conventional cameras that operated within the visible light portion of the EM spectrum. Today, we have multi-imaging systems that utilize other bands of the spectrum, including infrared, ultraviolet, and radio, as well as the visible light portion. This enables the imagery analysis specialists to see their target environment in ways that overcome shadows and other distortions that are inherent in single-imaging systems.

Satellites which are placed into orbit over the earth are now able to record changing phenomena and then broadcast video images to receiving stations on the ground. These pixels may then be inputted into computers and converted to usable images. The latter information can then be analyzed by the ground personnel and/or uplinked to ultimate users such as military aviation platforms. By the same token, cameras and video devices can be placed on spacecraft or deep-space probes to record data in space, or on the surface of planets and other heavenly bodies. The data received can then be transmitted, in practically real time, to ground receiving stations. There they can be stored on magnetic storage devices and processed to enhance the visible images, which then can be made part of the total data base, or transmitted to other final users.

In more practical dimensions, the continuing development of multi-spectral telescopes, as well as that of the wide-scan, high-resolution imaging systems, which have been carried by the many series of flybys and orbiting missions into the planetary regions of the solar system. This has enabled humans to turn their attention to the development of practical space navigational charts for the solar system and the celestial bodies that inhabit it. After a half-century of space operations, therefore, NASA and the other national space agencies have built a significant database of imagery and data to construct useful space navigation charts that can be used by all space-farers. The centuries of work by astronomers of all types has also provided the cartographical expertise to generate these charts for any given space mission.

The Earth-bound geographer has become accustomed to thinking of one's point of origin and a target destination on the Earth as being immutable; that is to say, fixed on a particular juncture of latitude and longitude on the surface of the globe. The imaginary grid that we have superimposed over the surface of the Earth, and the surface itself, also seems to be fixed at short intervals of distance. However, as soon as humans have begun to operate at longer distances, beginning with the oceans and other large bodies of water, it has become apparent that the surface of the Earth is actually moving beneath us. This relative movement of the surface of the Earth beneath our flight paths in the atmosphere has become more apparent, especially when we fly over greater distances.

Thus, geographers and other Earth scientists have begun to take notice of the effects of this global movement on the trueness of our trajectories; one of these being the so-called Coriolis Effect. Now, as we travel in space we have learned that the apparent movement of the celestial sphere with respect to our spacecraft also produces a kind of deviation that is effectively similar to the Coriolis Effect on the Earth (another spinning planet). Another force that causes deviations and requires mid-course corrections in a flight path in space is gravity. On Earth or another planet, the main effect of gravitational pull is downwards; in space, however, gravitational effects on a spacecraft can be applied from any direction.

What we have come to realize, in effect, is that the planetary grid system of latitudes and longitudes is also becoming "relative," in the same way that an absolute location on a grid system is always relative to other such locations. Thus, we see that the grid system of latitude and longitude can be utilized anywhere in the solar system, and even the Milky Way galaxy, with some extensions, to determine and communicate absolute locations throughout the region as a whole, or with respect to planets and other celestial bodies as well. Once again, the geographer arrives at the realization that "scale" is an overall determining variable of absolute and relative locations. However, the plotting of any line between two points in this much larger celestial sphere invariably requires that more attention be paid to the deviations from "true course" that are caused by the movements throughout the solar system of the celestial bodies.

The nearest and most familiar of the celestial bodies in our solar region is the Moon. It has been the subject of study by humans, even before the time that the telescope's invention in the 17th century. And almost immediately thereafter, humans began the process of measuring and otherwise quantifying their observations of this body. Unlike our own planet Earth, the Moon could be comprehended (at least its bright side) in its entirety and thus allowed the construction of a global scheme for assigning locations on its surface. The next step in the study of the Moon would then be the development of an atlas of its features. The main part of the atlas that has been developed consists of a detailed map of the near-side of the Moon, which is subdivided into 76 sections. This a traditional map based on an orthographic projection. It depicts the Moon as it appears from Earth at zero (static) libration (oscillation). A network of lunar co-ordinates is then superimposed on all map sections. The parallels of latitude are drawn as segments of straight lines; the meridians of longitude as ellipses (incomplete depictions). North is always uppermost; and east is to the right. This is the same view as would be observed through a non-inverting telescope. This is the algorithm that has been followed in the mapping of all other bodies within and beyond our solar system. It is how humans, including military aerospace aviators, are able to determine the "absolute location" and the "relative location" of any phenomenon on the surface of the Moon, and any other sphere in the solar system.

This concept of "relative" latitude-longitude coordinate systems now becomes even more important in the context of space reconnaissance operations. More specifically, reconnaissance systems of all types now benefit from computer-controlled imagery-interpretation systems that are based on the surface of the Earth. But the orbiting reconnaissance platforms, the satellites, are now the "big picture" air reconnaissance technology of the 21st century. It provides the most comprehensive and, at the same time, the most detailed source of geographic data for use in military air operations (and many other human endeavors as well). Orbiting satellites (or platforms) also can be seen as "relay stations" in terms of air reconnaissance. They permit the study of celestial objects and radiation from above the atmosphere. They also allow observer stations on the surface of the earth to "see" portions of the electromagnetic spectrum before they are absorbed by the Earth's atmosphere, including visible light and some of the infrared radiation and radio waves. Thus, the instruments on the orbiting platforms in space open up all of the regions of the spectrum to observers on the surface of the Earth.

Another important element of the orbiting reconnaissance platforms is that they provide enhanced "loiter time" over a specific target. This refers to the amount of time which an attack aircraft can spend over the target and continue to provide close air support to the friendly troops on the ground. In the same way, the orbiting reconnaissance platforms, which often work in groups, are now able to "loiter," 24-hours, 7-days per week over any particular target area of a planet. However, the greatest significance of "loiter-time," in the present context, is that "real- time," continuous, digital data is now being relayed to computer-based intelligence systems on the ground. Thus, the communication or relay of observed data to imagery-processing operations, and the operational end-user, no longer requires the physical transportation of imagery film to the processor operations on the ground, for analysis and transformation into usable intelligence. This, in effect, compresses the input-process-output process into a nearly "speed of light" cycle for air operational intelligence and military air operations. It is not quite instantaneous, but it's pretty close to it. The output of this system is now available in "real time" to combat aviators. It is also inputted into onboard computer systems, processed, and then outputted onto visual displays or to the controls of the various airborne systems, including navigation, flight-control, and weapon systems.

The space-based reconnaissance function within the solar system is now being carried out by the flyby space vehicles and the longer-lasting orbiters that travel past and around the planets, moons, and asteroids of the solar region. These operate in coordination with an extensive network of telescopes that are deployed both on Earth and in orbit, around the Sun, the planets and other bodies. The space-based observatory is a manifestation of the fact that, even though technological advances have been made to make possible the construction of larger and more powerful telescope systems on the surface of the Earth, it is nevertheless the case that the only solution to some astronomical observational challenges is to make some of these observations from vantage points that are located above the atmosphere of the Earth. An example of these vantage points are the Lagrange points of the astronomer, which are points at which a small body, under the gravitational influence of two large ones, will remain approximately in suspended animation relative to them.

Planetary surface reconnaissance is another area where NASA and USAF operations dovetail. The Curiosity rover that is now carrying out "reconnaissance" operations on the surface of Mars is an example of what future military reconnaissance on the planets and other objects in the solar system might be like. Consider that a specialized camera records color images and video of the Martian terrain. Then, a hand-held imagery device provides Earth-bound scientists with close-up view of the minerals, textures, and structures of rocks on the Red Planet. A Mars Descent Imager (MARDI) then films the landing site that has been pre-selected to help mission planners choose an optimum path of exploration for the robotic scout. On the surface of Mars, an Article Particle X-Ray Spectrometer

(APXS) measures the chemical elements in rocks and soil; a ChemCan fires a laser beam of energy at rocks and soil and then analyzes the composition of the vaporized material; meanwhile, a Chemistry and Mineralogy Instrument (CheMin) identifies and measures the abundance of various minerals. Also, a sample analysis of Mars searches for compounds of carbon that are associated with life; a Radiation Assessment of Detector (RAD) identifies and measures all high-energy surface radiation; another device attempts to detect subsurface supplies of water ice and molecules of water in minerals; and a Rover Environmental Monitoring Station (REMS) provides daily and seasonal reports of weather. (Air & Space, August 2012). All of these "reconnaissance" functions are now taking place on Mars, as the rover vehicle, Curiosity, joins two other rovers and begins its exploration of the Martian surface. This would appear to be the epitome of "situational awareness" for the space commander who is planning an operation on the surface of any celestial body in the solar system.

Many of the systems that NASA has put into place to support the orbiting satellites and spacecraft, including their hardware and software subsystems, parallel the support functions that are so familiar to the U.S. Air Force maintenance specialists on Earth. There are also sophisticated research and development, logistics, propulsion, and other systems that we identify with the aircraft and ICBMs that were the mainstay of the Cold War air forces. In this book, I argue that virtually all of this knowledge and skill base which is being developed by NASA can be converted toward military space operations. Consider the following support functions that are being utilized by NASA to keep its Space Shuttle fleet in operation. The focus of this fleet of reusable spacecraft is the International Space Station. A model for the maintenance of such a space station, through the use of reusable space vehicles, has developed over the past two decades. As part of this maintenance and support system, astronauts have made repeated "house calls" to keep the ISS healthy, to upgrade its capabilities, and to extend its life span. These maintenance sorties have included many hours of extra-vehicular activity (space walks) to repair and upgrade the station. Also, during this period of time, astronauts have produced new generations of architecture and capability for both the ISS and the Space Shuttle system itself. Meanwhile, the International Space Station experience has furthered the technology and methodologies that will enhance future military space operations relative to military space bases in the solar system.

Similarly, in order to provide maintenance to the Hubble Space Telescope, the NASA Space Shuttles have been launched to provide maintenance and system upgrades in order to extend the service life of the telescopic system. The first of these "field maintenance" missions installed new gyros and transmitters. Space Shuttle missions, over the years, have installed new gyros, transmitter systems, as well as new memory for the Hubble Station's computer system. Later space shuttle missions replaced the Faint Object Cameras and the Advanced Camera for Surveys (ACS) that extended the HST's field of vision. Several iterations of space shuttle missions then continued to repair and augment systems on the Hubble Space Station, and extended its life span to at least 2014.

———

Another support function for any sort of aerospace operations is that of logistics. Theory and practice continue to drive system designs for operability, and for the management of the flow of materiel services and information that is needed throughout the space system lifecycles. Once again, future military space systems will be able to utilize the experience of NASA during the past half-century of space operations. It is interesting to observe the philosophy and operational strategy that has been followed to create, what amounts to, a template for any future military-oriented logistical system to support space operations. On the one hand, it is obvious to this observer that many aspects of the U.S.A.F. Military Aerospace Transportation system on Earth have been incorporated into the NASA space program. Thus, the main portion of the logistics system that has been developed by NASA has been devoted to the support for the Space Shuttle system (formally, the Space Transport System). The heart of the STS system has been the Space Shuttle spacecraft/space plane (platform), which has functioned and probably will continue to function in different guises – very much like the C-130 Hercules tactical transport platform on Earth. In other words, the Space Shuttle is a reliable and versatile workhorse for the transportation of materiel and humans from one point to another within the solar system.

It is interesting to note that the Space Shuttle, like all other space vehicles, operates on an orbital system of space highways and connecting trajectory segments. So far, its point of launch has always been some point on the surface of the Earth, which itself is revolving around the Sun. To get from that moving point of origin on the surface of the Earth to a moving destination point on another solar orbit, mission planers and engineers have to plot a trajectory (a sub-orbit) which will rendezvous precisely with the rendezvous supply destination point. To date, these destination points have been artificial satellites, including the Hubble Space Station, or space stations like the International Space Station or Skylab. On viewing this pattern of nodes and linkages, some geographers who specialize in transportation geography would probably recognize this as another opportunity to apply their transportation models that have developed on Earth.

The logistics personnel who operate the U.S.A.F. logistical system would also see many opportunities to expand their conceptual horizons by doing a comparison analysis of the Earth-based and space-oriented logistical systems. Imagine space-based caching system whose nodes include: an orbiting logistics depot; orbiting modular spacecraft; an orbiting logistics depot composed of clusters of supply units deployed to an orbit off Mars. Then think of prepackaged supply units; Individual Supply Unit (ISU) pods being transported from an orbital position to a planetary base. Such a space-oriented system is now being planned by Spacehab, Inc., a NASA contractor. Its main objective would be to transport large quantities of cargo to and from the ISS. Spacehab, Inc. already has processed more than 290 spacecraft in preparation for launchings. And it has developed modules and integrated cargo carriers (ICC) to maximize the efficiency of the payload loading process. It has developed an almost routine Earth-Mars logistical network with established optimum links. It provides services to near-Earth missions in low-Earth geosynchronous orbit, as well as lunar and highly-elliptical orbits (nasa.gov).

The first instance of aerial refueling reportedly occurred in 1921, when stunt pilot leapt from a wing of a Curtiss JN-4 Jenny with a gasoline can strapped to his back and poured five gallons of fuel into the Jenny's tank. By the time of the Cold War era, the U.S. Air Force was using airborne refueling techniques to extend the mission range of the heavy bombers, such as the B-47 and, later, the B-52. The first of the airborne refueling tankers that I came into contact with was the KC-97 and its follow-on, the KC-135, which had much greater capacity for fuel payload, and enhanced speed capabilities to enable it to keep up, as it were, with the jet aircraft that it was refueling in the air. The same imperative to extend the range of the spacecraft has produced another refueling tactic that has proven to be effective in a variety of NASA space missions. In this case, the refueling technology involves a battery-like fuel cell which carries hydrogen as the main fuel ingredient. This technology was used in the Apollo series of missions as a means of extending the life of each mission without the need for heavier boosters to propel the rocket during its initial launch phase, as it escaped the gravitational pull of the Earth. This refueling technology and technique that is used in space is most analogous to the air refueling operation in the terrestrial atmosphere which involves a flying fuel station and a mid-air refueling rendezvous.

Another approach to space-based refueling involves the use of the fuel cell. This refers to any device that converts the chemical energy from a fuel (e.g., hydrogen) to electricity by means of a chemical reaction with oxygen or some other oxidizing agent. Fuel cells differ from batteries in that they require a continuous source of fuel to operate. On the other hand, they can produce electricity continuously, as long as the input fuel is supplied. Another practical advantage of the hydrogen fuel cell that has been realized is that the emissions of the system can be reutilized and thus increase the total efficiency of a given fuel cell by as much as 85%. A practical example of how fuel cell technology might be used on permanent military space bases can be seen in Iceland, where one public hydrogen refueling station produces all the hydrogen that is needed to generate all the electricity and water for the community which it serves.

Among the space refueling strategies that are being discussed is the deployment of caches that are placed in orbit about a planet, or an asteroid, to allow spacecraft to rendezvous for refueling. Another approach that is being considered would be to launch a "tanker" spacecraft separately from the spacecraft that bears the main payload. Ultimately, the decision as to which strategy to follow probably will boil down to the classic fuel-load versus pay-load ratio equation that is faced in all such transportation situations. NASA engineers have concluded that one aspect of having propellant depots is in determining the total mass that must be lifted into orbit for a given mission by multiple launches, rather than one large-mass lifting into orbit. Given the fact that NASA routinely utilizes the multiple-launch approach, in conjunction with final assembly in orbit, with respect to the International Space Station, it would appear that the multiple-launch strategy can be economically feasible.

———

Since the turn of the 20th century and into the middle of the century, most military air operations have been occurring within the atmosphere of the Earth. This is the medium within which humans have learned about the need for sufficient thrust to achieve the required lift and to move through this ocean of air. We have also learned much about how the properties of air and the characteristics of the atmosphere interact with an artificial platform in this medium. Our technological solution to the thrust problem has been twofold: The propeller which operates like a screw the medium of the atmosphere; and the gas jet that takes in air, then compresses and forces air through a rear outlet. Thus, we have developed the movement of air and gases past propellers and through compressors to achieve the propulsion that is needed to enable our platforms to move about the medium in three dimensions.

Through trial and error, we have become acutely aware of the differing properties and characteristics of the horizontal bands of gaseous material that we call air. We know, for example, that as we ascend through the lowest band, which we call the troposphere, and into the upper bands, the mesosphere and the stratosphere, that temperature tends to decrease and that the atmospheric pressure tends to decrease. Also, we have learned from practical experience that our bodies (and the air-breathing propulsion systems that we have made) begin to suffer from oxygen deprivation, as we fly at higher levels of the atmosphere. Then, it seems that our bodies also have difficulty keeping our organs "together" in the low pressure environment of the upper atmosphere. The forces of gravity that tends to resist the lifting that we need to escape the surface of the Earth continues to exert a downward pull at all times as we fly in the atmospheric medium. On the other hand, we have discovered that any sudden moves within the three dimensions of flight cause the forces of gravity to exert stresses on both the aircraft and the human crew aboard it.

The clouds that humans used to look up to in poetic terms, or in the more utilitarian manner of the practitioner of agriculture or sea-fare, are now seen as potential hazards and obstacles to movement by our air planes. On the other hand, as on the surface of Earth, the precipitation "events" that derive from clouds, such as thunderstorm and the turbulence that is associated with them, are seen as a threat to atmospheric flight in man-made aircraft. Clouds are also seen as impediments to "visibility," both in terms of navigation by visual flight rules, and as geographic obfuscators of a potential target on the ground, or in the air. In the latter circumstance, the clouds can be as effective as mountains in impeding the view of oncoming surface-to-air and air-to-air missiles, for example – even with the assistance of the electromagnetic technology.

The air in the atmosphere, which we generally consider to be in a state of rest, is anything but that. It moves in the form of winds in three-dimensional angles, and affects our navigation and ballistics in this medium. The winds or moving air regions, or "fluxes," if you will, are a constant but internally dynamic factor that greatly affects military air operations, even more than gravitational forces perhaps. We also have found that heat and cold serves as a catalyst that causes the air to move in various ways and to change properties that are related to density of the molecules; all of which have a particular effect on military operations in the various horizontal zones of the atmosphere. Thus, we soon learned that like the weather, all atmospheric military operations are local and dynamic. The sum of all these insights about the conduct of military air operations in our atmosphere has also provided us with a template for carrying out such operations within the atmospheres of the other planets of our solar system in the future.

With the beginning the age of aviation at the turn of the 20th century, it soon became obvious that the obstacles that are due to the "relief" in the atmosphere are as "concrete" as any mountain or ridge; valley or hollow, on the land surface of the planet. The other "topographical" properties of the atmosphere that affect military air operations are those that are related to what is called "weather." Among these are the clouds, water vapor, particles of soil and ice, winds, atmospheric pressure, turbulence, electrical activity, and oxygen-content, to name a few of the most immediate factors. Cold, especially when it is accompanied by precipitation, in all its forms, can also prove to be as much an obstacle in the air as it is on the ground, or on the water. By the same token, the clouds of ash and sand which are produced by volcanoes and the desert lands – both of which are lifted and blown by the winds – can be as significant to human activities in the atmosphere as on the surface of the planet. On another dimension, gravity and drag can create the same negative effect on the movement of objects through the atmosphere as the flora on the land.

The milieu of the terrestrial atmosphere is somewhat similar to the oceans of the Earth. In both of these media, there is natural buoyancy which enables humans to ameliorate the pull of gravity from the center of the planet. Unlike the ocean, which provides natural buoyancy, in the atmosphere, humans have always had to employ technology in order keep their aircraft aloft and to maneuver in that three-dimensional medium. Therefore, the first requirement toward achieving even the most basic military air operations within the atmosphere has always been to design and construct an efficacious aerodynamic "plane" or "foil" to which a superstructure could be integrated. One of the first these aerodynamic platforms, the heavier-than-air flying machine which was developed by the Wright brothers in 1903, became the proto-type for many other aircraft that would be developed during the 20th and 21st centuries. The genius of the Wright brothers' achievement appears to have been the development of the internal-combustion engine as the source of the propulsion for their airplane, which was thereby able to overcome the negative effects of Earth's gravity to remain aloft.

As long as human flight remained in the relatively comfortable zone of altitude above the ground (somewhere below 12,000 feet) the human physiology and senses were quite adequate to the task. It was within the distance from the surface of the Earth, where the human eye could still maintain "situational awareness," and a modicum of muscle and coordination could exert control over the forces of gravity and the wind. However, when the envelope" of human flight was pushed upward in altitude, or vertically in terms of speed or distance, it became necessary to employ additional technological assistance to achieve the objectives of the sortie. Initially, most of the research and development was focused on developing a more powerful airframe and engine to achieve the altitudes and speeds that were desired. Thus, to be able to climb to (and operate at) altitudes above 12,000 feet or so, ever more powerful internal combustion engines would be needed to turn increasingly sophisticated propellers, in order to increase the thrust-to-weight ratio. This, in turn, would require more efficient fuels and aerodynamics to increase the fuel-to-thrust ratio. Then, to achieve greater velocities in order to overcome the obstacles of spatial distance, more powerful engines would have to be developed. And, at some point, the achievement of greater altitudes and speeds would impose greater stresses on the airframe, and it would have to be strengthened.

———

Ironically, however, the advances in aircraft performance tended to increase the environmental stresses and performance demands on the human crewmembers inside the airframes. The "technology" has been developed to enable humans to operate effectively in the face of the stresses imposed on the human physiology is now embodied in the specialized field of medical science known as aviation medicine in the atmosphere, and space medicine outside the atmosphere. The profile of the "stresses" of aerospace flight in the atmosphere includes, among others: extreme temperatures, low atmospheric pressure, noise, vibrations, oxygen deprivation, radiation, and the dual forces of acceleration and deceleration. More generally, the flight surgeon is always vigilant for any signs or symptoms, or even any predisposition of the body or mind to react negatively to the effects of these stresses. Thus, the overall evaluation process for maintaining one's "flight status" also included, in my own experience, semi-annual tests of physical conditioning. All flight personnel had to run a mile and a half in no more than 18 minutes, and they had to perform other feats of strength and endurance. These were all part of the rigorous cardiac stress tests that were administered by flight-surgeon personnel. Careful attention was also given to vision and auditory abilities, as well as a review of what was called "mental hygiene" to ascertain the rated personnel's holistic level of resistance to the combined stresses of "flying and fighting."

Another aspect of the technologies that were developed to enable humans to cope with the stresses of high-performance flying and fighting was a comprehensive program of scientific experimentation, technology, education and intensive training. Toward this end, sophisticated simulators were developed to provide controlled-exposure to the stresses caused by lowered oxygen levels and atmospheric pressures. Other simulators prepared the aviators to deal effectively with the various "g" forces (acceleration or deceleration of gravity); which derive from sudden and extreme movements within the atmosphere of the Earth. Even the experience of being catapulted from their platform by a rocket-powered emergency ejector system was presented in a simulated fashion. All this, and more, became necessary as the platforms which were being developed towards the end of WWII began to be jet and rocket-propelled. With this added thrust capabilities, the platforms were now being pushed to higher service altitudes, greater velocities, and the possibility of quicker, sudden movements in the space of the atmosphere.

A human's normal body temperature is about 98 degrees Fahrenheit. It also happens that liquid water will boil and vaporize at that same temperature, at approximately 60,000 feet, where the atmospheric pressure is significantly diminished, as compared to even 12,000 feet above sea level. At these high altitudes, without the assistance of the pressurized suit or capsule, human aviators suffer many kinds of damage to the circulatory system, the lungs, and then the brain – all due to oxygen starvation. Ultimately, death will ensue. So, when military aviators began to operate at those stratospheric altitudes, it was found that they needed to have some form of technology to prevent what is called, "ebullism," which is the formation of bubbles in the bodily fluids due to the extreme reduction in the surrounding pressure. One important example of these technologies was the pressurized flight suit, which was designed to keep body and mind together, as it were.

Actually, technology to overcome the hazards of high-altitude air operations began to be developed even before the advent of air operations. Humans had already discovered, long before the beginning of human flight, that the environment at higher elevations, as on mountains, can be as cold as it is at the higher latitudes of our planet. Also, humans have long known that once the body temperature drops below 90 degrees (F), the normal shivering reaction ceases and, at that point, emergency medical treatment is required. In the case of hypothermia due to exposure on the ground or on water – as in the case of a downed aviator – the situation can be made worse if there are any underlying medical conditions, such as cerebral-vascular disease or hypothyroidism, for example; another reason why preventative care by the flight surgeon is so important.

This limit to "naked" human activity in the upper atmosphere began to be empirically tested and scientifically analyzed almost as soon as the Second World War had ended in 1945. The imperative for this activity was the inexorable movement towards higher altitudes and greater speeds in military aviation. The B-29 Superfortress had been put into production as early as 1942; its design and construction was done according to specifications set by the U.S. Air Corps. It was to be the heaviest bomber to date, able to carry great loads of bombs, and designed to cruise above 30,000 feet of altitude. To carry out its mission, it became the first heavy bomber to have a pressurized cabin for the crewmembers. But many questions surrounding the effects of high-altitude aerospace operations still remained open.

A more intensive search for some answers to these questions began in the period following World War II, when Colonel Joe Kittinger flew in a balloon which ascended to 103,000 feet altitude above the Earth's atmosphere. At that point, Kittinger bailed out of the balloon, partly in order to gather empirical data on the effects of such an aerospace maneuver at such high altitudes, and to gather continuous data as he descended to the surface of the Earth. The purpose of his "sortie" was to gather data about the effects of altitude above 63,000 feet on the human body, on the way up and on the way down, as well. Utilizing only the existing balloon technology and a pressurized flight suit, Colonel Kittinger ascended to a region several thousand feet above the safety of the comfort of the protective blanket of air that makes human life possible.

Throughout the "Cold War," there was a further definition of this high-altitude operational region. In addition to the high-altitude balloons, this was accomplished through the use of specially-designed high-altitude aircraft, such as the U-2 high-altitude jet reconnaissance and research aircraft. Its successor platform, the SR-71 Blackbird, could fly even higher. This generation of aircraft had a service ceiling of approximately 70,000 feet. This was an altitude that was far above the cruising altitude of any commercial airliner; high enough to see the curvature of the Earth. The benefits of such high-altitude operations were quite significant. Greater amounts of in- situ data regarding the aerospace environment at the upper regions of the atmosphere were able to be captured for subsequent analysis and incorporation into the more ambitious manned space missions, for example. Furthermore, the high-altitude flights within the atmosphere also increased our knowledge and skill bases, which would contribute to the space-based "reconnaissance" missions that would be conducted by NASA that utilized the space probe and the planetary orbiting spacecraft. Among these, was the knowledge and experience that was gained from operation of high-altitude and strategic reconnaissance of the Earth's surface on a global basis. Included in the overall amount of valuable experience that has been gained is the series of advances in terms of the ability for wider spatial coverage and the development of higher-resolution imaging.

But flying at these heights, even within the atmosphere, imposed costs to human crewmembers. There was, for example, problem of low atmospheric pressure, which was the flip-side of the type of decompression problems which deep-sea divers have felt and reported when they ascend to the surface too quickly. At very high altitudes, a forced bailout or loss of cabin pressure can create a significant health issues. These include intense headaches, nausea, extreme fatigue, and even hallucination, which have come to be recognized as the syndrome of symptoms that is now known as "DCS" (decompression sickness). U-2 pilots have also reported symptoms such as severe "itchy dots" on their bodies. The best "fix" to this problem has proven to be the pressurization of the cockpit or cabin. In the case of the U-2 platforms, where the cockpit was only partially pressurized, another countermeasure was to have the pilot breathe pure oxygen for an hour or so before the takeoff, and to equip them with a kind of pressurized flight suit.

Like the "red dots," many symptoms of DCS do not manifest themselves until the pilot has landed. These can be as minor as a headache or fatigue, such as one might feel as the result of caffeine withdrawal. In any case, Air Force pilots are now required to maintain a vigorous exercise regime to help them to deal with the harmful retention of nitrogen gas, as the increasing of the heart rate helps the body to rid itself of it, much faster. So, the problem appeared to have been resolved decades ago. However, according to an article in the magazine, Air & Space of April/May 2012, when the U-2s began flying more and more sorties during the war in Afghanistan, the symptoms of DCS were coming back, with a more virulence, and one pilot was nearly lost due to this syndrome. After that harrowing incident, it was ascertained that the pilot had removed his pressure-sealed helmet, after he had vomited into it, so he could see. Finally, he became so disoriented that he could not fly the aircraft and essentially blacked-out for a few precious minutes. Somehow, he was able to land the U-2 safely, but he had to be dragged out of cockpit. After having been placed in a hyperbaric chamber for several days, he was ultimately cleared for flight again.

Now, since the beginning of space flight the flight surgeons and other medical experts have recognized that the human body is adversely affected by weightlessness, mainly due to the post-mission reports of the astronauts. With the launching of the Skylab space station in the 1970s, however, medical professionals, including medical doctors and human biologists were able to monitor the effects of weightlessness on the human body in great detail and over a lengthy period of time. Body fluid sampling was done throughout the course of one mission and the effects on the legs and torso were tested. The results of these tests have been valuable in the design and planning of future long-duration manned space missions to the International Space Station and, eventually, to Mars.

Ultimately, as will be seen later in this book, it has become almost axiomatic humans must create an "artificial atmosphere" whenever they venture beyond the comfort zone of our oxygen-rich comfort zone. By the same token, we will need to replicate the protective shield which we call the geomagnetosphere; the envelope of charged particles and gases that deflects and otherwise mitigates the harmful effects of solar and cosmic radiation to humans on Earth. These "artificial atmospheres" can take the form of a space suit or a spacecraft, or even an artificial magnetic field that is propagated around the spacecraft. The space suit, for example, is vital to the conduct of any "extravehicular" human operations in the vacuum of space. As has been found in the Mercury and Apollo series of space missions, as well as the operation of the International Space Station, it is imperative that astronauts be able to work outside the capsule, in space or on the surface of other celestial bodies in the solar system. The space suit provides vital oxygen and it maintains the pressure necessary to keep the astronaut's blood from boiling, as the molecules try to escape the bond of the human skin. But it also must have a system for removing excess products of respiration, such as carbon dioxide and water vapor. Also, in an environment where temperatures can quickly swing from +250 degrees (F) to -250 degrees (F), the suit must protect from these conditions of extreme heat and cold. Then, there is the need for protection against the ubiquitous radiation of charged particles, the solar wind, magnetospheres, and plasmas which negatively affect astronauts and spacecraft.

The radiation and heat shields of the spacecraft and the space suit of 2012 are not the same as those that were first developed in the early 1960s, when the United States and the Soviet Union began their contest for putting humans into Earth orbit, and then, to the Moon. The later iterations of design and construction have produced a much more flexible suit that can provide the protection the astronaut needs in the vacuum of space, where there is zero gravity, no atmospheric pressure, and no oxygen, as well as extremely low temperature. The latest generations of space suit is constructed to include three layers. The innermost layer is a tight-fitting suit which provides a measure of creature comfort in an environment of extreme cold. The intermediate layer is somewhat less flexible, but hardy, pressurized endothermic layer which provides the atmospheric-like environment that the astronaut requires, including the oxygen and pressure requirements. The outer layer consists of a tough material that is made of chrome threads, but it also has the flexibility of a rubber-like material that provides the astronaut with the flexibility and freedom of movement that resemble what would be experienced on the surface of the Earth. There are also special boots and gloves to complete the effect. This utilization of layered architecture in the development of spacecraft and equipment continues today.

The state-of-the-art generation of the spacesuit represents a continuous effort to develop a spacesuit that can be used virtually independently of the mother-ship's life-support systems. In reality, the only remaining connection to the mother ship is a slender strip of material, a tether that prevents the astronaut from slipping away into the void of space. In any case, the cumulative benefits of these advances in spacesuit engineering have provided astronauts with the ability to conduct construction and maintenance operations in space. An example, astronauts are, almost as a matter of routine, conducting maintenance and renovation work outside the International Space Station, with the maximum freedom of movement and from unduly strict time constraints. The key piece of engineering in this situation was the development of lightweight, integral backpack for the astronaut, which contains a canister of compressed oxygen, a battery unit for electrical power, and a system of air conditioning to provide the optimum mix of heat and cooling to keep the astronaut at peak efficiency during the external operation. The spacesuit, in effect, had become the astronaut's own spaceship – at least for a few hours at a time.

So, the Earth's atmosphere is the medium and the environmental system in which virtually all military aerospace operations have occurred throughout the 20th century and into the 21st century. Then, in the middle of the 20th century, the U.S. Air Force began expanding its operations into the near-Earth region of outer space that lies just outside, relatively speaking, the our atmosphere. This represents the beginnings of military space operations. And, as is the case with many adjacent regions, the boundary between the two theaters of military aerospace operations is a frontier zone, rather than a definitive line – hence the term, "aerospace operations." Military atmospheric operations have occurred for about a century. Through these years, there have been periods of relative peace, during which the nation-states have, in effect prepared for the next air war. One way to track these episodes of military air preparation and combat operations is to analyze the various threads of technology that are interwoven into the overall matrix.

———

Many of the strands of this matrix of technology have been developed simultaneously; others have been coursed in stages and phases. Today, military aerospace operations technology now includes "enhancers" to the human senses; these are essential because our own eyes and ears are simply inadequate to maintain "situational awareness" and ballistic intelligence in the atmosphere and in and other optical sensors to add range and acuity to our own eyes. Another set of artificial sensory enhancers that have devolved in military aerospace operations is related to the human ear. The answer to some of the problems of natural hearing in the noisy environment of military flying and fighting was the discovery that sounds could be transmitted as waves that could be directed between specialized transmitter and receiver devices, with a minimum of external "noise." The control of the "waves" of electrical energy in order to send sound messages between stations, in space, the atmosphere, on the ground, and even under water has created a revolution in military air operations. Another aspect of such remotely-controlled actions or events is how electromagnetic waves now can be harnessed to control robotic versions of the human system, anywhere in the solar system. So now an operator that is located anywhere in the solar system can cause actions to occur at other point in the medium.

Whether military aerospace operations occur within the Earth's atmosphere, or in outer space, the basic imperative for such operations remains essentially the same: to achieve control of the air or space domain – in the furtherance of the objectives that are assigned by national authority. So, it seems that control of the battle domain is the essential overall objective in any set of military operations. From this imperative flows the subset of military activities that are commonly referred to as air reconnaissance, air attack, air interdiction, and close support of ground military operations. But there are also several sets of activities that enable and sustain the primary set of military activities (or, operations) – all together, and working together – this set "support" activities keep the combat fleet "flying and fighting."

Using the Vietnam War event as a model in which both strategic and tactical military air operations, we see all the elements of air power occurring within the Earth's atmosphere. Even in the subset of military air operations that occurred in the Laotian Plainne des Jarres region of the event, we see examples of virtually all of these specialized activities. So, in a typical 24-hour cycle, there would be air reconnaissance sorties being conducted over the battlefield. Some of these would be longer-cycled, in which an aircraft platform would overfly large swaths of the battlefield surface and return the film to be processed and analyzed by the imagery specialists on the ground. Other reconnaissance sorties would be carried out by smaller platforms which would cover smaller areas and report their findings by radio to orbiting command and control aircraft, which would then relay the target information to the attack airplanes.

There is now appearing in the military-related literature the term "war fighter" as a blanket term for all the various types of "platforms" that carry out combat missions in a military aerospace region. In the 21st century, the distinctions between bombing and air attack or close support functions are likely to be blurred as platforms become more multi-faceted and self-sustaining. Even the concept of the fighter-interceptor platform, which traditionally has provided "force security" for other platforms, has been blended into the multi-function platforms of today, except in certain "special operations" environments which, in many ways, is a kind of throwback to the conventional operational environments of the 20th century. Another variation in atmospheric military operations is the rise of the "UAV" (Unmanned Aerial Vehicles), which are performing reconnaissance and attack functions in places like Afghanistan in the first years of the 21st century.

To summarize: in my experience, the raison d'etre for air power, whether in the Earth's atmosphere or in outer space, is to gain control of the battlefield space. That degree of air superiority means that the opposing air force is incapable of effective interference with one's air operations over a given battle-space. This condition of air supremacy has only been achieved in certain wars, such as the German invasion of Poland which began the Second World War, or in the First Gulf War of 1991. However, with the evolution of ground-based air defenses, such as was faced by U.S. Air Force and the U.S. Naval aircraft over the Haiphong-Hanoi battle space, the old equation of aircraft against aircraft became much more complicated than ever, due to through the evolution of the AAA (Anti-Aircraft Artillery) and SAM (Surface-to-Air Missile) weapon systems. More precisely, to the increase in precision that have resulted from advances that have been made in fire-control systems technologies, and the increasing lethality of air defense ordnance, has once again reset the balance between the air attack and the air defense sides of the equation.

Given the advances in nano-technology and computer-related electronics, it is unlikely that one side of a given military aerospace event will be able to achieve the level of space control that provides complete freedom of action and security of force; such as was achieved by the United States, for example, in the two Gulf Wars. The most realistic objective of military aerospace forces in terrestrial and space operations in the 21st century might be to strive for situational space control over a particular battle space. This would be manifested by a lower level of air space control, the condition of "air denial" or "air parity," which involves the ability of maintaining one's military air operations, even while conceding some of the control of space at certain times or places. Another aspect of maintaining space control is the ability to maintain force security, primarily through the operations of the air-to-air platforms that provide "force security" for the attack platforms and other force assets. This is very much like the scenario that is now evolving in military space operations; where advances in technology and lowering cost of entry to appropriate weapons systems are diminishing the level of "permissiveness" in the near-Earth orbital zone. In such a scenario, it appears to me that the nuclear-powered naval aircraft carrier, the nuclear-powered submarine, as well as the long-range stealth B-2 bomber of the 21st century, are the most promising models for platforms that can be utilized in space warfare.

One military strategy that might be embraced in space is the based on what I call the "Strategic Aerospace Command," or "SAC" model. This refers to the aerospace system that was developed by the United States to deal with the long-term "threat" environment that was created by the development of a strategic nuclear capability by the Soviet Union in the years following World War

II. This sustained nuclear standoff in which the adversaries possessed the capability to destroy each other and the rest of the globe is referred to as the Cold War.

The strategic paradigm that was adopted by the United States in response to this continuous existential threat was a deterrence and massive retaliation system that became known as the U.S. Air Force Strategic Aerospace Command (SAC). An interesting aspect of the overall "SAC" system, as I see it, is that it was effectively a collective security arrangement that was based on the unthinkable consequences of any attempt to significantly threaten the global peace. In other words, it appears that there was a kind of global peace on Earth, even though there were many relatively small skirmishes occurring within the substrata of the Cold War. That is to say, it was an era of "limited wars" that were designed to avoid a nuclear destruction of the human species.

The key concepts in the SAC strategy can be summed up in a few words: concentrated power, continuous planning, ongoing preparation and training, constant vigilance, and precise execution when needed. In other words, the SAC strategy presented a "realistic and viable threat" to any nation that would threaten the global peace. To accomplish this state of being, it became necessary to construct a physical infrastructure that would support all of the military systems (or security systems) that would enforce the peace.

The infrastructure for this SAC system was constructed along a region that lay between the 30th and 50th degree latitudes in the northern hemisphere, and spanning the entire line of longitude around the globe. The main nodes of this system were located at Offutt AFB, Nebraska (Command and Control Center) and Cheyenne Mountain in Colorado (Air Defense Center). Along the northern perimeter of this spatial system there was a deployed a series of early warning radar sites facing the North Pole, which was the horizon over which the main antagonists of the Cold War were most likely to travel toward their respective targets. The other nodes of the system were the many SAC bases on which platforms like the B-52 and the Minuteman II ICBM with their nuclear payloads were located. This sketchy outline of the SAC system possessed all of the elements that might be replicated, in some form, in the outer space sometime in the future.

Such a strategy reflects an appreciation of the fact that future military space operations will necessarily be long-term in duration because of the vast distances and time that will be involved. Such a strategy has to do with flexibility, particularly in terms of platform configuration, propulsion system design, and payload, to name certain aspects. The strategy of flexibility in space operations also calls for careful mission planning and design; no less complex than the planning of long-range bombing missions during the Cold War. Future space operations will require a strategy that can maintain forward momentum and continue the mission, even if one component runs into problems or delays.

One of the most important questions that likely will arise in such an event is whether to deploy manned or unmanned platforms, or a mix of the two. As was realized during the Cold War, there are comparative advantages and disadvantages inherent in both types of platform. A major consideration for human military operations in space is the replication of the Earth's atmospheric environment on military bases in space and within spacecraft. It begins with technologies and methods that are necessary to deal with the stresses on the human system, such as: extreme temperatures; low atmospheric pressure; cosmic and solar radiation; noise; vibrations; oxygen deprivation; and the strong forces of acceleration and deceleration. Then there are the disruptions to the human physiology and psyche caused by weightlessness in space. Boredom, disrupted sleep patterns, wasting of muscles and bones, and even psychological breakdowns are also problematic factors that have to be dealt with. From experience in space operations, we know that the there have been problems of the wasting of muscles and the problems related to osteoporoses; problems that are usually associated with aging on Earth. There also have been problems related to motion sickness in an environment without the customary boundaries of the Earth's atmosphere. Finally, there have been problems of fatigue and hunger, as well as sleepiness due to the absence of the Earth's day-and-night cycle. It is as though astronauts were put into solitary confinement and sensory deprivation in space. All of these are issues that have been encountered on Earth with long-term submarine operations and the lessons that have been learned from these experiences will prove useful to the human engineering part of any strategic military space system in the future.

The set of issues related to the isolation of the human in space is likely to be a major problem that has to be resolved if terrestrials are going to be able to live and work effectively in space, over extended periods of time. Mission reports and studies by NASA provide a preliminary understanding of the effect that isolation and sensory deprivation can have on the human system in space. For one thing, the sensory deprivation that occurs during an extended space flight can cause disturbances in perception to the human system, in extreme cases, even hallucinations. Another negative effect on human operation in space has to do with the disorganizing effect of sleep deprivation. This can cause degradation in the ability to carry out necessary operations, especially in case of emergency.

The matter of how to deal with the problems that are associated with long-term human operations in space is being dealt with empirically on the International Space Station (ISS). This mission, which has been ongoing for several years, is providing a continuous and rigorous stream of knowledge as to how humans can live and function in space for extended periods of time and, perhaps, even on a permanent basis. This system of modules has a pressurized component and environmental systems for the humans. It also has its own power plant, and supplies for an extended period of time. However, whatever it cannot contain integrally, it receives from a steady stream of shuttle spacecraft from Earth. In many ways, the ISS experience is serving as a proto-type, not just for orbiting spacecraft, but for planetary bases throughout the solar system – including military space bases.

There also are inherent advantages and disadvantages associated with unmanned platforms. Mainly, these revolve around matters of the location of the decision-making with respect to trajectory adjustments and when to make a final attack on a target. Generally speaking, an unmanned platform is relatively immune to problems of life-support and, therefore, can operate in the most hazardous environments. On the other hand, guidance and control, as well as the decision to fire the weapon, becomes more tenuous and problematic with distance from the launch point. It appears, however, that these are questions that can be resolved by continuing advances in robotic and telemetry technologies, among others.

Actually, there already is a considerable reservoir of knowledge and experience that has been garnered about unarmed platforms – both within the Earth's atmosphere and in outer space. Thus, various artificial satellites, space vehicles, space probes, and space rovers are now performing many of the tasks that have been done by manned systems. As an instance, these unmanned platforms are now providing valuable intelligence to the U.S. Space Command functions on the surface of the Earth. Thus, for example, the array of Earth-orbiting satellites provide vital information about nuclear explosions and rocket launches from anywhere on the surface of the Earth. This information is being provided to appropriate military establishments of the U.S. and other nations, which enable them to monitor and possibly respond to any military threats emanating from any point within the geospace domain. Thus, commanders in such a future area of operations will likely want to send "scouts" into space to detect and report on other "threats" from other sectors of space. A likely model for these military space scouts is the series of space probes that are being deployed by NASA today. These are defined as unmanned spacecraft that are equipped with exploratory devices such as mass spectrometers and other observational technologies. These specialized sensors initially have revealed much about the surface and atmosphere of the planets and other celestial bodies of the solar system. These probes generally study the atmospheres and surfaces of the planets, moons, and other phenomena, such as the asteroid belt and the comets of the Kuiper zone.

———

In any case, as soon as the SAC paradigm is superimposed on the region of orbiting artificial satellites, a whole range of military "support missions" becomes necessary. What happens is that the orbiting artificial satellites – whether non-military or military – now become "assets" to be protected. Once military (and non-military) assets have been deployed in space, they also will require some sort of asset security and maintenance. In the case of the SAC system on Earth, this requirement resulted in a network of land-based airbases, such as the U.S. Air Force base at Grand Forks, North Dakota. In the same way, a system of space bases will have to be constructed to provide security for the artificial satellite system. Thus, it is likely that a command and control base – like SAC headquarters at Offutt AFB, Nebraska will be established, either on the surface of the Earth or on some other body in the geospace region. A model for such a military facility might be the NASA "command and control center" in Houston, Texas. And, continuing with the SAC model, there would also be several operational military space bases that would be equipped with "bomber" spacecraft which would have the capability to operate independently, in the manner of the atmospheric B-2 bomber today. Assuming that the geospace region will be the main area of military operations, some of these space "security" bases could be located on the surface of the Earth. Others could be deployed on the Moon or some asteroid. This is by no means meant to be a comprehensive description of what such a space security force might look like. Rather, it is a way of illustrating how the strategic objective of maintaining "peace through power" might be translated into a functioning military force in the solar system.

There already is a natural geographic "security system" operating in the geospace region. It consists of the geomagnetosphere, which protects the atmosphere and the surface of the Earth against most of the hazardous solar and cosmic radiation that reaches our planet. Therefore, we can say that in a physical and metaphorical sense, the security system for such military space assets begins with the magnetosphere and the atmosphere, which guard against cosmic and solar electromagnetic intrusions, disruptions, and even existential threats. But now, with the construction of the artificial satellite belt (ASB), security of the geospace domain must also deal with "cultural" threats from the Earth's surface against the ASB, or from within the ASB itself. The situation is similar to the one that is faced by the U.S. naval forces on Earth. They too must provide security for regions of the oceans, including the established shipping lanes (which might be seen as being analogous to the artificial satellite belt). The U.S. Navy responds to this kind of threat environment by essentially maintaining a "force presence" which can respond to any particular emergency within its area of security. In the case of the geospace domain, it is likely that the overall SAC model will be combined with the U.S. Navy "force presence" model to produce an overall strategy for maintaining the security of space domains, beginning with the geospace domain.

Given the above, within the geospace, there will be deployed military bases, both on the surface of the Earth and in outer space, whose main mission will be to maintain the peace and security within that military space domain. The structure of the geospace security forces may also include elements whose main mission will be analogous to naval "fleet security" on Earth. Specific platforms will be designed to operate in all areas of the geospace, including such systems as surface-based spacecraft (including on Earth and other bodies) and orbiting spacecraft that will be deployed to so as to provide quick-reaction response to any threat event.

It is at least arguable that the development of such a geospace military force will still require a system of supporting elements in space. Thus, we can expect that the artificial satellites which perform reconnaissance, navigational guidance and communications facilitation today will, at some point, also provide the geospace SAC system with another layer of protection against natural and cultural threats in the solar system. At the present time, there appear to be two platforms that might perform this function, the space shuttle and the emerging space attack vehicles. One indication of such platforms' ability to provide "Mig-Cap" (Combat Air Patrol) protection of assets in space is the fact that the Soviet Union/Russians have suspected that the space shuttles could operate as a sort of Stuka fighter-bomber from orbit into the Earth's atmosphere. It all depends on how these platforms are configured.

Given a scenario in which artificial satellites (including manned space stations) are deployed as military assets in space, there will almost surely appear the development of space platforms which will operate under the "sortie" model. That is to say, there will be space vehicles that will launch from a base on a planetary surface or a "mother ship" in orbit, complete an assigned mission, and then return to base. An example of such a platform is the NASA Space Shuttle. These are reusable platforms that are rocket-launched from the surface of the Earth, enter in an orbit to accomplish a mission, then re-enter a trajectory that will lead them to a re-entry point above the Earth's atmosphere and, finally, glide back to a base.

The application of the SAC model of military space operations will have to result in the development of an equally powerful and complex system within the geospace domain. That is to say, it will be able to operate on the same level of power and complexity as the solar system itself; it must match the solar system environment in terms of the parameters of space-time in space. This will require a shift in conceptual thinking and application from the macro to the nano or the micro scale of reality. The nominal speed of light will be one of these parameters of the military space system of the future. At the present time, the speed of light is understood to dominate all communications in space. As a practical matter, the dimensions of any military space operations in the solar systems will be defined by the time it takes for an electromagnetic signal to move from a point of origin to a reception point.

In order for these various space vehicles and machines to be used to their optimum effect, they will have to be integrated into such an overall system. That is to say, they must be able to interact and interrelate dynamically under a central command and control entity which is calibrated at the enhanced velocities of the space environment. Theoretically, the speed of light is the basic parameter of speed in the Universe, but there are other practical parameters that have been discerned as a result of human space operations. Orbiting satellites and other spacecraft, for example, generally travel at a rate of about 20,000 miles per hour in the solar system. The outlines of such a transport problem in space had been discussed and planned as early as 1952 by Werner von Braun and others, at least with respect to the propulsion systems of spacecraft. Later, the project was taken up NASA after it was given its charter for space exploration in 1958. The efforts in this area have been focused on the development more powerful propulsion fuels, as well as that of lighter materials to improve the thrust/load ratio. The parameters of the problem also are defined by the transport of natural objects in the solar system. Thus, a Near Earth Object (NEO) is an asteroid or comet with an orbit that is close to that of Earth; in which perihelion (or nearest point to the Sun) is less than 1.3 astronomical units (1.3 times the distance from the Earth to the Sun). Potentially hazardous NEOs are 500 ft. or so in diameter and follow orbital paths that come within 4.65 million miles of Earth.

———

So, it is clear that the strategies for defending against natural objects – asteroids and comets – are also valid for defending against man-made objects, such as missiles. Therefore, in this book I draw an abstract notion of near-Earth objects to include both natural and artificial phenomena. Near-Earth object impact avoidance comprises a number of methods by which near-Earth objects could be diverted, thus preventing potentially catastrophic impact events. A sufficiently large impact would likely cause massive tsunamis or (by placing large quantities of dust into the stratosphere, blocking sunlight) an "impact winter," or both. The "large-impact" object could be either a natural body, such as an asteroid or comet, or it could be a man-made object which would combine kinetic and explosive energy to create the catastrophic impact event. Thus, I will use the asteroid impact event as the conceptual vehicle for the analysis of these objects in terms of military geographic terms.

Although they are considered to be a statistically rare, low-probability event, asteroid impacts are seen as potential global catastrophes, nevertheless. This is becoming more the case as humans continue to extend their activities into the areas of the solar system that lie outside our planet Earth. Now, with more than 1,000 asteroids currently listed as potentially hazardous to Earth and with about 100 or more new potential impacts currently flagged each month by the NASA Jet Propulsion Laboratory's automated collision-monitoring system, the threat is being taken increasingly seriously by governments and space agencies. According to NASA, as of November, 2011, over 8,000 NEOs have been discovered, of which 830 are asteroids with a diameter of approximately 1 mile or larger.

The fact is, however, that there are almost surely many more natural objects within the geospace domain are still to be discovered. While roughly 94% of the largest NEOs are believed to have been located, there are some 60% still not detected in 300-meter size. Detection numbers are even lower for smaller NEOs between 100 and 300 meters in diameter with only 10% of the estimated population accounted for, while only about 1% of the smallest ones – like the approximately 50-meter asteroid that burst over Tunguska, Siberia, in 1908 – have been discovered. When it comes to objects that can do serious damage, it is obvious that our existing detection and survey systems, especially radars, are in need of upgrading and enhancement. Options include new land-based telescope projects like the Atlas (Asteroid Terrestrial-impact Last Alert System) and Large Synoptic Survey Telescope as well as space-based systems. These could include hosted payload-type concepts in which a staring array would be mounted on the "backside of a commercial payload," scanning as it orbits the Earth. Such schemes are less capable than a dedicated survey telescope, but much more affordable.

Countermeasures that have been proposed to deflect an asteroid or comet include the use of focused solar energy onto the surface of the incoming object in order to create thrust from the resulting vaporization in order to either speed up or slow down, or otherwise perturb the original trajectory of the object. This would not be a single "zapping" of the body, but rather a cumulative, longer-term process that take place over the course of months or years. Other such strategies to cause perturbation in the trajectory of the object include the use of: nuclear weapons, kinetic impact weapons, ion beam "shepherds," and even a conventional rocket motor. These methods and technologies, it should be noted, could also be used in military warfare in space, against a hostile man-made object, like a space station or other spacecraft.

Near-Earth object avoidance also comprises a number of other methods for detecting these objects at distances far enough where they might be somehow diverted. Among these are: (1) the use of nuclear weapons to either destroy the body or, at least to divert it from its initial trajectory; (2) the use of kinetic impact to divert the body and; (3) the use of ion beams, to name but a few. One thing all these proposed techniques have in common is that they propose to take a natural force and to convert it to purposes that help ensure the survival of the human species (or a subset of it), in one way or another. Thus, we see the natural force of the fission or fusion of the atom being used to divert a threatening body of outer space, rather than the destructive purposes for which it was first harnessed by humans. Again, there is the utilization of electromagnetic radiation as possibly being the archetype of weaponry and tactics in space. This is a major theme that I will stress in various parts of this book; which is the utilization of natural "hazards" for intelligent military purposes.

Some military thinkers have concluded that military strategy in the space age can no longer simply defined as the science of military victory. In the first place, as humans venture out beyond the comfort zone of our atmosphere and geomagnetic field, a totally different set of natural and cultural circumstances argue against the old terrestrial model of warfare between subsets of the human species, in which the territorial imperative, of one kind or another, induces military conflict.

Instead, it is argued that military strategy is now equally, if not more, the art of coercion, of intimidation and deterrence against any element that would threaten the order and wellbeing of the human species. Thus, warfare in space will be bound by the same motivations and constraints, mostly of the political nature, as was the case during the bipolar Cold War era. However, the art of deterrence and coercion, as well as accommodation, will now be practiced by a greater variety of the participants, not all of which will be utilizing nuclear weaponry to achieve their aims. Rather, the "weaponry" will likely occur along a wide spectrum of tools, from "terrorism" to the use of the so-called "weapons of mass destruction" on Earth, to the use of cyberwarfare in space. What we are seeing develop in the space age is the combining of electromagnetic radiation with computer technology in a seemingly infinite variations. The other technological trend that will "democratize" space conflict has to do with the military technologies that will flow from nanotechnology, which will effectively bring down the cost of operating in space generally.

———

It can be reasonably said the present military state in space is one of "watchfulness and preparedness." Some have referred to it as the beginning phase of a new kind of "cold war," with similar dynamics as the Cold War that occurred in the 20th century. As with the first such prolonged period of global military conflict that existed between the United States and the Soviet Union, this "cold war" that is occurring in geospace is began with the initial construction of detection, early warning, and tracking systems which are orbiting the Earth. These are hallmarks of the state of "watchfulness" that characterizes this latter iteration of cold war in the 21st century. One dimension of this global conflict that distinguishes it from the earlier cold war is the addition of the outer space threat environment. Thus, whereas the earlier period of "controlled readiness" for explicit warfare was confined to the stratosphere of the Earth, since 1957, the new threat environment has been expanded to include altitudes of at least 22,000 miles above the Earth. Indeed, the center of gravity of this cold war can be said to have shifted upwards to about the 200-mile altitudinal zone of the low-orbiting artificial satellites. Also, as the boundaries of this threat environment have been shifted outwardly, the complexity of the natural and man-made systems has also increased.

———

The dictum: "know your enemy..." is perhaps the best short-hand term for the military aerospace operations that are usually called "intelligence, surveillance and reconnaissance." In the end, regardless of time and place, these functions are performed by some kind of platform. It may be in the form of an air balloon, aircraft, spacecraft or an orbiting artificial satellite. Or, the "platform" may be a telescope or other sensor system that is deployed on the surface of some planetary body within the solar system. Ultimately, all these ISR systems are designed to perform more or less the same general function: to maintain situational awareness of the relevant military region or space domain. One can see that the United States military already has been developing the basic infrastructure for carrying out these situational awareness objectives. These systems are now in place and are working to provide the early-warning and surveillance intelligence that is needed in an environment that is presenting increasing levels of military and non-military activities within the geospace domain.

One manifestation of this effort has been the creation of a practical system for scanning the skies for possible threats from objects that might be coming down from outer space and into the Earth's atmosphere. The United States Space Surveillance Network is an example of such a system. Its mission is to detect and track artificial objects that are orbiting the Earth. These include active or inactive satellites, spent rocket bodies, or fragmentation debris from previous space missions. But the system's sensors and trackers also maintain a detection-and-tracking vigil on natural objects, such as asteroids, comets, and their meteorites. In either case, when a possible threatening or unknown object appears on their scopes, the ground operatives pass notifications on to the elements of the military that can respond as necessary to deal with the event.

The following narrative summary of the tasks that have been assigned to this early-warning and "space traffic control" system provides specific examples of the "job description" of this system. One such task is to predict the most likely location and time when a decaying space object will re-enter the Earth's atmosphere. Another is to maintain surveillance over all man-made objects whose trajectory or orbit characteristics indicate they might pose a threat to our national interests; a corollary of this is the responsibility to prevent false alarms from spurious missile threats that might trigger a response from our ground-based (or possibly space-based) weapon systems that are designed to operate in much the same fashion as NORAD (North American Aerospace Defense) did during the Cold War. Then, there is the mandate to maintain a catalog or database record for every country (or agency) that owns a re-entering space object; and to inform NASA whether or not objects may interfere with satellites and ISS orbits.

The "tip of the spear" in military space operations of the future will probably involve some aspects of interdiction operations. It is possible that the growing dependence on orbiting satellites will produce the conditions for similar military operations within the region of these artificial near-Earth objects. A likely reason for interdiction operations in this space domain will involve the imperative to protect against one's own lines of communications, or to maintain the security of some military or non-military asset system in space, such as the artificial satellites that nation-states, and now private enterprises, are placing in orbit about the Earth.

Interestingly, another term for interdiction in military space terms might be a "rendezvous" with a certain point along an orbit. The most well-known of this example of this type of maneuver in space is the docking operations of the Space Shuttle spacecraft with the International Space Station. This is has become a rather common event in the 21st century, as the NASA Space Shuttles continue to provide logistical and maintenance support to the ISS. And, more recently, a private corporation has successfully docked its own shuttle platform to this multi-national space station as well. The various technological and operational challenges that are facing such rendezvous operations in space also bear a resemblance to those that were faced in carrying out long-range interdiction of naval vessels on the high seas by enemy submarines during World War II.

However, it was the naval carrier that ultimately became the primary system for carrying out interdiction missions on the high seas. The main reason was that the carrier could project much greater military power at greater ranges than the submarine. It could launch attack aircraft to deliver a concerted attack at 200 miles or more from the interdiction point, whereas submarines had to approach their target at distances of a few hundred feet. The foremost tactical question with respect to interdiction operations during the early years of the 20th century was whether aircraft could deliver enough destructive power to supersede the submarine. By the end of the Second World War, it was a moot point; the naval carrier proved to be the most effective military system for carrying out interdiction operations at sea. This, in turn, was due to such advances as the development of engines that could carry adequate payloads, dive-bombers and torpedo-plane designs had matured, carrier-arresting gear, and associated flight-deck handling facilities were up to the tasks, and proficient strike tactics had been well practiced.

So now, in the 21st century, there is another trend occurring in space which will affect military space operations in the coming years. In the current era, where the move is toward the development of lighter and faster military spacecraft systems, the imperative to maintain a forward deployment posture, along with a visible preparation for conflict of many kinds is paramount. All the U.S. military services demonstrate an understanding of this fact, and are taking the steps to develop the force structure and systems that are needed to carry out their derived missions. Perhaps not so surprising, the U.S. Navy, with its nuclear-powered aircraft carrier, appears to have the most experience with long-range projection of massive power in a three-dimensional medium, whether in the Earth's atmosphere or in outer space. Its doctrinal literature gives recognition to the fact that the sea battle space is no longer a self-contained battlefield in which adversarial sea platforms fight each other until one side submits in a definitive manner. Instead, I see references to the ocean in the same vein as the terrestrial atmosphere and outer space might be seen by aerospace forces. Thus, today, the sea, the atmosphere, the planetary gaseous envelopes, and the inter-planetary space, are all media within which armed conflict may occur. This reference to the "medium" for military operations reminds me of the abstract plane of the ideal "region" of the geographer. In the real-world, however, the conflict medium is a three-dimensional space that is organized by humans into a distributed array of nodes that are connected by perceived linkages. In the language of the naval planners, the main objective of the [space] force is to protect the freedom of movement via these linkages.

Based on the history of NASA space missions that have taken place during the past half-century, it would seem that "interdiction" in the context of outer space basically involves a series of orbital maneuvers. One which has become almost common-place is the rendezvous maneuvers between the International Space Station and the NASA Space Shuttle spacecraft. But there are many other examples, such as the "interdiction" of an ascending landing capsule which is docking with a command module, as in the case of the Apollo lunar landing missions. Indeed, every mission that seeks to gain access to an orbital path around another planet or body in the solar system can be seen as an interdiction. These have all been examples of peaceful interdictions; but from these, one can extrapolate what military interdiction operations in space might be like.

So, when spacecraft are equipped with weaponized payloads, then they will be capable of the classic "interception" of enemy spacecraft, or the "interdiction" of the enemy's lines of communication in space. Just imagine what it would be like if spacecraft like the Space Shuttle were to be armed with kinetic or directed energy weapons. Now we would have an effective military space interceptor or the equivalent of an air-breathing attack aircraft on the surface of the Earth. Of course the space interceptor or line-of-communication attack function could also be performed by a "ground-based" weapon system that would be located on the surface of a planet or other body, such as an asteroid, for example. In the latter scenario, the rovers which have been so useful in planetary exploration could easily be "weaponized" with surface-to-space missiles and, thereby, become surface-based defensive systems. Once again we see that "civilian" space systems can easily be converted to military space systems.

———

The first of the major operational objectives of a hypothetical military space force likely would be the gathering of as much intelligence about the potential areas of operation in the solar system. Such "military exploration" would resemble the NASA space exploration programs of the past 50 years. In a sense, the early strategic military reconnaissance activities in space might be likened to the Lewis and Clark mission; which set about to explore and catalogue the geography of the new territories that the young United States had purchased from Napoleon at the turn of the 19th century. Thus, the Lewis and Clark mission had as its main objective to make empirical and careful scientific observations of the vast new territories in North America that had been acquired as the result of the so-called Louisiana Purchase. This acquisition of territory essentially doubled the lands that were under the sovereign control of the young United States of America. The main architect of the deal had been President Thomas Jefferson, who promptly requested an appropriation of $2500 from Congress to finance the venture. Upon receiving approval from Congress, Jefferson then assigned to Meriwether Clark (his secretary) the task of organizing, planning, and putting together the needed resources of men, guns, boats, and other equipment and supplies for the planned mission. This was to be a mission of "space exploration" across a portion of the surface of the Earth that was hardly known to anyone but the native Americans who lived on it, and a relatively few European fur traders. Incidentally, both of the latter would prove to be vitally instrumental in the ultimate success of this mission.

Somewhat like the case with our solar system, detailed maps and charts of this vast unknown territory were practically non-existent, except for the general "maps" that had been created by fur traders. Therefore, one of the most important objectives of the Lewis & Clark mission was to fill in the blanks by careful observation, recording, and cataloging as much of the visible geography through which the explorers would pass. The mapping of the Missouri River and its tributaries as well as the passage to the Pacific Ocean, was of primary concern; it was foreseen by President Jefferson that it would be the main passage route for new settlers of those new lands. The search for relatively easy passes through the various mountain chains that were thought to block the ultimate passage to the Pacific Ocean was also one the main objectives of the expedition.

This narrative of the "Corps of Discovery" exploration mission contains many analogies to the manned space exploration missions of the 20th century. The science and technology of the Lewis & Clark mission may have been low-tech by comparison, but the essential elements of this 19th century operation are very much like those of the space exploration missions of the 20th century. Both enterprises involved a long early phase of organizing, training, and equipment prior to the launching of the operation. Both missions employed the latest technologies of the time, and they borrowed some innovations from military institutions and academic centers, which they then adapted to the needs of their respective operations. Two examples of the cutting-edge innovations that were developed by Lewis & Clark included the standardization of their weaponry and ordnance, and the production of gun locks with strict tolerances, so that they could be used interchangeably in the field. The other was the construction of a metal-frame boat, covered with animal skins, which could be disassembled and reassembled as needed. Two other things that the Lewis & Clark expedition did (which are now being planned in modern space operations) was to cache equipment, parts, weapons, and food supplies along their planned return route, so as to obviate the need for carrying all their resources with them.

In contrast to the Lewis and Clark mandate, the specific strategic mission that was assigned to NASA by President J.F. Kennedy and the Congress, was essentially to do what was necessary to get an American astronaut on the Moon by no later than 1969. Despite the fact that NASA was explicitly defined as a civilian government agency, whose mission was to conduct peaceful exploration of the solar system, the methods and organizational concepts that were developed contained a great deal of the Department of Defense "DNA." As an example, the rocket-launching capabilities were transferred in lot from the U.S. Army. Moreover, the first astronauts had either been military pilots or test pilots for the military-industrial complex prior to their assignment to NASA. For this reason, it is wholly logical to view the exploration of space in terms of military aerospace operations. In this conceptual context, it can be said that the first steps in the process of "strategic reconnaissance" of the potential military space region have been ongoing since the early 1960s.

The initial phase of the strategic reconnaissance of outer space as a potential military space region can be said to have been done by surface-based telescopes, as opposed to the airborne platforms which were the mainstay of such reconnaissance of the Earth. The reason for this configuration of strategic surveillance and intelligence gathering in the 20th century was due to the fact that all the potential targets and air defenses were on the surface of our planet. Now, however, in potential military space operations, the target complexes can be deployed anywhere in the area that lies between the surface of the Earth and its atmosphere, into the transition zone between the terrestrial atmosphere and near-space, and also in low and high Earth-orbit zones in space. Even now, a vital component of the strategic reconnaissance system is the network of Earth-based telescopic systems, primarily because of the power and stability that very large platforms, and because they can be capture extremely wide scans and very-high resolution power. However, with the advent of sustained operational space capabilities that are provided by orbiting platforms, we now have the ability to deploy telescopes and other sensors in space itself.

If we can define space exploration as an exercise in observation and map making, then it appears that this type of human activity began at least some thousands of years ago. Unfortunately, without the technology to view the skies in any kind of detail or resolution, much of the "map" that was drawn relied on myths and legends to fill in the areas that could not be comprehended by the human eye and some logical deductions. However, by the 19th century, the advances in optics and the telescope began to provide more detailed information about what was out there in the skies above us. Since the middle of the 20th century, space-flight has made possible more scientific exploration of the solar system, and beyond. Earthlings now have a greater (and more empirical) understanding of the many objects and other geographic phenomena in space. Now that humans have actually gone into space, either via manned flights or unmanned probes, they have also gained a much better perspective of the geographic phenomena on the surface of the Earth, as well as the other planets of the solar system. And, we are gaining a deeper understanding of the galaxies and star systems in our corner of the Universe. All of these activities: discovery, scientific understanding, and the application of this understanding to serve human purposes, are elements of space exploration. Indeed, this represents the dawning of the end of much the romanticism with respect to outer space.

Within the context of the United States Space Exploration Program, NASA has been the primary agency of the government which has been tasked with the responsibility to carry out space exploration activities. Thus, since 1959, they have conducted a total of more than 200 manned missions and a score of unmanned missions to undertake the exploration of the solar system, and to conduct science testing, both on the surface of the Earth and in space itself. Some of these have been "test flight" type of operations to develop the human and non-human systems that are needed to carry out the overall mission objective that has been assigned to NASA and other agencies. Among the more prominent of these have been the Mercury sorties which were flown in support of the ultimate lunar-landing objectives of the Apollo Program during the 1960s. These were designed to answer questions about a human's ability to live and work in the yet "alien" environment of outer space, which lies beyond the comfort zone of the terrestrial atmosphere.

During the first sixty years of space exploration and in situ analysis, the high cost of such operations has made it necessary for national governments to take the lead in that endeavor. The private sector has now entered the arena in the past few years, mainly by being involved in the placing of satellites in orbit and, now, even launching their own rockets and rendezvousing with the International Space Station. Now, the exploration and scientific study of the solar system has begun to form "clusters" of scientific studies, technological development, and research and development. Some of these make up what might be called the "NASA-Industrial-Academic" complex. However, it is mainly the national governments that are willing to take on the overall risks involved in space exploration as of now. So, the space exploration by the United States NASA and other national space agencies has taken on a certain maturity that usually marks such massive and long-term enterprises. The first indication of this was the conscious decision of NASA to pursue more scientific studies in its missions, as opposed to simple discovery of new worlds and the development of technologies and maneuvers in the pursuit of its objectives. Now, in 2012 there is more serious discussion of conserving money and budgeting for the most "bang for the buck." Mars is still the most favored object for study (see www.sciencemag.org, July 27, 2012), but other "flagship missions" are also competing for attention and resources. Jupiter's moon Europa has many supporters in the planetary science committee. So too is there strong consideration of Uranus orbiter mission.

One of the first missions of the NASA space exploration program to be devoted to explicit scientific studies was the twin Voyager 1 and 2 missions, both of which were launched in 1977. They set out to exploring areas of the solar system beyond Mars, which no other humans have ventured to do before. This twin space probe mission is now in its 34th year (as of 2012) and has gone beyond Pluto. Now in the second phase of its mission, and now called the Voyager Interstellar Mission, it will go beyond the outer limits of the solar system, to the very ends of the Sun's sphere of influence, and possibly beyond. The extended mission is continuing to describe the outer environs of the solar system, in search of the heliosphere boundary, and into the outer limits of the Sun's magnetic field and the solar wind. Out there, it will attempt to conduct measurements of the interstellar environment, including extra-solar wind interstellar fields, particles, and waves. Both probes are still sending back data about their surroundings via the Deep Space Network (DSN).

If we review the NASA space missions that have occurred since the lunar landing event, they may be characterized as taking two tracks: one was the quest for greater scientific knowledge and understanding, in order to develop ever more sophisticated and efficacious technologies with which to expand the exploration of the inner solar system; the second has been to begin our understanding of the outer regions of the solar system, including the Jovian subsystem and the frontier zone that lies between the heliosphere and the Milky Way galaxy. The second track was aimed at peering and moving into the outer reaches of the solar system, including the giant planets: Jupiter, Uranus, Saturn and Neptune, as well as their moons and other suburbs. There were also the missions to explore the zones of the asteroids and the comets, with the objective of learning more about the early formation of the solar system itself and, later, to understand the intrinsic composition of these most ancient bodies. Along the way, much has been learned about the precise mechanisms that control the location and movement of the bodies; and thereby often adding to or subtracting from the various theories and models that had been constructed during the past two millennia prior to these in situ empirical observations.

Although the crowning achievement of the Apollo program – and for which it will most recognized – is the lunar landings by human astronauts, the most significant thing about this program in terms of military space operations, was the great body of knowledge and skills that it created for "flying and fighting" outside our terrestrial atmosphere. The overall program involved a step-wise process, in which precursor sub-programs flowed into an overall developmental time-line that eventually was culminated in the Lunar Landing of 1969. One of the most important of these precursor operations was the Mercury program, in which a single astronaut rode a Mercury space capsule on a 300-mile flight into space, which lasted for 15 minutes, and attained an altitude of 116 miles. The first successful U.S. manned flight into orbit around the Earth completed three orbits and then landed in the Atlantic Ocean, near the Bahamas. The last mission made 22 orbits and also was successfully recovered in the ocean, some 34 hours later.

Then, there are the so-called Gemini missions. These were a series of 12 two-man spacecraft which were placed into orbit around the Earth in the period from 1964 to 1967. They were preceded by the Mercury Project which provided much of the base-line data and technological experience with respect to orbiting operations that the Gemini Program would utilize to do the preparatory work for the Apollo series of sorties that would follow. In this transitional period, the spacecraft crew would grow from one to three astronauts. So Gemini was primarily intended to test the ability of astronauts to maneuver their spacecraft by manual control, thus developing much of the precursor knowledge and skill sets for such later missions as that of the Space Shuttle program. It also helped to develop the techniques for orbit rendezvous and docking maneuvers which are still used in the support operations for the International Space Station. Another outcome of the Gemini Program was the continued improvement of environmental control and electrical power systems of spacecraft by NASA engineers. In one of the later sorties, the first "space walk" outside a spaceship was accomplished. Other achievements of space exploration included the longest space flight to date, and the first automatically-controlled reentry into the Earth's atmosphere. It was also shown that useful observations of the Earth, both visually and photographically, could be made. It was perhaps the first occasion in which humans were able to comprehend our home planet from a truly global perspective.

The concurrent Apollo program consisted of a series of space sorties that are best known for the culminating missions that landed humans on the Moon. However, there is another lesser known event that has had equally significant consequences in determining the tactics and systems that have been developed during the entire space exploration and science program. After an intense internal debate among the administrators and the engineers of NASA, it was finally decided to follow an architecture strategy wherein the spacecraft that would be used by Apollo would be launched in several discrete, smaller components, which would then finally be assembled through a series of rendezvous in Earth orbit. This would be the strategy that would also be followed in the construction of space stations, such as the International Space Station later on. This decision to send smaller payloads which could be assembled later at a site in space would ultimately return great savings in the cost of launching and assembling these systems. The Apollo spacecraft would, therefore, consist of modules. The Service Module would be utilized to transport the equipment and the rocket engine that would be needed to guide the spacecraft into lunar orbit and then send back to Earth. Meanwhile, the Lunar Module would serve as a shuttle vehicle between the Command Module in lunar orbit, and the surface of the Moon.

The most technically significant capability that was built into the Lunar Module was that of being able to perform a docking maneuver with the Command Module while in lunar orbit. There also were other efforts to develop the capacity of the human astronaut to work more independently of the space vehicle through the development of more effective EVAs (Extravehicular Activities) during the operation of the Space Shuttle Program. The first space-walks of the shuttle era demonstrated that astronauts could operate independently of the space vehicle through the development of improved space suits. Nevertheless, there were still plenty of hazards which faced the astronaut during an EVA. Among these were the reported ionization layers that built up on both the vehicle and the astronaut's space suit.

On another track, in the period following the completion of the lunar-based Apollo Program, NASA turned to the challenge of developing a comprehensive exploration and study of the other parts of the solar system. This next phase of the NASA space program would widen its scope of operations into the domain of the inner planets, including Mars and Venus. To prepare – or complement – this new phase of exploration, however, NASA set about developing a strong infrastructure of manned space vehicles and stations, as well as the capabilities to do real science work in space. This would require space vehicles that would incorporate habitat and laboratory modules into the rocket architecture. The upshot has been the formation of a fleet of manned and unmanned space vehicles and surface vehicles for doing work on the surfaces of the planets and other bodies.

In the years following the successful lunar landings, NASA turned to the job of developing a practical logistical support system for the next generation of space missions, such as the

International Space Station, in which humans would remain to live and work in space, for extended periods of time. These "STS" platforms can be characterized as the first transport spacecraft to be utilized in space. They are partially-reusable, rocket-launched vehicles that are designed to get into Earth-orbit, deliver cargo and people to an orbiting space station or other orbiting platform, and then to return to the Earth's atmosphere and, finally, glide down to recovery on a runway on the Earth's surface. The fleet of NASA space shuttles was inaugurated in 1981, primarily to transport the artificial satellites into orbit around the Earth. Latter, the logistical and training support function for the International Space Station was added to its mission assignments. Aside from delivering cargo and people, the space shuttles have also served in support of the maintenance and updating of the various orbiting assets of NASA and other clients. Another of its capabilities is to serve as a relatively longer-term platform in space for conducting scientific experiments under the weightless conditions of space, as it is able to loiter on station for as long as two weeks.

The first Spacelab mission was designed to conduct tests of scientific modules which were to focus on five major areas of study: Earth studies, space physics, astronomy, life science, and materials science. The latter has become perhaps the most important to the conduct of space operations, especially for longer-term sorties, and has spurred intensive research into the development of materials that can deal with the extremes of temperature and the radiation environment in space. Another of the most productive of the systems that were deployed into orbit around the Earth to conduct astronomical studies was the International Ultraviolet Explorer (IUE) satellite, which was launched in 1978. This operation was a joint venture among the U.S. NASA, the European Space Agency, and other European members. Equipped with an 18 inch reflecting telescope, two spectrographs which were linked to television cameras, and other instruments, the IUE has sent back over 100,000 spectrographic images and data that have helped construct the overall data base of space operations until 1996, when it was shut down. Its main assets revolved around its ability to function like any modern astronomy station on Earth. It was made accessible to all astronomers and provided a venue for true academic work which resulted in valuable peer reviews of the great number of scientific papers that were produced during its operation. And, it also was able to function like the leading ground-based observatories in its capability for responding quickly to transient targets such as comets and supernovas.

The Skylab Program represented the first U.S. attempt to deploy a long-term space station into orbit around the Earth, which occurred in 1973. The impetus for this was the desire to match the technological achievements of the Soviet Soyuz station program which launched its first space station in 1967. There was also the practical decision that was made, following the successful Moon landings of the Apollo Program, to begin sustained scientific operations in Earth orbit. The first U.S. space station did indeed utilize much of the technical knowledge and skill-sets that had been developed by the lunar-landing missions. Thus, the Apollo spacecraft and systems which were developed for the Moon landings incorporated and adapted in the development of the specialized rocket system and the modules for use in a variety of scientific missions that would follow. The overall platform architecture would be designed with the goal of developing a space laboratory which could operate in orbit for long durations, much like the nuclear submarines on Earth. Another of the primary missions, aside from conducting scientific studies in space, was to begin the study of the Sun in greater detail and with increased clarity from the vantage point of outer space.

At the end of the period of the Cold War, in 1993, NASA and Russia agreed to merge, what originally were their own orbiting space stations, into a single facility. The approach that was agreed upon was to integrate their respective modules. Since then, there would be an incorporation of scientific and engineering contributions from the European Space Agency and Japan too. During the 1990s, the Russian control module Zarya and the American Unity connecting node were added to the basic configuration. Notably, this task of linking the two nodes was done by U.S. Space Shuttle astronauts in orbit. In 2000, the module Zvezda, which is a habitat and control center facility, was added to the ISS architecture. In the same year, it received its first on-board crew. Then, a NASA microgravity laboratory nodule, called Density, was added. This complex of laboratories and habitats were crossed by a long truss supporting four large solar-power arrays and thermal radiators. Ultimately, a complex of telemetry units, computer batteries, and a pair of power-producing solar panels were added.

One of the more important initial missions of the ISS was to perform scientific experiments in the weightless environment there, and this was accomplished by the addition of a microgravity system to the overall complex of habitats and laboratories onboard the system. The continued operation of this platform in orbit for over a decade has in itself been an important milestone in the exploration of space. Among the more notable advances in the technology of maintaining a "mother ship" in space, has been the continued improvements in specialized maneuvers, such as the docking of other craft to the ISS, which has become almost commonplace during the past several years. This has further been demonstrated by the fact that commercial spaceships are now able to perform docking maneuvers as well. However, one can see a movement toward a situation in which many of the space operations may ultimately be conducted almost entirely in space orbit. Meanwhile, the "Low-Earth Orbit" (LOE) space operations region will continue to comprehend the surface of the Earth, the upper reaches of the terrestrial atmosphere, and the area that is sometimes referred to as near space.

This LOE region is already being defined as an aerospace area of operations (the nomenclature that is being used today is "space domain" by U.S. military planners). At least part of this space domain already is being militarized by the United States and other nation-states as well. Thus, the U.S. Air Force Space Command now has been given the responsibility for operating many of the military satellites and for developing capabilities to defend them against any type of hostile move by other nation-states or rogue groups, such as Al Qaida. The latter mainly poses a threat in the cyberwarfare area of operations, at the present time. This first military space domain has been largely created and operated by manned systems, but the next developing one, here called the geospace domain, has been steadily relying more on unmanned systems as the need for sustained operations there has increased. Here is a summary of some of the capabilities for space operations that have been developing in the more recent decades of space exploration activities.

————

These first generation of unmanned spacecraft were sent into orbit around the Moon to perform what, in essence, was reconnoitering of a potential area of operations. They obtained thousands of photographs of much of the surface of the Moon, which enabled planners of subsequent missions to determine the best landing sites for the Apollo lunar-landing sorties that would follow. The payloads that were carried by these lunar orbiters included sophisticated cameras which enabled them to obtain wide-angle and high-resolution photographs of many areas of the Moon (an activity that is not unlike "air reconnaissance"). Some of these photographs were taken as close as 28 miles from the surface of the Moon in order to achieve high-resolution, while still acquiring a wide field of observation. Thus, they also provided the broad view and high-resolution detail of a potential target region that would be recognizable to military imagery analysts. At any rate, these lunar orbiters made possible the construction of detailed lunar maps; as much as 100 times the detail that was available from the Earth-based telescopes of the 1960s.

By 1975, unmanned spacecraft were being sent into orbit around Mars. Among these were the twin Viking orbiters whose mission was to map large expanses of the planet's surface. Other objectives included the observation of weather patterns and to photograph the two small moons of Mars: Deimos and Phobos. These "reconnaissance" platforms also carried an instrumented orbiter and lander vehicle (the latter was designed for "in situ" reconnaissance). After surveying possible landing sites from orbit, the landers were to be sent down to the surface, where they would continue to map and analyze large expanses of the planet's surface from their ground vantage points. The data and imagery that was acquired by the landers was relayed back to the orbiter, and subsequently retransmitted to ground stations for further analysis. The orbiter itself was equipped to do some preliminary onboard analysis of the acquired soil samples. In this way, experiments could be carried onboard, before the samples had an opportunity to become degraded during a return flight to Earth. As it turned out, these onboard experiments were unable to detect any signs of life in the soil that was collected. The final data from Viking was received by ground stations in November of 1982.

At the same time, another series of unmanned U.S. space probes were tasked to carry out studies of planetary bodies, other than the Moon. They were designed to measure various interplanetary particles and magnetic fields with the aim of determining their respective effects on the conduct of space operations. Among these were the Pioneer series of unmanned space probes. The first of these (Pioneer, 1965) was placed into solar orbit in order to determine what the space conditions would be like in a journey from Earth to Venus. A later version of this series of unmanned probes (Pioneer 10, 1972) did a flyby of Jupiter in 1973. This sortie was followed by another probe (Pioneer 11, 1973) which passed by Jupiter and continued on to within 20,000 miles of the planet Saturn in 1979. Meanwhile, Pioneer 12 and 13 (1978) conducted studies of the clouds and atmosphere of Venus; it successfully mapped about 90% of its surface through the use of penetrating radar. The second spacecraft inserted several instrument packages into the planet's atmosphere at various points in order to measure the various physical and chemical properties of its surface.

From the above narrative about the history of unmanned spacecraft that have been utilized by NASA, we can infer that a conscious cost-benefit and risk analysis probably prompted the decision to use robotic systems, as opposed to manned ones, in the exploration of the solar system beyond the relative "comfort zone" of the Earth-Moon system. Another factor in the decision-making process had to do with the tremendous advances in technology which made the unmanned sorties to the other planets much more feasible. One of the most important of these advances was in the areas of computer and telemetry technology. In a now familiar paradigm, computers began to get more powerful and able to store much more programming code and data than ever before – a trend which continues to this day in the era of nanotechnology. Also, the computers continued to get smaller and lighter, which meant that they could be placed, at lower cost, onboard the spacecraft – as well as the landers and rovers which would operate on the surface of the planets themselves. Still another revolution that began at about the same time was characterized by the advances in artificial intelligence and robotics. These served to enhance the technological and economic feasibility of relying on increasingly autonomous equipment to do the job of space reconnaissance and surface reconnoitering on the planets of the solar system. By the same token, during the period of the Vietnam War, I noticed the development of more powerful and sensitive remote sensors which were being used to do air reconnaissance and even weapons guidance and control in many different battlefield environments, both in the air and on the surface of the Earth.

The robotic space probes are a type of unmanned spacecraft that are being sent out on scientific exploration missions into space, beyond the Earth's gravitational attraction. They may approach the Moon, enter the interplanetary space medium, and flyby, orbit, or land on other planetary bodies. They even approach or enter interstellar space, beyond the limits of the heliosphere. Examples of this type of spacecraft are NASA Voyagers 1 and 2. These are twins that have been sent out to explore the outer solar system, and beyond. More particularly, these were a pair of robotic interplanetary probes that were launched to observe and transmit information back to Earth about the so-called giant planets of the outer solar system, as well as the farthest reaches of the Sun's sphere of influence, as measured by the ultimate reach of the solar wind. Mariner 2 flew past Venus in 1962 and returned information about its atmosphere and extremely hot temperatures. Meanwhile, the Soviet Venera series of probes entered the sulfurous atmosphere of the planet Venus, and one of these, the Venera 7, was the first spacecraft to land on another planet's surface. And, enroute to Mercury in 1974, Mariner 10 photographed the atmosphere of Venus in ultraviolet light and returned information about its circulation.

One thing that all these space missions have in common is they occurred within a spatial region, and the basic process, or transaction, for carrying out the mission was encompassed in a "sortie." That is, a loop that began at a launch point, proceeded enroute to a target area and point, and performed the essential activity to complete the assigned mission.

THE SORTIE CELL REGION

The sortie can be said to be the basic transaction of military space operations, just as it has been in military air operations within the atmosphere of Earth. In a sense, it is analogous to the basic "transaction" that underlies all business activities. In much the same way, the sortie defines the most basic operation that underlies all military aerospace operations. In its most basic form in military space operations, the sortie consists of a launch or takeoff (sometimes a spacecraft is launched from an air-breathing aircraft) into an initial orbit, a cruise orbit toward the target area, followed by the delivery of the payload, and a return to the original launch region. On Earth, the first sorties are fairly straightforward; the main elements consist of one platform (aircraft), one flight path, and one recovery point. Later, the military sortie would become more complex. Consider, for example, the Doolittle raid on Tokyo, Japan for example. In this case, none of the aircraft returned to the original launch point. Instead, the "recovery" points of the sortie transaction were scattered throughout the Chinese landscape and the Pacific Ocean.

In space operations, the same level of complexity has evolved. Once again, it begins with a single launch of a single rocket, and its payload. It then proceeds to a point along an orbit around the Earth, and it ends with a return to one of several points on the surface of the Earth, one of which might be a "splash-down" point in an ocean. Later, the sortie would become even more complex and sophisticated. There still might be a single launch point, but the "target" orbital point in space might involve any one of several intermediary points on the way to multiple targets. On the other hand, some sorties involve one-way missions wherein the rocket and the payload would remain in orbit, either until their batteries expire or they complete their mission. Thereafter they might simply become "space junk." Another addition to the complexity of the sortie in space would be the addition of multiple-level orbits. At first, the target-orbit path would be the "low-Earth orbit" that lay between 200 and 22,000 miles above the surface of our planet. Then, the target orbits were extended in order to reach the Moon; and even farther out to the region of the inner planets, such as Mercury, Venus and Mars, as well as into the Main Asteroid Belt. Later, "slingshot" or gravity assistance maneuvers would be utilized to reach the orbits of Jupiter, Saturn, Uranus, and Neptune, as well as the outer rim of the solar system, where Pluto and the Comets of the Kuiper Belt are located.

Almost immediately following a successful launch event from the surface of Earth (at an altitude beyond the 62 miles) the profile of the natural environmental factors which interact with the sortie change drastically. These include such factors as the atmospheric ambience of oxygen, relatively high atmospheric pressures, gravitational influences, water vapor, and heat, as well as protection from the most of the harmful cosmic and solar radiation. Within the interplanetary medium the sortie environment now presents almost zero gravity, no oxygen, extreme cold, very little light, and exposure to very high levels of electromagnetic radiation, from many natural sources. There also is no protection from a benevolent atmosphere from the meteorites and charged particles that can damage or destroy a spacesuit or a spacecraft.

Another difference between the terrestrial sortie and the space sortie is in the navigation environment. Whereas terrestrial platforms rely on Global Positioning Satellite Systems as their "external" navigation point, in space, the spacecraft utilizes the stars and their internal inertial systems to navigate from point to point, while maintaining the correct "attitude" at all times. The Earthly grid of latitudes and longitudes remain the same throughout the solar system, however. The imaginary lines that pass through the terrestrial poles and Equator are simply extended outward throughout the "celestial sphere." The orbit in space replaces the "great circle" routes of the planetary globes. The location of the spacecraft is now plotted along an elliptic orbit around the Earth, or some other body in the solar system. It is referred to as being at some point along the orbit with respect to the intersecting line that originates from the observer's location in the solar system. So, in effect, the traditional astronomer's method for tracking the location of the bodies of the solar system has now been adopted in military space operations, except that the "body" in question is now a spacecraft.

There are several conceptual approaches to the study of spatial phenomena in the science of geography. One that I find especially useful is to see military aerospace operations as simply one more type of human activity, very much like agriculture or a construction site, for instance. Thus, as in all human endeavors, it is a constant that natural and cultural geographic factors interact with military aerospace operations just as they do with other human activities that occur in the natural-cultural environment. It is also an invariable truth that all such relationships will be dynamic; always changing in terms of space and time. Furthermore, the changes in the relationship between sequences of human activities, including those we call military aerospace operations, will rarely occur in a straight-line sequence; rather they will trace continuous changes in trajectory, and sometimes loop back. These variations in the nature of the "man-land" relationship will also tend to occur both in the short-term and the longer terms as well. This relationship and its inherent variations are also serial in time, but not always sequential or consequential. There is seldom a continuing inherent "order" in the relationship; instead, the tendency toward disorder is always present. Finally, there is no inherent order of importance in these relationships; there is only a continual reordering of these relationships that are dependent on the particular situation in which they occur. In short, the interrelationship is complicated in each instance of space-time.

So, the geographer will quickly understand that the activity which is here called the "sortie" has to occur within the confines of the concept of the geographic region – which is a well-understood conceptual vehicle for analyzing any activity or phenomenon that occurs on Earth, or even throughout the Universe. Thus, since the beginning of the space age, it has become ever more obvious that the "space-time event" can be understood as being a region, especially in the case of space operations that occur outside the planetary gravitational spheres; into the realm of the inter-planetary and inter-stellar media. One of the reasons for this transmutation of the Earthly region is the fact that objects in space are traveling at speeds that are somewhat closer to the speed of light than ever before. Another is the greater distances in the outer solar system, such that a beam of energy that is traveling at the speed of light following emission from the Sun takes about 8 minutes to reach the Earth.

Returning to the analysis of the sortie, we see that the "region" in which it takes place can assume many forms, and is highly-changeable during transit, depending on the particular geography in which it occurs, and the kind of activity that is happening. We will see an example of this phenomenon in our discussion of sortie-cell regions later on. Moreover, regions can be nested within larger regions, or they can be conjoined and, therefore, have an overlapping subregion of commonality. Sortie regions can also be either functional and/or situational. The overriding characteristic of a sortie region seems to depend on whether the functions that take place within a three-dimensional space are elements of an integrated and interrelated operational system, which is dedicated towards an objective or a set of objectives. In other words, a sortie region is an "area of operations," whose configuration "depends" on many variables.

I find that it is beneficial to any analysis of the region(s) in which the definitive military aerospace activity – the sortie – takes place, to see them as being a time-space series of three-dimensional areas of atmosphere, or of the interplanetary space. Aviation and space operations literature is filled with references to orbits, circular turns, and circular climbs and descents by air platforms, and now space platforms. The trajectory of the sortie invariably is plotted as a curvilinear line on a two-dimensional matrix, or in a three-dimensional portrayal by modern computer modeling. It appears that every platform, including vertical takeoff and landing aircraft and rockets, eventually transcribe a curvilinear parabola in a three-dimensional space. In the case of the sortie itself, a round-trip is implied; an elliptical path, rather than a perfect circular one. The great circle route itself is curvilinear along a path that extends beyond a few hundred feet. The concept of the radius is used to define the distance from one's own platform and the relative location of another object in the three-dimensional space about them. Normally, the target of a given sortie is referred to as an area, rather than a point, and the precision of the delivery of the ordnance on the target is called the CEP (Circular Error Probable). The axis of the cylindrical region, in the context of the sortie, is usually the platform which is carrying out the operation. Therefore, in this discussion of the various phases of the sortie region, I will utilize the cylinder as the "space" within which humans operate their machines, both inside and outside the terrestrial atmosphere.

As we have seen earlier, the notion of "situational" relationships that occur between human activities and the natural-cultural environment can be expressed in terms of the "cell regions" that occur sequentially along the trajectory of a sortie. Furthermore, within each space-time cell there is a certain combination (system) of properties and characteristics that change as the cell moves through a medium. One aspect of the biological cell that especially interests me as a geographer, is that it as it moves through its medium, it is affected variably by its surrounding environment, and makes changes in response to the effects of its environment. Finally, as the cell moves in time and space, its internal composition and physiology seems to adapt to the effects of its environment.

So, in applying the living cell model of biology and the idea of integration from calculus to the analysis of the military aerospace sortie, we can elicit many insights into the dynamic nature of a sortie as it moves through its particular medium. Thus, as a platform travels along its trajectory in a particular medium (atmosphere or space) the nature of its total environment is continually changing. Calculus provides a framework for describing the notions of position, speed, and acceleration through a medium. So, by combining ideas from the fields of biology, calculus and geography, we get the ability to track and measure what goes on within a "cell-region" as a platform moves (and accelerates or decelerates) as it moves toward the target point of the sortie.

In this circumstance, it is useful to view the platform as moving through a series of three-dimensional environmental space-time events. Such a phenomenon is becoming more obvious as military operations continue to occur over ever greater distances and time-durations, as well as variable accelerations or decelerations in outer space. Also, our ability to measure both space and time at the sub-atomic level of scale, and the nano-second time interval also allows us to deal with "slices" of time and space (or cell-regions) at these levels. In the terminology of calculus, acceleration is "integrated" (that is, summed a little at a time) to derive velocity; then velocity is integrated into "position." Thus, the inertia of the mass (platform) causes it to tend to remain stationary (for a nano-second), but acceleration of the platform tends to displace the environment relative to the mass. At this level of appreciation, there is also a recognition that military aerospace operations occur as series of dynamic time-space events. I also envision this phenomenon as a series of time-lapse photographs in which the "reality" is always transitory. This reality, I believe, will become more dominant in space, where time and space are accelerated to the point where Einstein's theory of special relationships becomes an actuality. This can be seen in the tracking of the movement in space and time by NASA space missions to the outer areas of the solar system, in which the trajectory is depicted as a sequential series of events in discrete time-space cells.

Like all geographic regions, this time-space region of military area of operations has a series of defined boundaries. Also, the "nodes" (places and other phenomena) that are contained within them continue to be interconnected in some dynamic, systemic order that is imposed by electromagnetic and gravitational forces. The time-space region (or, cell-region) also continues to be defined by the number and intensity of the linkages among its nodes, which are often conceptualized as vectors and fields. The linkages can be in the form of "information" or manifested in the movement of packets of light and matter. These linkages also develop patterns of interaction whose direction and intensity can provide a "contours" which depict the core (or cores), as well as the boundaries of the region. "Linkages" in the case of military aviation are actuated by the movements by aircraft or other platforms that carry out their operations in terms of the "sortie."

I also see the concept of the military "sortie" as a useful conceptual framework for analyzing the activities that continue to occur within the platform, even as the series of space-time events are occurring outside it. So, one can see that the use of the concept of "sortie" (takeoff-operation-landing) cycle enables one to do a detailed and dynamic analysis of the ways in which the relevant "geography" affects military aerospace operations at every point in time and space – and within the platform itself. Thus, the notion of the cell-region and the integration of phenomena during acceleration (deceleration) through a medium can be a very useful to framework for analyzing the nature of the interrelationship between military aerospace activities and the relevant natural-cultural environment at every space-time point in the process of the sortie. The sortie concept lends itself well to many ways of measurement during the course of a particular mission event. Like the photon in the movement of electromagnetic energy, it is a defined "packet" of activities that can be measured according to a variety of "metrics" during its cycle. Moreover, the concept of the sortie in military aerospace operations is useful in the analysis of the many different tasks and in the measurement of their efficacy in meeting predetermined task objectives, within the scope of the overall missions that they are designed to support.

———

The concept of the "cell-region" represents an attempt to relate military aerospace operations with the "space-time" of Einstein's theory of relativity, in which time is the fourth dimension. As I envision it, regions can be seen as a momentary snapshot of a spatial reality, like the series of three-dimension images that created by computers today. Thus, in military aerospace operations, the sortie can be understood to be a series of moving time-space events. Each of these slices of time-space can be analyzed in terms of their interaction with its particular and temporal "geographic environment." So, in the case of the military aerospace sortie, we see that within each "slice" of time-space there is a set of weather conditions, for example. The following analysis of a hypothetical space mission is undertaken within the context of the "slices" of reality that occur during a "sortie" in space.

The initial cell-region of a given sortie is the "takeoff cell-region." Within this cell-region, some kind of rocket propulsion system for achieving "escape velocity" is occurring. This term refers to the rate of acceleration that is necessary to escape the gravitational attraction of a planet or other body in the solar system. During this period of space-time, the platform is interacting with a particular environmental regime, that is to say, the weather. The most important of these is the lightning and wind phenomena which can cause a fuel explosion, as well as the disruption of the electronic-based systems, and the compromise the initial trajectory of the rocket within the narrow window of opportunity for achieving the appropriate initial point in orbit. At a later phase of the sortie, the movement of a spacecraft through the series of cell-regions in space presents a constantly-changing series of environmental conditions to the platform as it continues in its sortie trajectories.

Within the current thinking in the science of physics, the trajectory is now seen as the position of a moving object in space, over a period of time, along a vector. Other scientists view a trajectory as a time-ordered set of "states" within a dynamic system. The reality is that all objects, such as platforms, are always moving in some direction, at a given velocity. This means that the "commuter path" of the aerospace platform is always transmuting. And, as often occurs in the modern era, the systems within the platform also are undergoing a dynamic transformation. The general response of military scientists and engineers to the constantly-changing operational cell-region, at any point in time, is to utilize more technology – both in terms of equipment and people – to overcome any encumbrances these might present to the success of the mission. Therefore, the response of the human or robotic "aviator" function is to go through a "sense-process-react" cycle in order to respond to any syndrome of challenges that are thrown up against them, in the actual moment. This interaction of environment and humans and robots is the subject of a great many scientific studies by geographers. This methodology is now applied to the interaction between platform systems and their environments in general.

Returning to the cruise phase of the sortie, we see that the overall operational environment continues to evolve through a series of "environments," which present themselves along the way toward the target point of the sortie. To a military geographer, this series of time-space events seem to be like the series of images that are recorded by the cameras and other remote sensors of the reconnaissance platform. In terms of regional analysis, the military geographer will see the value of treating each event both as a single-occurrence and as a flowing set of occurrences. The modern computer, with its powerful ability to store vast quantities of data and imagery, and the processing power to integrate the additional information within a dynamic model, now makes possible such "real-time" geographic analyses. Thus, what we are seeing, in effect, is the advent of artificial self-awareness and situational awareness during a sortie, which might also be described as a form of artificial intelligence.

This brings up the question of whether humans will continue to play a part in future military space operations and, if so, how and to what extent. The concept of the cell-region enables that particular decision to be made during each time-space event that occurs during a sortie. That is, the sortie can now be analyzed in "real-time" with respect to the particular micro operations environment. And furthermore, a decision can be made about the nature of the particular human-robotic relationship. So now we are provided with a conceptual vehicle and the technology to observe and analyze the almost instantaneous sequences of "man-land" relationships as they occur throughout a sortie. The next question then becomes: how do we intervene in order to respond these micro-environment conditions during the sortie?

Actually, the issue of the role of the human factor during a sortie has become as much one of "intervention," as opposed to integration. I say this because it appears that the technologies for performing the sensor-actuator process during the course of a sortie, has been fairly-well worked out. So, the main focus of the debate now has to do with the decision of where to place the locus of control during the sortie. This applies to both manned and unmanned sorties. Even in the case of the manned sortie, the locus of control may not necessarily be onboard the spacecraft, but on the surface of the Earth at Houston Control, for example. Furthermore, the locus of control may involve a shared arrangement, depending on the phase or environmental conditions in space, including unforeseen conditions. On another dimension, the locus of control, at every stage of the sortie, may either be preprogrammed or assigned to the "astronaut," which can be either a human or a robot (like the Robonaut 2 onboard the International Space Station). Or, the locus of control can be assigned dynamically during the course of the sortie, in response to actual conditions.

————

Even in the case of "unmanned" sorties, such as the NASA probes, there is an ongoing debate with respect to the optimum site for the locus of control. One option is to front-load the programming software, with the assumption that all of the possible values of the variables have been identified and dealt with satisfactorily with the appropriate algorithms. Another approach is to only partially pre-program the sortie, with the assumption that only a percentage of the probabilities can be identified and handled by coded instructions to an on-board computer. This also implies that a human subsystem will be available to assume control when it is not practical for the robotic subsystem maintain operational control. Even in that case, the location of the human (locus of control) may be either on-board the spacecraft, or in a control room on the surface of the Earth. In any case, the sortie continues toward the target and the cell-regions continue to serve as a conceptual vehicle for analyzing the ongoing and dynamic relationship between the environment and the platform.

Indeed, during the turn of the 21st century, there has been a quiet revolution in the definition of the "pilot" function with respect to the platform involved in a sortie. Prior to the year 2008, and during the later years of the Afghanistan Conflict, the dominant sort of platform which was used to carry out air-to-air or air-to-ground combat operations had been the air-breathing aircraft. The configuration of the air forces in that area of operations was still a mixture of helicopters and fixed-wing aircraft that was still reminiscent of the Vietnam War era. Then, in that same year, reports of "drone" or remotely-piloted vehicles began to appear in the military literature and in the popular media. The name of a new "remotely-piloted vehicle" (RPV) platform that was performing air reconnaissance missions and, later, ground strikes against selected "high-value" targets over Afghanistan and parts of Pakistan is called the Predator. The locus of control of air platforms was shifting away from the cockpit to the flight control room on the ground.

One interesting fact about the shifting of the locus of control of a platform, from the cockpit to a remote location (either on the ground or on another platform), is that the locus of control is becoming more remote. This seems to be occurring even as the sortie environment is becoming increasingly complex and the required reaction times are becoming increasingly short. Even the intermediate or "cruise" phase of the sortie is becoming more complicated and requires more split-second decision-making on the part of the human/computer interface. And with the advent of space operations occurring beyond the Jovian planets and into the outer reaches of the heliosphere, we are seeing the time-space distance between the locus of control and the immediate environment of the sortie extending ever farther.

One approach to dealing with this problem by NASA has been to develop technologies and engineering within the areas of artificial intelligence and automated software development. The art and science of software programming has been perhaps the most crucial element in dealing with distance and response time in space operations. Actually, it is the software side of computer science that has presented the greatest challenge in military aerospace operations. A review of the post-mission reports of the NASA sorties reveals that there have been more problems with software than the computer-system hardware itself (The USA in Space). Much of this stems from the problem that faces all computer programmers, which essentially is to think of every step of every procedure that is embedded in the computer instructions to all of the electromechanical systems of a platform during its sortie. This is the nature of the task that faces the NASA flight engineers for every given mission, and continues to evolve as the variables of a sortie become more complex.

One effort to solve the problem of writing "perfect" preprogrammed instructions for a mission has been to rely more on "real-time" software code as the situation evolves from cell-region to cell-region. So now, the main computer that controls the platform systems and their payload systems is becoming more remote from the actual location of the programmer. This means that the programming instruction to the platform systems is more dependent on the electromagnetic spectrum than ever before. It also implies that the electromagnetic signals between the flight control center on the ground and the platform systems will be more greatly affected by the cosmic and solar radiation within its cell-region environments.

Another approach to the programming problem has been to enhance the capabilities of the on-board computers so that they could, in effect, program themselves in response to actual conditions. In other words, this approach would involve the enhancement of the artificial intelligence capabilities of the on-board computers. The implications of this alternative were clear: one of the on-board computers would have to learn to program the other processors in real-time fashion. Now the locus of control would shift back to the platform itself, and the flight engineers would again be faced with the challenge of "thinking of everything" prior to the launch of the mission. In fact, what has happened is that the on-board computer with the artificial intelligence has been pre-programmed with certain instructions that are more related to the overall doctrine and strategic mission objectives of the sortie. This brings up images to me of the on-board computer "HAL" on the 2001: A Space Odyssey, of Stanley Kubrick. Another consequence of this distancing of control is that the military aerospace sortie will have to depend more on remote-sensors, remote actuators, and a shorter input-processing-output loop in order to execute each of the steps in any algorithm. It is this emerging condition that is driving the development of more powerful technologies related to command and control and communications.

Another factor that is adding to the complexity of the sortie, especially in outer space, is that the essential task of moving a platform from point A to a target point B is becoming more eventful and difficult. Indeed, as the distance to point B has increased, many more cell-regions will be encountered by the aviator. Furthermore, the complexity or difficulty factor of such a movement will increase as well. For example, the most hazardous phenomena in space are cosmic and solar radiation, and meteorite storms. But, in the future, it is likely that there will also be "cultural" hazards as well. These could include everything from hostile spacecraft to satellite "mines" that might be encountered. Thus, generally speaking, the longer the platform is in space, the greater the probabilities of encountering these natural and cultural environmental hazards.

In this book, I have taken note of these complexities of space navigation by turning to the sciences of geometry, mathematics and physics for conceptual tools for analyzing the "man-land" relationship that occurs throughout the sortie. The concept of the vector is employed to represent direction, distance, and "magnitude" (e.g., energy) of a moving body or force. In the case of military aerospace operations, I have found the vector to be a powerful way to describe a movement from one cell-region to another during the sortie. Thus, a platform in these three-dimensional spaces can be considered to be free to move in any direction. However, within any cell-region, is bound to be met by many physical and cultural forces. Some of the forces in this "force-field" can act on the moving platform at the same time, thus producing a combined and cumulative force on that platform.

As we have seen, the movements of vectors within a real space will invariably face obstacles to a successful arrival at the destination point. We can now appreciate the value of using the concept of a vector, which describes not only direction and distance, but also magnitude. The latter attribute implies that variances in the number of platforms, their combined payload, or in the effective destructive (explosive) power of the ordnance that is being carried will affect the "magnitude" of a given sortie. Again, the fuel that is being carried onboard, or which will become available during the sortie will affect the distance of the vector of the sortie. By the same token, the effective horsepower of the power-plants (engines) will work in a cumulative sense to affect the actual vector of the sortie.

———

So, now we consider the "man-land" interrelationship that operates in serial fashion, as a platform moves through the three-dimensional cell-regions. Beginning with the aeronautical navigation challenge, we see that since the development of the advanced inertial-guidance systems and the orbiting satellite-based positioning systems to maintain "situational and positional awareness." The constellation of GPS (Global Positioning System) satellites, which are presently orbiting the Earth synchronously, now provides the navigation subsystem with the external reference data to ascertain its absolute location in space. The onboard computer system then makes navigation decisions based on directions from pre-written computer software, or in response to signals from internal sensors and actuators which rely on "real-time" environmental data to make autonomous decisions.

One aspect of this "situational awareness" has to do with the nature of the commuter cell-region within which the mission-objective is to be accomplished. What in fact is being encountered is a series of geographic regions, occurring within what can be described as corridor; a corridor filled with a series of states of combinations of natural and cultural variables, which can negatively impact the success of the sortie. Unlike the set of challenges that are present at the takeoff point, the cell-region is less controlled. A dramatic example of such a situation is the "Doolittle Raid" of WWII; a sortie whose objective was to bomb Tokyo. In that case, once the squadron of B-25s had launched from their aircraft carrier runway, it was imperative that they continue on with the sortie. The focus then was on the ordnance drop point and, hopefully, recovery on mainland China.

In any case, within each commuter cell-region, the platform will encounter many geographic factors, such as the emissions of electromagnetic radiation, which can be natural or cultural in origin. Within the atmosphere, these are most likely to be emitted by cultural phenomena, such as the radar sites. Some of the emissions will be from friends, but others will be from foes; some emitters will be stationed on the surface of the Earth or some other celestial body within the area of operations, while others may be located within the interplanetary medium. On the friendly side of the ledger, there will be those specialized airborne platforms which will provide electronic-intelligence to the sortie platform. The "foe" side of the equation will be manifested by enemy weapons-control radar or infrared sensors. The platform which is the target of an enemy's electromagnetic beam of energy is said to have been "painted" by the radiated energy.

So, at some point along the flight path towards the target area, the military platform will enter the domain of the enemy's air defenses. Long before approaching the target zone, the first indication, may be in the form of a heightened-state of electromagnetic energy environment. Perhaps an enemy orbiting satellite will have detected the encroachment of the military platform. On the other hand, friendly satellite might have been deployed to conduct electronic countermeasure operations in order to facilitate the success of the friendly aircraft. However, by then, the enemy air interceptors may already have been alerted; and even scrambled, and the interceptors would begin scanning for the radar signature of the friendly aircraft. The sortie is now entering its "time over target."

———

So, consider what most target-systems will look like in such a military space-domain. Assuming that the imperative for control of the space is paramount, the aggressive force will first target the space defense systems for the orbiting satellites. The specific targets will be those that perform the function of electromagnetic fire control systems; the command and control systems that acquire and tracked the trajectory of the attacking platforms, and then direct the fire of the missile and anti-aircraft artillery that defended the main target systems. These will be the equivalent of the SAM (surface-to-air missiles) that confronted the B-52 bombers over the Haiphong-Hanoi centers of North Vietnam during the Vietnam War. In the geospace domain of the 21st century, such command and control nodes will be the primary targets in the military space context, as well. These primary targets will include the components and linkages within an adversary's military space infrastructure. Initially, this refers to those elements of the adversary's ability to "see" the attacking forces and to coordinate the defensive forces. Consider the command and control functions that could be located on the surface of the Earth or in a parallel orbit about our planet. These may be surface bases or orbiting spacecraft in geospace. Other elements of an adversary's command and control system that might be targeted would be their early-warning sensors (like the radar sites of the 20th century) or their communications relay sites. Perhaps the most lucrative target in the military warfare of the 21st century, however, is the computer, and more especially, the operating systems and software that control all the electrically-based systems. This implies a whole new constellation of targets, which can only be attacked through the use of cyberwarfare. Indeed, the essence of military aerospace warfare during the course of the 21st century might be characterized as being another generation of what was called electronic warfare (EW) during the 20th century.

The next logical set of targets in a geospace environment of the 21st century would most likely comprise the anti-aircraft weapons themselves. These constitute the immediate hazard to any form of military operations in that domain. Some of these will be improved versions of the lethal surface-to-air missiles and, maybe even the "anti-aircraft artillery" of the 21st century. The latter could consist of either directed energy weapons (e.g., laser beams) or kinetic energy projectiles. There is even the possibility that the modern descendents of the tethered barrage balloons of the Second World War might be deployed in the space above the air defense systems on the surfaces of other bodies in the solar system. In the geospace environment of the future, these could be "mines" that are connected to the Earth's surface by miles-long cables made of nano-technology material.

The main point with respect to space defense systems, which would have to be degraded or destroyed in order to attain "domain supremacy," is that the target cell-region air defense environment will be extremely complex. Therefore, the problem that will be presented to the attacking space force will be multivariate on many levels and within the space-time dimensions. So, we are once again returning to the reality that both the offense and defense in military space operations will be more and more reliant on telemetry and computers in order to exert command

and control discipline in such operations. In turn, this will require more application of technologies in the area of passive electronic warfare countermeasures; these could include what is referred to as "stealth" technology that essentially "bends" light waves so they do not strike their target or, if they do, distort the signal that returns to the emitter.

The final series of cell-regions within a given sortie ultimately lead to what we might call the target cell-region. Here again, the target cell-region represents a classic geographic region and, in this case, a functional region. Typically, it encompasses a spatial system whose function may be political, economic, or strictly military. In any case, the target spatial system will present one or more nodes which are the likely objective of the sortie payload. One way to provide context to this analysis of the "target cell-region" in the military space operations is to first review some examples of such cell-regions that have occurred on Earth. The target systems with which I am personally familiar are those of the Vietnam War, but they also may be found in any combat aerospace campaign. As always, the types of targets that are emphasized in a conflict are derived from the strategic and tactical mission objectives that were set down by the policy-makers and the major commanders.

———

In the case of the Vietnam War, the basic strategic objective was framed as an effort on the part of the United States and its allies to stymie the Communist Bloc attempts to take control of Southeast Asia. This geo-political region mainly comprised the territory which lies south of China. During the turn of the 20th century, it was largely a colonial possession of France, which named it "Indo-China" (Cambodia, Laos, and Vietnam). Later, following the initial war for independence, the region came to be known as Southeast Asia. It also became a military aerospace region (MAR) when the U.S. Air Force began conducting operations there. Although legally not combatants, Laos and Cambodia – as well as Thailand – were part of the MAR, in practice. It must be stressed here, that the Southeast Asia MAR was not a homogeneous or invariable one, neither in terms of space nor time. More to the point, the kinds of targets that had to be planned for attack varied considerably in terms of the ordnance and fusing that had to be employed, as well as the delivery tactics. Overall, the so-called Vietnam War actually consisted of at least two main areas of fixed-target or recurring-target systems: Haiphong harbor-Hanoi city in North Vietnam; and the Ho Chi Minh Trail, which ran from north to south, generally along the a low, but rugged, mountain chain that separated the Vietnams from Laos and Cambodia.

In a fundamental sense, the strategy that drove military air operations in the Haiphong-Hanoi region mostly had to do with interdiction of military materiel that flowed into the country from the Communist world. Part of that overall strategic objective held true for the Ho Chi Minh Trail region, except that the flow of materiel was from staging bases in North Vietnam which were then sent down the line-of-communication system to allied fighters in South Vietnam. So, as can be appreciated, the target systems in North Vietnam had more mass than those within the Ho Chi Minh pipeline. The kinds of targets were generally of the same genre because they both were aspects of the same line of communication (LOC). However, the aerospace ordnance systems were somewhat different because of the differences in massiveness of the individual targets, and their areal concentration. In any case, the most significant target-type in these military aerospace regions was represented by the "choke points" of the lines-of-communication, where the return per unit of ordnance was greater. This has often been the case in any blockade or interdiction operation in warfare. Thus, the most significant target-types in North Vietnam and the Ho Chi Minh were the input points and the output points, rather than the flow-points, although the latter were important too. Regardless of the location and configuration of these nodal points, however, the more important attribute of the targets, in terms of military geography, as I see it, is the specific "signature" of each type of target.

I first became familiar with the notion that each type of target has a specific electromagnetic and chemical "signature" during my tour as an air intelligence officer in the Laos theatre of the Vietnam War. This was the time in air warfare history when the "smart bomb" technology was really taking hold. The "dumb" iron bombs of the World War II era were now being fitted with sensor devices which could detect such signatures and, thereby, provide a more intelligent guidance and control, and fusing process. The detonation event was now being initiated by sensory data, either about changes in the atmospheric environment or based on the electromagnetic or chemical "signature" of the intended target. In the same vein, the patterns of phenomena, or their "footprints," were used as a basis for detecting possible targets even before the Vietnam War era. As an example photo-interpreters used "visible-light" photographs to detect the distinctive star-shape of the Soviet missiles on the ground in Cuba, which precipitated the Cuban Missile Crisis in 1962. This is a case where "P-Is" (photo-interpreters) were able to deduce the existence of the Soviet missile sites in Cuba by their distinctive "star-shaped" configuration on air reconnaissance film that was produced by the U-2s.

It was when aerospace reconnaissance began to use film that was sensitive to waves-lengths in the infrared portion of the electromagnetic spectrum that the concept of a target-type's "heat signature" became more apparent. Now, "imagery-interpreters" began to learn how to "see" heat patterns in their attempt to ferret out potential enemy targets. Ultimately, the technology of "remote sensing" took hold in the 1980s and the idea of the "signature" of various kinds of objects, throughout the solar system, and beyond, came to be appreciated. Another specialized sensor system that made it possible to actually "see" the components of particular and varied signature wavelengths has been the spectroscope. This technology essentially makes it possible to discern the most basic properties of any object (target), the arrangement of its atoms and their movements within a molecule, which determine the quantity and composition of energy that is emitted. It is this unique energy emission profile that gives a target its "signature," which can be specifically and precisely used for identification and measurement of differences. Then, all that is needed is a "smart" sensor-detonation system which can either be integral to the ordnance, or used by an external system which communicates data to the attack platform. This technology will likely be even more important in the conduct of military space operations. Another similar technique is used to measure the apparent difference between the frequency at which sound or light waves, from the time they leave the source, and when they reach the receiver. A classic example of is the shift in frequency that occurs from the time a train whistle approaches the observer, to the point where the train passes, and the sound of the whistle fades away. The observation of the Doppler shifts in atomic spectral lines is a powerful tool with which to measure relative motion in the process of target detection, identification, and tracking.

Regardless of how we detect, identify and track the target, there will always be two basic types of targets on one sense: they either will be passive and active. The difference between the two, in short, is that the first type cannot shoot back, but the second type can. In reality, the additional layer of target nodes, the air defense system, serves to make the original target much more inaccessible and difficult to hit. Now, not only is the overall target system more complex; it is also more lethal for the attacking platforms. So, now the attacking platform either must be escorted by other specialized platforms whose job it is to "quell" air defenses, or the original attack platform itself must be equipped with integral subsystems to ensure its survivability in the now more "lethal" target environment. But it is not just a case of there now being more kinds of targets, which require an additional layer of cost in terms of specialized platforms and on-board equipment. There is also an entirely new dimension to the targeting problem: the so-called electronic warfare that is waged on the virtual electromagnetic spectrum battlefield. Thus, what had begun as a rather straight-forward targeting problem which centered on factors related to cloud cover, gravity and wind, as well as the Coriolis-effect, has now become much more complex.

By the time of the major wars in the last two decades of the 20th century, the electronic battle had become so dominant that specialized platforms had to be developed to deal with the air defense weapon systems which were controlled by electromagnetic spectrum emissions. Thus, in Bosnia and in Iraq, the "shock and awe" effect had to begin with a "blinding" phase of the aerospace attack in which the electromagnetic sensors and control systems were made inoperable. Aside from the weapon control systems, this precursor phase was also designed to degrade communications systems of the radio portion of the EM spectrum. Only when the enemy's air defense systems were rendered electronically "deaf and blind," could the attacks on the other, "passive" targets proceed with the expectation of acceptable cost.

It seems to this observer that the same model is likely to be followed in the hypothetical space wars of the 21st century. Under current international agreement, one of the most important types of passive target systems would most likely be the constellations of artificial satellites that have been placed into orbit around the Earth. Thus, in this hypothetical scenario, the initial targeting problem for attack in space basically will be one of calculating the correct rendezvous or interception point along the target's orbit. In this case, the actual attack on the target would have to utilize ordnance or directed energy that would minimize any collateral damage on friendly assets in nearby space. In the end, however, the problem basically will be of the kind that had been long-solved by the engineers of the inter-continental ballistic missiles on Earth.

In space however, the main force that influences the path or trajectory of the missile or projectile is gravity; often the relative gravitational influence that causes a particular "deformity" in the fabric of space-time. As it happens, such gravitational deformities are the main contributors to the kind of problem that NASA mission engineers must solve any time a rocket is launched to reach a particular point in space. Just consider the precision with which the Mars rover, Curiosity, had to hit a particular orbit insertion point, in order to reach its target orbit point around Mars. The same holds true for the mid-course corrections that must be made. The solution of the targeting in the intra-planetary medium also involves such matters as determining the appropriate "window of opportunity" when one's launch point is oriented toward the target in space, because of the rotations and orbital movements of the bodies there.

Indeed, the very success that has been achieved in the few attacks that have been made on artificial satellites in orbit has provoked a series of countermeasures by the nation-states that have artificial satellites in orbit. One of these has been more political and diplomatic than of the military kind. Thus, there was formulated the Outer Space Treaty of 1967, whose signatories included the United States and the Soviet Union. Among other provisions, the treaty makes nations responsible for their activities in space and liable for any damage which is caused by objects launched into space from their respective territories. This treaty was designed to at least discourage Earth-based missile attacks on the orbiting artificial satellites, but it added one more "cost" to such actions. Unfortunately, one notable nation that did not sign the treaty was China. More effective than the legal proscription, is the practical reality is that, as with chemical weapons on Earth, the risk of collateral damage on friendly assets is simply too great. What is serving as a more poignant deterrent in the near-Earth orbit region of space is the real likelihood that the nation-states which have orbiting assets will deploy anti-missile systems, both passive and active, which will greatly raise the cost of such attacks.

In summary, it appears that the concept of the "cell-region," which is borrowed from the sciences of biology, geography, and physics, provides a useful vehicle for the dynamic analysis of the sortie in the atmosphere or in space. The idea of the cell is a way to encapsulate a single environment at a time during the entire sequence of a sortie. Think of a "black box" of an aircraft that is continuously recording selected aspects of a flight mission. With the technology that is available today, each space-time "slice" data throughout the entire sortie can be acquired and analyzed, and the appropriate actuator or feedback response action can be taken in real-time fashion. This kind of microsecond input-processing-output loop is now possible because of the advances that have been made with respect to sensors, telecommunications, microprocessors and actuators – all operating at virtually the speed of light in at a micro-field level. The next part of this book analyzes these fields as geographic regions.

THE DYNAMIC GEOGRAPHY OF THE SOLAR SYSTEM

———

We have established that the solar system is a geographic region, but it also can be seen to be a field, in the language of physics. In this respect, the whole of the solar system consists of an energy field which is manifested as a complex of electrical, magnetic, plasma, cosmic and solar sub-fields. The physical phenomenon which is known as a solar nebula was the proto-field within the context of this conceptual model. Within this context, the following is an exploration of the various fields of the heliosphere as geographic "uniform" regions (as opposed to the functional region, or spatial system). Because of this, the field can be analyzed and understood in the same way as any natural or cultural geographic region in the solar system.

On another level, we can identify the various nodes of the solar system and describe their characteristics and properties. Within the solar system, these would be the planets and their moons; the asteroids and comets; and the smaller planetisms that inhabit it. Then there are the moving fields of energy. They are flowing, dispersing and propagating phenomena, composed of both particles and energy. These fields include those that are formed by the interaction of electrical and magnetic fields, cosmic particles and energy, the solar winds, and plasmas of charged particles that radiate and flow through the solar system amongst the celestial objects. These "natural" fields of the solar system have more recently been added to by the "cultural" fields of directed energy by man-made emitters in the radar, infrared, and the electro-optical phases of the electromagnetic spectrum.

The largest geographic field in the solar system is the heliosphere itself. Even prior to the establishment of the solar system itself, which occurred some 4.5 billion years ago, there was a precursor force, which we call the solar nebula – a gaseous field from which many scientists believe the Sun and the planets of the solar system were formed, as a result of condensation within the cloud. Now, our star, the Sun, is the dominant emitter of the particles and energy fluxes within this geographic field; known more commonly as the "solar wind," it affects the entire range of military space operations. The Sun is the source of the propagation of the solar wind, which behaves like much like the atmospheric winds that occur on some of the planets and celestial bodies, and which creates the solar field. The solar wind is relatively constant throughout the solar field, but it varies in terms of intensity, from place to place. However, there are other elements of what might be called solar weather that can have a more immediate effect on spacecraft. More particularly, these are the inconsistent elements of solar weather, which include such occasional phenomena as solar flares, coronal holes, and coronal mass ejections (CMEs). Each of these has become familiar to space system designers and their characteristic behavior has been taken into consideration when conducting mission design and planning. Aside from design countermeasures, the strategy that is being taken with respect to these erratic phenomena is to develop systems and procedures for to general surveillance, detection, and warning in sufficient time to be able to implement countermeasures.

Because the Sun is the dominant mass in the solar system, it causes the deepest and widest "depression" in the space-time field that we mostly perceive as gravitational attraction throughout its domain. Seen from another perspective, the gravitational field is also the basic mechanism that regulates the process of accretion and orbit positioning of the matter in the solar system, regardless of the mass that is involved. Thus, through a series of iterative applications of this force, the location of the Sun and the planets, their moons and rings, as well as the other celestial objects within the system has been dynamically established – and is being established. That is, although the basic relative positions of the major planets have been determined during the past 3 billion years or so, there are still perturbations and collisions from wandering asteroids and comets that tend to continue to modify the structure of the system. And the gravitational forces continue to cause tectonic changes inside the cores and the crusts of the various celestial bodies.

The Sun dominates the solar system, not only through it gravitational influence, but also through the effect of the solar wind of charged particles, which propagates out to about 100 AU, towards the farthest reaches of its stellar domain. The Sun has both positive and negative effects on everything in its domain. It provides the Earth and the rest of the planets and bodies within the solar system with light and heat, with variable intensity. Perhaps the most direct effect of the Sun on military space operations is the light energy that can be utilized as power for spacecraft and other space systems. On the other hand, the solar wind, which is a flux of plasma propagating from the Sun, creates a bubble in the interstellar medium. This phenomenon is called the heliosphere, which extends outward to the edge of the solar domain.

The solar wind itself is relatively low in energy in its constant state, but there are spikes in energy manifested by solar flares, CMEs and coronal holes. These are more powerful fluxes; they can extend beyond the orbit of Mars, and they can be damaging to space mission operations. However, on the positive side of the equation, the unfiltered solar radiation can be harnessed more efficiently and can be converted to electricity for use by military space systems. The intermediate medium which actuates the conversion is the solar panel of arrayed photo-electric sensors. One example of this technology is the solar panel on a space-based observatory. These solar panels can extend outward, as much as 56-foot in length. On such a panel, there are arrays of individual photo-voltaic cells, which receive the Sun's energy and then convert it into electrical power, which can then be stored in batteries for utilization as needed. This onboard source of electricity then provides the power to operate the various systems of space operations.

On the other hand, the space-based observatories are continuously being bombarded by flows of solar energy – and occasional bursts of more extreme radiation – which can be detrimental to both humans and their space infrastructures. More specifically, these bursts of radiation pose a hazard to any electronically-based subsystems on the space vehicle. These are the two faces of the Sun as it relates to military space operations. The first is beneficial to the successful completion of the mission; the second is clearly a potential hazard to it. In both cases, the solar energy that is being received and falling on the observatory is more acute than it would have been on Earth, because it no longer has the protective and filtering attributes of the terrestrial atmosphere to ameliorate it. On balance, however, it is nevertheless a fact that the amount of this unfiltered solar energy that is incident on objects in space vehicles is more than adequate to provide electrical power to any number of space stations and even space colonies in theory.

Solar radiation has always affected traditional military air operations, even when it has been restricted to the areas under the protective blankets of the atmosphere and geomagnetic field on Earth. Within our atmosphere, the main influence of the Sun on human activities – including military aerospace operations – has been a function of emitted light and heat that actually reaches through our atmosphere and onto the surface of our planet. Thus, for example, solar heat is instrumental in the heating of the air which, when it rises, causes some of the lift on the airfoil that enables prolonged flight. On the other hand, the heat derived from the Sun also generates the turbulence in the storms that pose a hazard to air flight. Solar radiation also heats objects and the reradiation can then be utilized by infrared sensors in detecting, tracking and directing ordnance in military air operations. And now, increasingly, solar radiation is being utilized to generate secondary light via such mechanisms as the photovoltaic cell.

I have come to realize that the solar system consists of moving fields and, within these fields, there are moving nodes. The former refers to the field that is defined by the meandering orbits of the planets and other celestial objects. The movements of the orbital fields vary widely, both in terms of their orbit around the Sun, and within the range of orbital paths that occur during varying segments of time. Perhaps the most vivid analog on Earth would be the larger river systems, such as the Amazon River or the Mississippi River. These vast riverine systems may be depicted as a line on a map, but the reality of their movements must be seen as a flux, much like the solar wind. Thus, there is a dynamic three-dimensional flow of matter and energy that propagates and re-propagates throughout the width and length of its boundaries. Another analog might be the ocean currents on Earth, whose "orbits" within that medium move in three dimensions along the axis of their respective trajectories. In a similar fashion, the orbit paths that are defined by the planetary bodies in the solar system also contain many internal horizontal and vertical variations and perturbations caused by the friction of relative gravitational attractions in different parts of the solar system. Consequently, any geographic analysis of the solar system must be done from the perspective of an observer of "relative geography." By this, I mean that like Einstein's observer of relative time and space, one's observation point is constantly moving – as are the geographic orbits and nodes that are being observed.

————

Another point to be made with respect to military geography that the outer solar system has now become an active part of the "aerospace" in which military space sorties will take place from now on. As Isaac Asimov might have put it in his book, Asimov's Guide to Science , if I might paraphrase: "No matter how glorious the unimaginable depths of the universe are, and however puny by comparison it may be, it is on earth [the solar system] that we live..." Therefore, the military geographer must acknowledge the greater proportional influence of cosmic, solar, and electromagnetic radiation on military space operations in the region of the solar system that lies outside the Earth's protective magnetic field and atmosphere. Another difference between the environment within the Earth's electromagnetic and chemical "bubbles," and without, is that the availability of oxygen and the forces of gravity become exotic to humans. One of the most interesting difference between conducting aerospace activities in the two environments is that the "weather" conditions on Earth refers to phenomena such as cloud cover and precipitation. Outside the Earth's atmosphere, "space weather" consists of solar winds and micro-meteorites. The degree of exoticness, however, varies throughout the solar system outside our atmosphere. Thus, each of the major planets contains its own oxygen-gravity environment.

It is well-understood that planets of the solar system, including Mercury, Venus, Mars, Jupiter, Saturn, Uranus, Neptune, and our own Earth, orbit the Sun. These, as well as their satellites, can be seen as being moving fields within this regional system. These bodies revolve about the Sun, each in its own orbital path. Within these orbital paths, there is also variability in the movement, motion, and even the direction (in at least one case) of these planets, as they move around the Sun at different speeds. Moreover, each of the planets and their subsidiary nodes are more or less "erratic" in their rotation about their respective axis; and their "attitude" or inclinations vary as well. Finally, these nodes tend to "wobble" a lot; this behavior is derived from the secondary gravitational influences of close neighbor bodies. All in all, it is like watching a bumper car scene at a carnival, and when humans send their artificial bodies, such as the orbiting satellites, into this moderate chaos, they too are subject to the perturbations that are experienced by the natural bodies.

The farther from the Sun a planet is, the longer its "Earth-year" will be. Thus, Pluto, which is farthest from the Sun, has a year that is about 250 years longer than that of Earth. Some of the planets have satellites (including moons and "rings"), which revolve around the planets in the same way that planets revolve around the Sun. Since the start of the age of space exploration in the latter half of the 20th century, there have also been an ever increasing number of "artificial moons" or satellites that are orbiting around the Earth. Among these most recent discoveries is that of Eris, a large (diameter of 1550 miles) and distant node of the solar system. This dwarf planet revolves around the Sun, well beyond the orbits of Pluto, which has recently been recognized as a dwarf planet too. Farther out, there lies the realm of the Oort cloud – an immense, roughly spherical cloud of small, icy bodies that are thought to revolve around the Sun at distances of more than 1,000 times that of the orbit of the farthest planet, Neptune. So great is the distance of the Oort cloud, that it takes the so-called long-period comets more than 200 Earth-years to orbit the Sun, at distances of 40,000 to 50,000 AU. Another group of comets, the short-period ones take less time to complete a solar orbit and reside in the so-called Kuiper belt.

At some level of scale, the solar system also can be seen as a "uniform" region that is characterized by fields of gravitational forces and electromagnetic radiation, for instance. Like the uniform regions on Earth (such as the Corn Belt) the diffusion and distribution of these phenomena is not absolutely uniform; rather, there are local variations (micro-regions) that will affect military space operations in particular ways. One so-called "uniform" phenomenon within the solar system – temperature – may be seen to be somewhat uniform on a macro-scale. However, at a more local level there appear extremes and subtle variations. Thus, on Earth, we speak globally of the extremes of temperature as occurring at approximately 140 degrees (F) at the hotter level of the temperature scale; and about -70 degrees (F) at the colder end of the scale. Within this macro-spectrum, there are many micro-spectrums on the planet, which loosely are related to latitudinal and altitudinal location. But, even within the micro-spectrums, there are shorter spectrums that are the result of diurnal cycles, as well as particular "niche" locations which are a function of variations of solar radiation, winds, and precipitation, among others. These niche micro-spectrums will prove to be greatly significant to military aerospace operations in the future as the trend toward precision and high-resolution progresses; especially with respect to target detection and ordnance delivery.

Within the relative order that exists among the planets and their satellites, there are "celestial herds" called the Asteroid Belt and the Oort cloud; their populations consist of most of the asteroids and comets, respectively. Although these objects tend to follow certain generalized migration with respect to the preponderant gravitational influence of the Sun, they are sometimes captured or at least perturbed by the "gravitational influence of opportunity" of a body of greater mass which they happen to encounter as they move about the solar system. These competing influences sometimes cause them to go apart from the herd, and to go off on their own more erratic paths. When they do so, they sometimes collide with the more stable orbiting bodies, thus causing both good and bad things to happen. On a longer geologic time-scale, these asteroids and comets do generally superficial damage and create cosmetic changes to the outer surfaces of the objects they strike. However, especially during the adolescent years of the solar system, these strikes could generate changes on a global scale, as they have done on Earth when they have caused species extinctions. Even today, even a modest meteorite could produce a cataclysmic event if it were to strike our planet. Further, as humans venture out more and more into the other bodies of the solar system, the strikes such as that which occurred recently on Jupiter will affect us more directly as well.

Based on the most advanced dating of the asteroids and meteorites that have landed on the surface of the Earth, the origin of the solar system has been established to have occurred 4.567 billion years ago – plus or minus a few million years. The most commonly-accepted model of the origin and evolution of the solar system has as its precursor situation as a cloud of interstellar gas and/or dust which is called the solar nebula. It is thought that, at some point, this nebula was disturbed by some particular event, and that it collapsed under its own gravity. According to this theory, as the cloud collapsed – in a period of less than 100,000 years – it heated up and compression occurred in its center. This caused the heat to build up to the point where the dust is vaporized. The center of the cloud then compressed to the extent that it became a protostar, and the rest of the gas propagated outward and was sorted into orbit around the center.

The process still continued, and some of the gas flowed back inward and added to the mass of the star that was forming; meanwhile the gas continued to rotate. The centrifugal force from that rotation prevented some of the gas from actually reaching the star. As the gas cooled down enough for the metal, rock, and ice which results from the process, the metals condensed, the very outset of the accretion process then followed. However, some of the dust particles did not join in the accretion; instead, they continued to collide with one another until some of the particles eventually grew to the size of boulders or small asteroids, or even become planets. In the meantime, other rocks continued to collide, but they only achieve the size of planetesimals, which then were sorted out by gravity and assigned their own orbits around the star.

Now, in the 21st century, the upshot of all these developmental activities is that there are three geographic regions of the solar system that are of major significance to military space operations: geospace, near-Earth space, and Jovian space. Then there is the region that we call the Asteroid Belt, which can be likened to a sandstorm on Earth, albeit on the mega-scale of the solar system. This is the natural geography of the solar system. In the last half-century, there has been developing another "layer" of cultural geography that is being integrated with the natural geography. The features of this overlay include the orbiting artificial satellites, space stations, and other human systems that are operating in the solar system today. Then there are the artificial "fields" of electromagnetic energy are being formed by the various emitters that have been constructed by humans in space. Finally, there is the emerging system of nodes and linkages that constitute the military geography of the solar system; these are being rationalized spatially into military space domains. The first of these is the geospace domain. Later, it is conceivable that there will be other military space domains that will encompass the orbit of the Asteroid Belt and the Jovian subsystem.

One of the most important geographic aspects of military operations in the geo-space has to do with the hazards posed by the asteroids and micro-meteorites. In the case of the asteroids, the probability of interacting with a member of the asteroid system is relatively low, probably on the order of being struck by lightning. Once military space operations are extended into near-Earth space, however, the potential for interaction with a member of the asteroid belt will increase considerably. Now the terrestrial analog would be a hail storm effect, in which the probability of interaction not only increases in terms of the likelihood of being struck by one asteroid, but the size of the asteroid varies from the diameter of a grain of sand, to that of a small moon. The higher probability of being struck by some kind of meteoroid was shown during the Apollo 9 mission of March 1969. Following their return to Earth, the astronaut, James A. McDivitt reported what he thought was being struck by a tiny element of an asteroid while performing extra-vehicular check-out operations outside his Lunar module. Fortunately, it did not penetrate his space suit.

Once again, we see how geographic bodies and fields can be both beneficial and hazardous to the conduct of military space operations. Thus, in the case of the smaller celestial bodies, such as asteroids, it is apparent that they will be available for utilization as military space bases for spacecraft. One such function of an asteroid could be as a supply center, a refueling site, or as a field maintenance facility in support of theater operations. As such, these space bases would not be so different from our present military bases in harsh environments, such as Thule, Greenland or Grand Forks AFB, North Dakota in the winter, for example. What makes the asteroid plausible as a permanent space base has to do with its lower mass and, therefore, the lower escape velocity that is needed to take off from it. The concomitant lower fuel requirements relative to payload capability at an asteroid means that routine take offs and landings could be done with current technologies. Another thing that makes an asteroid attractive as a potential military space base is its composition, which is essentially metallic and water ice, in many cases.

The solar system is also home to a number of subregions which are populated by smaller objects. One of these is the Main Asteroid belt; another is the Kuiper belt of the comets. There also are communities of linked populations of objects in the region of Neptune that are composed mainly of ices, including water, ammonia, and methane. Then there are five objects which are recognized as being large enough to have been rounded by their own gravity. These include Pluto, Haumea, Makemake, and Eris, which collectively are called "dwarf planets." Thousands of other small objects in the outer regions, including the comets, centaurs, and the interplanetary dust populate the solar system as well. Then there are the so-called "homeless" planets which have only recently been discovered. These solitary planets apparently wander aimlessly throughout space, without conforming to any apparent orbital discipline. Six of the planets and three of the dwarf planets have their own natural satellites in orbit. The outer planets are encircled by planetary rings of dust and particles as well.

With a geographer's eye, the structure of the solar system can be seen as being analogous to the planet Earth in terms of its aggregate physiography. Thus, there are the planets which seem to be constantly moving and "drifting" apart as the cosmos itself expands, somewhat like the continents on Earth – only a much faster rate. The inter-planetary area of energy, dust, and the "dark matter" that has recently been discovered by scientists seem to interact with the planets and other bodies like the oceans and the seas do on Earth. There are cosmic and solar storms of radiated particles, just as there are storms of liquid water on our planet. Similarly, the solar winds and move about the solar system and produce all sorts of disturbances and turbulence in the non-contiguous and variegated "atmosphere" of the solar system. In short, the solar system appears to be a larger-scaled version of its constituent planetary systems.

THE REGIONAL GEOGRAPHY OF THE SOLAR SYSTEM

If we view the solar system as a series of concentric circles of matter and energy that radiate outwards from the Sun, then the statistical population of this regional system would consist of the orbital paths that define these circles. Following this model, the planetary orbits would be the major "nodes." However, there would be many secondary nodes in this solar system, just as there are in most regional systems. Among the secondary orbital nodes there are the moons, or satellites, that orbit around the planets themselves, especially the so-called "gas giants" like Jupiter and Saturn. Many of these satellites are not discrete bodies but, rather, particle-sized formations that we call rings. Then, there are the "rogue" orbits that are defined by the movement of the asteroids and comets that have strayed from the home orbits, or belts.

In the inner region of the solar system, whose nodes include Venus, Mars, along with the Earth and its Moon, relatively high temperatures during the early development period made it impossible for materials other than high-melting-temperature compounds to survive. Among these are the metal oxides and the silicate materials which make up these planets. Farther from the Sun, the relatively lower temperatures allowed not only rocky materials, but also more volatile carbon compounds and ices to exist as solids. Aside from these major differences in the composition of the solid component of the solar nebula, there would be differences between the kinds of materials within each region. It is within the inner region that most military space operations are most likely to occur, at least in the short-term. One of the active military areas of operation in space today, centers on the belt of artificial satellites and other spacecraft that has been developing in Earth-orbit since 1957. In any case, it is safe to assume that by the end of the 21st century, there will be other military space regions (or domains), either on or about some of the planets and other bodies of the inner region or; more likely, along the lines of communication that are developing among them.

The terrestrial planets are generally warmer than the planets of the Jovian system because of their closer proximity to the Sun. The average temperature of Mercury, for instance, is about twice that of Mars. At temperatures in the 800 degree (F) range, the light elements, such as hydrogen and helium, possess too much kinetic energy for them to be retained by the relatively low gravity of this group of planets. This is consistent with the theory of planetary formulation where the hot protostar temperatures would have kept the volatile substances, such as hydrogen, helium, ammonia, and methane in the gaseous state and would have prevented them from condensing out during the process of accretion of material by the inner planets. However, in the colder outer regions of the solar system, where temperatures are in the -400 degree (F) range, these substances would have condensed out and become available to the planets.

At the present time, in the early decades of the 21st century, there are two "artificial orbits" above the Earth; these represent the beginnings of the "cultural military geography" of the solar system. One of these, the belt of artificial satellites is a reality, and it is growing in size and complexity – to the point where some experts have called for the establishment of a "satellite traffic control" system, like the air traffic control system on Earth. There are also discrete sub-orbits of artificial satellites that are occurring at various levels of altitude above the surface of the Earth, depending on the specific observational or communication requirements of each satellite system. For instance, there are satellites that inhabit a specific near-Earth orbit because of the requirement for global surveillance and intelligence gathering of the Earth's surface. Then there are other specialized satellites, like the International Space Station, which are placed on a given orbit in the solar system, partly to make them more easily accessible for logistical and maintenance support by Space Shuttles. Other satellites, such as the Hubble Space Telescope, occupy an orbit at higher altitudes above the Earth so they can encompass a wider global scope of a planet, or the Sun.

The geographic distribution of mass also plays a role in the variances among the structures of the planets in the solar system. Without significant gravity, for example, a planet will lose whatever atmosphere it might have gotten during the early formation of the planets. The amount of mass of the bodies is basically derived from their original location in the solar nebula. However, out-gassing from volcanic activity, which is associated with a molten interior and impaction of material, also continues to contribute to the integrity and vigor of atmospheres of the terrestrial planets. In the case of the planet Mercury, its small size and its closeness to the Sun causes it to have virtually no atmosphere. With the higher temperatures that exist close to the Sun, most of its gasses are sufficiently energetic to escape from its low gravitational pull. Mars, for its part, has been only slightly more successful in retaining an atmosphere, as tenuous as it is. Earth's atmosphere is the densest and most complex of the four inner planets. Its atmospheric composition has been further enriched by the gasses that have been produced by its seemingly unique biosphere.

On the other hand, the four outer planets of the solar system, which are collectively known as the gas giants, are comparatively huge relative to the inner planets, both in terms of their overall sizes and their atmospheres. These four planets can be further divided into gas (Jupiter and Saturn) and ice (Uranus and Neptune) giants. Jupiter and Saturn, in particular, are distinctive because of the rings that orbit around them. These bands consist mostly of hydrogen and helium gases. They grow increasingly dense with depth until they reach a rocky core of these planets. From observations that have been made, the gas giants appear to have no real surfaces. However, it must be noted that only the upper levels of the atmospheres have been observed, as of now.

It should also be stressed, at this point, that our analysis of the military geography of the solar system, in general, is still very much a work in progress. Indeed, it can be reasonably said that we are only beginning to formulate the methodologies for studying the complex interrelationship between the physical geography of the heliosphere and the conduct of military space operations. So far, the methodologies that geographers have utilized with respect to our own planet are proving to be valid throughout the solar system. So, we are constantly learning how to study the subject by applying what we have learned about it on our own planet. On the other hand, much has been learned from non-military exploration and scientific analyses that have been done – and are being done – by the NASA scientists, and others, who are focusing their attention to the solar system in all its aspects. Thus, for example, the effect of the absolute mass of a body and it relative mass with respect to other bodies on military space operations is just beginning to be understood.

————

One of the things that geographers find interesting when studying a region, like the solar system, are the patterns of movement within it that define subregions. Consider that the planets and the other celestial bodies of the solar system are engaged in constant movement. They trace orbits that vary as to their distance from the Sun, and within certain dynamic boundaries. I see the "orbit" as the essential manifestation of the geometry of the solar system and the most obvious expression of the regional boundaries within the solar system. Thus, from the perspective of the orbits, the heliosphere region can be seen as being organized in terms of concentric "rings" around the Sun, somewhat like the rings around Saturn; only the particles in these rings are much more massive. These subregions of the heliosphere also are quite varied as to their internal composition; some are rockier, others are more gaseous, and still others contain water ice. Then, there are the various fluxes of charged particles that propagate through the region, such as the solar wind. Finally, there are the "bubbles" of charged particles that form around planets when the solar wind interacts with the magnetic field of some celestial bodies.

The absolute position of an orbit within the heliosphere can be said to be the result of a "dynamic tension" between the relative gravitational attractions and centrifugal forces that are occurring within its spatial neighborhood. These celestial mechanics also are occurring within the solar system, which is itself is flying through space in the shape of a speeding bullet, at approximately 62,000 miles per hour, according to reports from NASA's Voyager spacecraft. All of this is occurring within the boundaries of the heliosphere, a bubble of charged particles that interacts with the interstellar magnetic field.

Most of the planetary orbits in the solar system describe an ellipse. An ellipse is a closed plane curve generated by a point, moving in such a way that the net sum of its distances from two fixed points is a constant. The ellipse also is the geometrical form that is the basic reference concept when describing and measuring the movement of the planets and their moons in the solar system. This form gives rise to the many points of relative location with respect to the controlling planet, or vis-a-vis other celestial bodies. Most planet-moon systems follow a path of movement that is described as closed; as opposed to an open orbit which is an open curvilinear path. They constantly travel around the Sun in what are called heliocentric orbits. In the case of the moons, they orbit around a parent planet. Then there are the clusters of asteroids and comets that also maintain a heliocentric orbit. Many "wandering" celestial bodies maintain somewhat erratic orbital movements throughout the solar system that complicate the basic pattern of orbits. Among these are the asteroids and comets that have left their home belts as they seek a final resolution of their own dynamic tension among forces. One of these is the turbulence or perturbations that are caused by the tidal forces among the various bodies of the solar system. Thus, we are left with an initial impression of a mix of discerned order and seeming chaos as we begin our "descriptive analysis" of our solar region.

Another, more detailed, analysis of the orbits of the major bodies in the solar system reveals additional information about the orbital characteristics shows that the eccentricity of the orbits of Mercury and, to a lesser degree, Mars, is more pronounced than the other planets and the Moon in the inner region of the solar system. One would also notice that, from a polar perception, where the observer is looking down on the Northern hemisphere of the Earth, the planets all appear to revolve in a counter-clockwise sense. If one were to project a plane outwards from the Earth's Equator, one would see that the some of the planetary orbits are above that plane; others are below it. From a "side perspective" of the solar system the tilt of the planetary orbits would also vary as to the elliptical plane. And, the movement and motion of the innermost planets would be much faster than of the outermost planets. Then there is the relatively erratic planetary migration which occurs when a planet or other stellar satellite interacts with a disk of gas or planetesimals, which results in the satellite's orbital parameters being modified.

―――――

At this point in analysis of the geography of the solar system, we should note that there are many "tools" that are available to the geographer with which to begin to develop more abstract generalizations about the phenomena that is being empirically observed. Once again, we find that there already are some basic "rules of the game" (or laws) that have been developed by other scientists in fields such as astronomy and physics that will be valuable to the geographer. Among the most prominent of these scientists, who derived their models from empirical observations, are Johannes Kepler and Isaac Newton.

Kepler essentially sought to formalize abstract generalizations about the empirical observations he made about the solar system. His laws, therefore, can be seen as an attempt to describe the geometry of the solar region, including the distances between the central node (the Sun) and the respective planets, or between the planets themselves. His first law states that the relationship between the "central node" of this region, the Sun, and the celestial objects that revolve around it, follows a definitive rule; that the orbit of each of the planets forms an ellipsis and that the orbits revolve around the Sun as their common focal point. The second law states that a line that is drawn from the Sun to a planet is equidistant at all times; and the third law is a rule that defines the distance of between a planet and the Sun, or from one planet to another – if one knows its orbital period.

Isaac Newton deals with the basic principles of motion with his three laws, which are the basic principles of modern physics. So, in physics, motion refers to changes in the position or orientation of a body, over a given period of time. The concepts that derive from his principles include: translation, rotation, velocity, and acceleration. Translation is the movement along a line or curve, the motion that changes the orientation of a body is called rotation, velocity refers to directed speed, and acceleration is the time rate of change of velocity. Newton's first law states that the momentum of an object remains constant unless an outside force acts on the body. According to the second law, there must be a force that is acting on a planet to cause its path to curve toward the Sun; this force is gravitational attraction. More generally, Newton, in his law of universal gravitation, also concluded that the force of attraction between two massive bodies was proportional to the inverse square of the distance between them, and the product of their masses. These models of the solar system are not unlike those that have been developed by geographers to explain the functional region on Earth. One that has proven useful in a variety of arenas is the Central Place Theory, which seeks to ascertain the controlling variable(s) that determine the absolute and relative location, and the movement of human "actors" within a functional region.

In the case of the solar region, it is the prevalent notion among physicists and other astro-scientists that the overall shape of a planet in the solar system is determined by the effects of external and internal gravitational forces, and the rotation of the planet on its axis. Gravity pulls any object of sufficient mass into the shape of a sphere; only smaller objects (like asteroids) in the solar system have non-spherical shapes. Generally, all of the planets exhibit a similar internal configuration. The classic model is that of a sphere which resembles an onion, with layer upon layer of material radiating outwards from a central core. The size and composition of the core is the result of a process in which the heavier elements drop to the center of the globe. Since this process is variable and not all cores are the same. Some are made of molten iron, while others are more liquid or gaseous in composition. Indeed, Venus may not have a central core at all. Similarly, variable tectonic processes either move or distort a planet's surface to produce several outcomes. We know that they have played a crucial role in the configuration of Earth's surface, and are now discovering that they also operate on other planets and moons in our solar system, but to varying degrees.

Another geographic feature that occurs on many planets and some other celestial objects is a magnetic bubble that envelopes these bodies. Again, the nature and vigor of this bubble of charged particles is a function of the interaction between the core of the planets and the solar wind that impacts them. It is actually the core of the planets that produces a more or less powerful magnetic field around them, and which then interacts with the solar wind and other cosmic radiation to form the magnetosphere. It is thought that the swirling motion of the liquefied outer core behaves like a spinning conductor in a generator of a bicycle. The process is driven by radioactive heat and convection currents which are rising through the outer core to the surface of the planet. The molten metallic material swirls around, and due to the fact that it is electrically-charged, a continuously changing electromagnetic field is propagated around the planet.

With the above in mind, geographers continue to operate on the basic assumption that all human endeavors can be seen in terms of their spatial manifestation, that is to say, as occurring within geographic regions. The overall region of military operations in the 21st century is the solar system. Within the solar system region, there are several subregions. In the context of military space operations, the natural region can be understood as being any place in the solar system (including the entire solar system), which includes the Interplanetary Medium (IPM), as well as the planets and other matter and energy phenomena that occur within the IPM. On another dimension, there are evolving certain "cultural regions" (such as the belt of artificial satellites that are orbiting the Earth) that are being inter-woven with the natural regions. All together, this is the "environment" which interacts with the human activity that we call military space operations.

The natural region that is called the solar system can also be seen as comprehending sub-systems that interact together to develop the overall operating system. These can be seen as the smaller components and processes that maintain the operating integrity of the whole system. In the broadest sense, the solar system comprises the thinly-scattered matter that exists within the planets and other bodies, and the bodies themselves. Two subregions of the solar system that have been better-understood during the past half-century subregions of the outer solar system, are the Main Asteroid Belt, the Kuiper Belt and the Oort cloud, which are composed of smaller objects in the solar system, including the asteroids, comets, and even smaller objects which we generally refer to as meteoroids or fragments. Then there are the quasi-gaseous regions; the plasmas and magnetic fields that are like the suburbs of the Sun and the planets. Another way to categorize the natural subregions is to view them as various configurations of material and energy. The individual components of these regional systems vary greatly as to size and mass. We are familiar with the relatively massive planets, moons, asteroids and comets. But now we know that there are also populations of micro-objects that are occurring and interacting within another scale-context. This is matter and energy that operates at a sub-atomic level. These too are affected by electromagnetic, gravitational, and the strong and weak forces that enforce the laws of physics – albeit under a different model than the larger matter in the Universe. This is the domain of quantum physics.

———

From the perspective of the geographer, the solar system also can be seen to be a multi-nodal region, like many that I have encountered in a multitude of regional studies on Earth. Like all functional regions, this one is a type of geographic spatial construct wherein a field of single nodes is interconnected by a web of linkages to form the whole interactive and interdependent system. This type of region also occurs throughout the solar system, at many levels of scale. Where ever one encounters it, the overall diameter or size of the nodal region seems to be a function of the number and extent of the relevant processes that give it its identity. The functionality is what distinguishes nodal region from an area of a uniform region, such as a corn belt or an asteroid belt.

In conducting the geographic analysis, one also attempts to quantify the features of the array of nodes (or population) that might occur within my conceptualized view of that particular space. Often the quantities and the intensity of some properties of the units of population provide some usable information. Also, in the eye of the geographer, it is the nature of the perceived relationships that appear to occur among nodes in space that convert an "area" into a functional region. Sometimes the pattern is stationary, but often it is dynamic in terms of its morphology and physiology. In the latter case, one seeks to discern the patterns of distribution of the populations in an effort to gain insight into the nature of the system. That is to say, being able to "connect the dots" then enables the observer to apprehend the relevant details of the nodes as well.

On the other hand, in conducting a regional analysis, it is sometimes also fruitful to "telescope" the nodes so as to study their internal properties and characteristics in larger, more detailed scale. Thus, using any one of the planets as an example of this geographic technique, we can develop a methodology for conducting "reconnaissance" of the various nodes within the regional system. NASA apparently has created a protocol for the primary iteration of analysis of the planets and moons in the solar system. It serves their purposes in terms of space exploration and the furtherance of space science. The military geographer may also begin an analysis of these celestial bodies for purposes of conducting military space operations in the solar system. In this regard, the protocol for studying the planets that is followed to first study the composition, structure, heat and pressure characteristics of the planetary atmosphere and surface.

As we have seen, one approach to the study of a functional geographic region is to focus on the linkages which connect the "dots," as it were. On the other hand, sometimes it is more productive if one is able to comprehend the nature of the individual nodes, or class of nodes, that make up the region one wishes to study. This, however, depends on one's field of vision which, in turn is a function of the point of observation, and the sensor systems that are available. Stated in the language of the psychology of perception, it depends on whether one wants to focus on the foreground or the background of a reality. In the case of the Earth scientist, including the geographer, it was not possible to comprehend the whole of the subject of study, simply because humans have historically been "too close to the tree to see the forest." What was required to overcome this narrow field of vision was to move our "sensor" outward from the subject of study; that is to say, to establish one's focal point at a distance of about 200 miles above the Earth.

But, simply changing the focal point alone is not the total answer either. There still has to be some kind of combination of a wide field of view with a greater appreciation of detail (the individual tree if you will). Thus, when humans finally landed on the Moon and were able to view and photograph the planet Earth from a distance that brought the entire globe into view, it was realized that the "astronomer" had only exchanged the location of the "observatory" from Earth to the Moon. So, even though humans could appreciate the home planet as a globe and a sphere, only a portion of it – the part of the sphere that presented itself to the eye or the camera at any particular point in time – could be seen in one field of vision. One solution to this problem of observation is to simply let the planet Earth rotate on its axis, and to continuously photograph the presented phases as they pass one's observatory. Indeed, a variant of this approach has turned out to be a process in which the observatory (or observatories) revolves around the Earth and photographs the surface of the sphere – one swath at a time.

This is reminiscent of the techniques employed by high-altitude air reconnaissance aircraft, such as the U-2, as they flew over the Soviet Union during the Cold War. Indeed, the early space-based reconnaissance of Earth utilized the same basic technique to "map" the sphere. That is, the "strips" of imagery were "pasted" together to form a kind of mosaic map. Another technique involved the streaming of imagery – like television images – to a receiver on Earth. So, in an almost ironic twist, the art and science of terrestrial astronomy is now being applied to the observation of all the planets of the solar system, including the Earth. The main difference is that the space age astronomer can now choose to place the observatory at any point in within the solar system.

Another advance in the technology of observation and imagery presentation has been the improvement in computer graphics systems. In the 1960s, vector graphics were used to reconstruct images out of straight-line segments, which were then combined for almost real-time display on specialized computer video monitors. This was better than having to, in effect, "cut and paste" live imagery onto video mosaics, but it still lacked the resolution and clarity which the space scientists and other end-users hoped for. However, by the 1980s, raster graphics (like that which had been developed for television) became available to space scientists. In addition, computer technology had progressed to the point where extremely large "bit" maps could be stored in computer memory and displayed onto video displays in real-time. The observational technology was now making possible for the geographer to comprehending regions as large as the solar system. So, humans are now able observe larger portions of the solar system at a time and, in the manner of the geographer with respect to the Earth, is able to utilize ever more powerful observational and analytical technologies in order to analyze the solar system as a region, in its entirety, or in its parts. However, observation alone is insufficient to understand the overall structure and workings of the solar system. As with any region that is the subject of analysis, there must be some type of logical order imposed on it. Some of this can be done by intuitive thinking, as has often been done in the science of physics and astronomy, but at some point, the scientific method will demand some precise measurement as well. Fortunately, there is a precedent for drawing conclusions about an entire population from direct analysis of a representative sample of that population. Humans have been applying the methodology of sample observation and statistical generalization, as well as the utilization of indirect or analog measurement to learn about the geography of our own planet, long before we acquired the vantage point of outer space for observation and the technology of the remote sensor and computer to enhance our powers in this regard.

———

Another approach to modeling in the analysis of regions is to depict an area as an abstract construct of a plane, nodes and linkages. This has proved to be quite powerful in the analysis of all manner of functional regions, including battlefields and the domain of the urban centers. One major advantage of this kind of spatial abstraction is that the nodes and linkages can be assigned quantitative values relating to dimension, intensity or speed, for instance. These numerical values can then be manipulated in various ways by computer-modeling to produce quite powerful information to the analyst. The central node and the subsidiary nodes of the region, and their interrelationship thus become apparent and can be analyzed as a system. Also, the region itself can be depicted as one node in an overall hierarchal network of nodes. At the same time, the frequency and timing of one's observations will affect the outcome of the analyst's interpretations. In the case of the solar system region, the use of abstraction highlights the fact that this spatial system is dominated by a hierarchy of nodes which are linked by a series of oval orbits to the central node, the Sun. The overall region, called the heliosphere, can then be further analyzed within the context of the system of galaxies.

Therefore, we can see that at our present point in human history, geographers and other scientists are now able to utilize the increasing powers of observation, measurement, processing and the derivation of true "information" in the study of our solar system, as never before. So, with their greater power and resolution, our telescopes and other sensors are discovering new celestial bodies and other phenomena on an almost monthly basis. Some of these telescopes are now deployed in orbit around the Earth, above the obfuscating clouds and the other distorting elements of our atmosphere. Other "in situ" (as opposed to remote) sensor systems are now gathering and relaying back data from the atmospheres and surfaces of the other planets, moons, and smaller bodies in our solar system. Perhaps most important, however, is the fact that these sensors are now able to operate in electromagnetic wavelengths other than visible light.

Our arrival at this point in the human understanding of our solar system also is the result of many precursor learning-events which have occurred over the past several thousand years. Perspectives have changed, as have the tools for observing and analyzing the structure and physiology of the region that we call the solar system. It is now almost universally-accepted that the planets and their satellites revolve around the central node: our Sun. The traditional focus on the Sun and the planets, also has given way to our modern understanding of six categories of objects. These include the planets, moons, asteroids, and comets, as well as magnetic fields and plasmas. Since the beginning of the space age of exploration, there is now also a greater understanding of the solar wind and its interaction with planetary magnetic fields. That is to say, we now see it as a continuous outflow of particles from the Sun which "blows" throughout the solar system, and which are only hindered by magnetic fields. We also have a more precise knowledge of how the solar wind also flows along a spiral path which is dictated by the magnetic fields in the solar system, and is propagated out from the Sun into the interplanetary medium.

In general, it can be said this solar system and its properties are no longer just a matter of logical reasoning and inferential knowledge. Now that humans are actually conducting operations (doing work) in outer space, beyond the comfort of our own atmosphere, the mechanics and the composition of this new geographic realm are a matter of practical necessity – particularly with respect to the conduct of military operations. So, now the realities of zero gravity and oxygen, as well as that of the cosmic and solar radiation are a matter of operational concern. NASA, the U.S. Space Agency has been leading the way in the scientific study of our solar region, since the 1960s. One such effort is the Dawn mission, a part of the NASA Discovery Program, whose mission statement involves the push to develop new technologies and to explore some of the areas of the solar system that were previously little-known or understood. In the case of the Dawn mission, the spacecraft has been traveling through the main asteroid belt, and orbiting Vesta and Ceres, two of the largest asteroids.

One subregion of the solar system is comprised of the group rocky-terrestrial planets: Mercury, Venus, Earth and Mars; as well as the Moon. These are considered a subregion mainly because of their relative nearness to the Sun and their unique rocky composition. Another subregion of planetary systems comprises the so-called "gas giant" planets that revolve farther out around the Sun, beyond the orbit of the main asteroid belt and the Jovian complex of planets and other bodies. This group of planets includes Jupiter, Saturn, Uranus, and Neptune. And since the middle of the 20th century, there has been developing a "cultural subregion" which is made up of the artificial satellites that are in orbit around the Earth. This zone of artificial satellites and spacecraft can also be seen as a military space domain that is currently operating within what might be called the "geospace military region." A subregion within the overall geospace military region is the one that has been developed on the surface of the Earth, including the command and control function, a network of telescopes for tracking near-Earth objects, and other sensor and even weapon systems that interact operationally with natural and artificial objects in orbit.

––––––

So, now we can analyze the U.S. military space operations that are occurring within this Military Space Domain, which is under the operational command of the Air Force Space Command. It has established military bases in Colorado and California, among others, to provide the physical infrastructure involves launch facilities, recovery bases, satellite constellations that perform a variety of missions, and the all-important cyberspace protection mission as well. Its geographic area of responsibility extends from the surface of the Earth, through the 200-mile altitude zone of the artificial satellites, and outward to about 22,000 miles (which is the altitude that is necessary to achieve a geosynchronous orbit with respect to the Earth). Its main weapon system configuration at the present time is in the form of surface-to-air missiles which can strike targets in the low-Earth orbit space. This military space region (domain) has been active since the 1960s, and its mission continues to evolve as the overall presence of the United States in that region continues to grow.

Based on the increasing space exploration activity – and the many precursor operations that point toward increasing commercial exploitation of space – there is another military space domain that is quite likely to take shape sometime in the middle of the 21st century. This region would encompass the orbits of the Moon, Mars, and some of the asteroids that lie within it. If the overall plans for the utilization of Mars and the other nodes within this regional system were to come to fruition by 2025, as many experts expect, it is likely that Mars would be the central node in this emerging military space domain. As for the other major nodes in this potential system, the Moon has already been visited several times by crewed NASA missions, as have at least two of the major asteroids (Ceres and Pallas). Now, in 2012, it seems that the tempo with respect to the exploration and scientific study of Mars is now increasing as well. Thus, the continuing mapping and surveying of the Red Planet by orbiters and the trio of rovers that are doing geological and chemical studies of the planet, all point to a seemingly logical outcome in the future that may even involve a manned base on Mars, or the possibility of mining operations on nearby asteroids, or Mars itself.

The third military space domain that I see forming in the latter decades of the 21st century is one that would coincide with the so-called Jovian System. This region would encompass the main asteroid belt, Jupiter, Saturn, Uranus, and Neptune. The most important nodes in this hypothetical military space region, however, could turn out to be the moons of the gas planets. I can foresee a time in the latter part of this century, in which Europa, Titan, Triton, and the other moons will be used as staging bases for any military space operations occurring in the outer edges of the solar system. ...which then might lead to the development of a military space domain that would be centered on the frontier zone where the heliosphere blends with the interstellar region.

The Orbits of the Solar System

Spatial patterns of movement always attract the attention of the geographer, and those that are seen in the solar system are no exception. The longest of these are the periods of revolution; the movement of the planets around the Sun. Within this major orbital movement, there are many subsidiary, minor orbital movements of the moons and other satellites around the planets themselves. Then there are the longer-range, often erratic movements of the asteroids and comets that have escaped from the more orderly movements of their colleagues in the main asteroid belt and the Kuiper Belt of the comets. Weaving through all these movements of more discrete objects are the fluxes of plasma and charged particles that are propelled by the solar wind and the interaction between electrical and magnetic fields. More recently, there are the human-made satellites and other spacecraft that move along their own orbits around the Earth and other planets and smaller bodies in the solar system. These can also be seen as subjects for analysis that have a spatial character and the analytical techniques of the geographer can be quite useful for the understanding these phenomena. The combination of facts, techniques and theories presented here establishes a geographic view of the solar system and its sub-systems. Learning to think in spatial terms is an important step in understanding spatial theories and analytical techniques which provide us with new insights about the overall Universe, as well.

Geographers have also developed a set of basic spatial concepts that can be useful in the analysis of any "place" within the solar system, the galaxies, or the Cosmos, for that matter. Beginning with the concept of the "point," we may then derive two additional elements so essential to geography that they are considered primitives in their own right. Thus, a series of points arranged sequentially creates a "line." A collection of adjacent points that are arranged in a nonlinear fashion, or a line which closes upon itself, defines an "orbit" which comprises an "area." Thus, point, line, and area are concepts that are important in any geographic analysis, including that of the solar system, for several reasons. Not only do they form the basic or primitive elements in various systems of geometry and topology, but they may also be used to describe all manner of phenomena in the real world. Another concept, "volume," is derived logically from area. Two-dimensional distributions are plotted along two axes, which are designated as x and y. The third dimension is plotted on a third, or z, axis. Geographic locations in the form of points, lines, and areas are always depicted by x and y coordinates. On any map of suitable "scale," points may be thought of as concentrations of foci. Lines represent a double function as either paths or movement, or boundaries. Areas show the extent of things and represent distributions or dispersals. Or, areas can be used to generalize and classify the reality around us.

The approach that seems most efficacious in understanding the geography of the solar system to me involves the placing emphasis on the orbit as the primary geographic primitive in the geography of solar system. This geometric form seems to occur throughout our Sun's gravitational hegemony over the objects there, as well as the subsidiary gravitational hegemony among the planets, moons, as well as communities of smaller bodies, such as the asteroids and comets. There are two main types of orbits: open and closed. As we have seen, the closed orbits can be either circular or elliptical in shape. In either case, however, a body on a closed orbit constantly moves about another, such as in the case of the planets orbiting the Sun. An open orbit follows a mathematical shape, either as a parabola or as a hyperbola. Both describe sweeping curves that never join together. Thus, objects on open orbits simply fly by other celestial objects – like some spacecrafts and the comets. It seems to me that all of the major planets of the solar system are best defined and analyzed spatially in terms of their orbital characteristics. This is true in the sense of describing the movements of these bodies, but it is even more fruitful in the analysis of the relative locations of the planets and their moons (as well as the belts of asteroids and the clouds of comets) in terms of the orbital behaviors. Any attempt to define the "absolute" location of these bodies in the solar system can only be transitory, both in space and time, however. It would be like trying to give a lasting definition of a moving NASCAR race car along an oval track.

From the various NASA mission reports, it appears that the planning for space operations always involves the orbital imperative for moving about in the solar system. NASA space missions, for example, always define their flight paths in terms of orbits or segments of an orbit, such a transit or spiral orbit. Only the shortest flight paths can take a straight-line path from one point in space to another. In a sense, this is an extension of the reality of long-range aerospace missions within the Earth's atmosphere that is based on the concept of the great circle over a sphere. The relative efficiency of the great circle trajectory or orbit in space operations has also been proven in the planning of missions to the various planets and areas of the solar system. In the particular case of military space operations, it appears that military space flight planning will also occur within the matrix of the orbital trajectory, not only because of the efficiency of the orbit in terms of fuel and other cost considerations, but also because that is where the adversary will also likely be operating in space.

Like the boundary and the zone with which all geographers are familiar in the study of regions, the orbit is a mental construct which exists only in the mind of the beholder. However, like the notion of the boundary, it is a geographic object in the sense that it can be plotted on a "map" and can be communicated in the abstract or concrete from one mind to another. Thus, in the case of the solar system, which is itself a mental construct, it can be organized in the mind as a series of open and closed orbits, whose parameters and other metrics can then be catalogued and stored in a data base of some kind or other for use in later mission planning. The orbit will always be an essential geographic concept in the practical applications of flight planning in space. Consider the reference to orbits in the various propulsion systems that are now being utilized in mission-planning by NASA engineers prior to the missions to the outer solar system. "Two modules are launched into low-orbit…" This reflects a realization that the ion-propulsion rockets are too weak to break out of Earth's gravitational pull for normal rocket liftoffs, but that the technology has proven useful for accelerating or decelerating in space. Also the ion-propulsion rockets have been used successfully to escape the gravitational pull of less massive bodies in the solar system. So, the practice has been to utilize this technology to slowly push a spacecraft in lower-gravity environments. Another tactic involves the use of ion-propulsion rockets to launch smaller manned-spacecraft a sufficiently high altitude. It could then rendezvous with the main spaceship, which would utilize heavier chemical rockets to escape the Earth's gravity and, eventually, be inserted into lunar orbit. (Scientific American (December, 2011).

The Hohmann Orbit is an elliptical trajectory, which is named for Walter Hohmann, a German scientist who was already working on the mechanics of space travel in the early decades of the 20th century. Also referred to as the "transfer orbit," it is considered to be the most efficient path for a spacecraft to take from one orbit to another, in terms of both fuel and energy. Such a transfer orbit just touches both the original orbit of the planet of departure and the orbit of planet of destination body. Therefore, it can be used for changing the orbit of an Earth satellite or for sending a probe to another planet. It involves two firings of the spacecraft's engine: one to break out of the original orbit and another to enter the destination orbit. The chief disadvantage is that it requires relatively long flight times. However, this can be overcome by the use of gravity assist maneuvers. The overall pattern of movement is essentially that of a spiral, in which the path of a spacecraft is moving around a central line, while continuously receding from or approaching it.

The actual location of a target or destination planet is in space can be determined by simply tracking the position in the sky and then using what is known about the kind of orbit describes in order to predict where it will be at some future point. This kind of prediction is fairly accurate over a period of more than 100 years, and it can therefore be treated as a known factor. When travelling among the planets, the objective is to minimize the propellant mass that is needed by the spacecraft and its launch vehicle. This also makes it possible, with current launch technologies and costs to make such a mission achievable. Incidentally, the orbit could very well become the Line-of-Communication (LOC) of the solar system military area of operations. In such a case, the strategy would be one of interdiction and instead of rendezvousing with an object in orbit the objective would be to damage or destroy the enemy object.

To launch a spacecraft from Earth to a planet such as Mars, and using the minimum of propellant, we must first take into consideration that the spacecraft is already in orbit around the Sun, even as it sits on the launch pad. The image that springs to mind is that of a person standing on a circularly moving conveyor belt in some airport. As a consequence, the existing solar orbit of the spacecraft on the launch pad must be adjusted in order to cause it to take the spacecraft to Mars. The desired orbit's perihelion (closest approach to the Sun) will begin at the distance of the Earth's orbit; the aphelion (farthest distance from the Sun) will arrive at the particular distance of the orbit of Mars. This, in other words, can be described as the Hohmann orbit for that particular set of spatial circumstances. So the spacecraft lifts off the launch pad and rises above the Earth's atmosphere, and uses additional energy from its rocket to accelerate in the direction of the revolution of the Earth around the Sun. If just the right amount of energy is added at perihelion, it will cause the new orbit of the spacecraft to have an aphelion equal to the orbit of Mars. The acceleration is tangential to the existing orbit. After this brief acceleration away from Earth, the spacecraft will have achieved its new orbit, and will simply coast the rest of the way toward it rendezvous point with the orbit of Mars. Ultimately, the Hohmann orbit to Mars requires about 260 days; therefore, it takes about 32 months in order to make the round trip, allowing for a waiting period of 455 days on Mars while the relevant planets realign themselves properly and thus enabling the returning spacecraft to successfully rendezvous with Earth.

As is the case with aeronautical navigation, location and direction of a particular orbit is determined through the use of a "geographic" frame of reference in space, which is called the celestial grid. At any given moment some celestial object is at the zenith of any particular location on the Earth's (or other planet, or platform) surface. This is called the "ground position" (GP) of the body. Ground position can be stated in terms of celestial coordinates, with the declination of the celestial object being equal to latitude, and the Greenwich hour angle being equal to longitude. These concepts of space navigation which guide the orbits in the solar system also provide a scheme for locating the positions (at any given time) of the planets and other celestial bodies that orbit around the Sun along the elliptical plane of Earth. There are also gravitational "regions" of dynamic attraction among the planets and the moons which have become more meaningful to space operations since the beginning of the modern space age.

One of these orbital systems, to which we commonly relate now, is the orbital relationship between the Earth and the Moon as a single orbit pattern, wherein the Moon orbits the Earth. However, it is more accurate to say that the two bodies orbit one another about a common center of mass, which is located within the Earth, about 2,900 miles from its center. Actually, it is this center of mass that follows an elliptical path around the Sun. In the course of the movement of the Moon around the Earth, the distance between the two bodies varies greatly, mainly due to the effects of the combined gravity of the Earth, the Sun, and the planets. Thus, during the last three decades of the 20th century, the apogee (farthest point) of the Moon ranged from 251,000 to 252,700 miles; the perigee (nearest point) ranged from about 221,000 and 230,200 miles. The eccentricity of the Moon's orbit causes it to travel faster in the nearest part of the orbit and slower in the farthest part of the orbit. This orbital eccentricity also affects solar eclipses, in which the Moon passes between the Sun and the Earth, thus casting a moving shadow across the Earth's sunlit surface. In an illustration of the fact that all the aspects of an orbit are generally dynamic, the time in which the Moon revolves around the Earth is slowly changing, as is the time between full moons. As a result, the Moon is slowly receding from the Earth and, consequently, the days and the months are getting longer over astronomical time. Until the beginning of the 1960s, the above facts were primarily a matter of intellectual interest to humans. This all changed in 1969 when the first humans landed on the Moon.

Ever since the start of the NASA Apollo program, humans have been gradually constructing what is, in effect, a line-of-communications regional overlay on this natural subregion, within the solar system. Aside from the zone of the orbiting artificial satellites, which is more properly a node within the geospace subregion, the nodes of the developing Near-Earth subregion have been steadily put into place by space agencies, like NASA, the European Space Agency (ESA) and others during the past 50 years or so. Actually, the central node(s) at this point in history are the launch facilities and tracking systems that are deployed on the surface of the Earth; "Houston" of the Apollo program is one classic example. The Moon became the next major node of this emerging space region even before the landing of the astronauts in 1969. That is to say, it had become a "target" node even in the early 1960s when NASA and the Soviets were conducting "reconnaissance and research and development operations in space in preparation for the "amphibious" landing of 1969.

Even during the planning and preparation phase of the lunar landing operation, the orbit had become the dominant geographic linkage feature in the system that was developing. It began with the so-called "low-Earth orbit" zone which lies about 200 miles in altitude above the surface of the Earth. This is the lowest altitude that spacecraft must achieve in order to orbit the Earth. At about 200 miles in altitude, spacecraft in this zone of orbits are able to circle the Earth once every 90 minutes or so. Thus, this altitudinal zone has become the manmade region of artificial satellites that has been created by humans since 1957. An orbital projection scheme also provides one with a construct for developing a "mental map" of the planets and the secondary bodies in the solar system. This is the now-familiar elliptical plane on which all matter in the solar system is plotted. Thus, the basic understanding of the planets and other bodies in terms of their distance from the Sun now became more complex and subtle as the orbits themselves have become primary objects of "geographic" analysis. Before the era of Sputnik I, which was placed into low-Earth orbit by the Soviet Union in 1957, the only objects that were orbiting the Earth were the Moon and some asteroids. Today, in the first decades of the 21st century, there are thousands of artificial satellites that are orbiting Earth; and there are even people constantly orbiting our planet in the International Space Station. All geostationary satellites orbit the Earth, west to east, along the Equator at a height of about 22,000 miles. At this height, they orbit the Earth at the same speed as the rotation of our planet; therefore, they appear to be stationary to a ground observer. One group of these, the Geostationary Operational Environment Satellites (GOES) are found in the Clarke belt. An advantage of the geostationary orbit is that the communication ground stations do not need to move in order to track the target satellite. Below the Clarke belt, at heights between 100 and 12,000 miles, are found some artificial satellites, such as the Hubble Space Telescope which maintains a height of about 375 miles. Intelligence satellites reach orbits as high as 60,000 miles, as do satellites that provide early-warning of any ICBM launches from the surface of Earth.

Farther out, away from the Sun, the next target node in another as yet hypothetical military space domain, lies the orbit of the planet Mars. Mars moves around the Sun at a mean distance of 140 million miles; about 1.4 AU. Because of its relatively elongated orbit, the distance between Mars and the Sun varies from 128 to 155 million miles. At its closes approach, Mars is less than 35 million miles from Earth, but it then recedes to about 250 million miles when the two planets are on opposite sides of the solar system. What all this seems to mean for space operations in the near solar system region is that the solution to any particular orbital path of a space mission requires a level of precision greater than that which is provided by nominal distance from the Earth and the Sun. That is to say, the planets and other bodies, including man-made satellites and space stations become moving targets, which then complicates the mission orbital planning. The outer regions of the solar system, beginning with the orbital regions of Jupiter, Uranus, Saturn, and Neptune add another level of complexity for the mission planner, due primarily to the eccentric behavior that is caused by competing gravitational force of the giant planets and their moons. The upshot of all these variable orbits that are occurring in the solar systems is a highly-dynamic system nodes and linkages. Gravity is the primary force that determines the nature of these orbits. The Sun is the major center of gravity overall, but giant planets, like Jupiter, exercise a regional hegemony within their smaller domains.

The orbits in the solar system are a human construct, but they are as real as any boundary or linkage in any spatial region. Thus, the separate "horizontal" orbits of the various bodies around the central node and the secondary nodes are now being linked by the "vertical insertions" of the man-made spacecraft. These can be seen as inter-orbit linkages that do not occur in nature – except for the occasional asteroid, comet, or one of their meteorites. So, now we see that rocket-propelled crewed spacecraft and unmanned probes and orbiters – some carrying landers and rovers – are adding another level of spatial complexity to the solar system. These "cultural" insertions begin as vertical trajectories, but they soon join in the spatial harmony of orbits as they reach their targets in the solar system.

The Dimensions of the Solar System

The first thing that strikes the observer about the solar system is how immense it is, even though it is only a small corner within Milky Way galaxy, and only one of billions of such stellar regions. One measure of it overall size is the effective reach of the propagation of solar wind, which has been considered to reach approximately 100 AU from its point of origin. On its outer frontier, the Oort cloud, which is a spherical cloud of small ice-bodies, has been calculated to revolve at a mean distance of about 4 AU. By comparison, the nearest mean orbit of the planet Mars around the Sun is 1.5 AU, and that of Venus is .7 AU. By international consensus, such great distances are generally expressed in terms of Astronomical Units (AU); where one AU is equal to the nominal distance of the Earth from the Sun. As can be appreciated, the geographic analysis of the solar region requires some changes in our working understanding of distance. Instead of thinking in terms of miles or kilometers, the "astronomical unit" (AU), which is defined as the distance from the Earth to the Sun, which is approximately 93 million miles, will be the most commonly-used measurement in space operations. This is also useful information if one wishes to comprehend the overall size and scope of the region. However, the more relevant measure of distance in terms of accomplishing military objectives in space, I think, would be an "operational distance." This geographer refers to this concept as "relative functional distance." This, I submit, is a more pragmatic term for mission planners to plug into their calculus of time-space for any given mission. Operational distance is not only a simple measure of a hypothetical line between point A and point B in space. It implies many variables, including the thrust-load ratio, material composition, and available trajectory, among others.

But, for purposes of practical military space operations in the solar system, "distance" again can be said to be determined but the time-length of the journey. At our present level of space propulsion technology, the travel distances from the Earth to the outer planets and other bodies of our inner solar system is measured in terms of days or months: the trip to the Moon takes about 4 days; the voyage to Mars is about 7 months. But the travel distances to the outer regions of the solar system is measured in terms of years; many years. Consider the odyssey of the space probe Galileo which was sent out to do flybys of one of the farthest planets, Jupiter. This probe was placed into Earth orbit by the space shuttle Atlantis in October, 1989. It conducted a flyby of Venus in February, 1990, flew through the asteroid belt during 1991-1993, and witnessed the collision of the Comet Shoemaker-Levy with Jupiter, as it approached the planet in July, 1994. One change in the calculus might be a change in the point of origin, from the Earth, to a forward base – such as the Moon, or asteroid of the Main Belt, or even Mars. This may seem like an impossible task to overcome in terms of military operations or commercial endeavors, but we must remember that the trip from Amsterdam or Palos (Spain) to the new world in the 15th century was a journey of about two months. Now, in the age of jet propulsion, the trip between London and New York takes less than a few hours. The critical factor in this "shrinking distance" was, of course, the propulsion technology and its associated technologies.

As far as is known today, the fastest mode of transportation in the Universe is via "light" or electromagnetic energy. The only things that can travel in that mode are the charged particles of radiation which humans have been able to harness for some applications – mainly for communication, navigation, tracking of objects, and determining distance. Through the manipulation of electromagnetic energy waves, we are able to send and receive data and information from any point in the solar system to another, or to broadcast it. The data and information can be transmitted and received as sound or as video, or a signal can be used to control electronics or mechanical systems throughout the solar system.

We are not yet able to particularize humans and hardware for shipment on a wave of electromagnetic energy (a la Star Trek) but we can utilize the energy waves to transmit data and information throughout the solar system and even to deep space. So, when it comes to transporting humans or their machines through the solar system, the practical speed of operation drops to a maximum of about 17,000 km/second (this the approximate speed at which our space probes have been able to move). NASA and the international community of scientists seems to have set the standard of the Astronomical Unit (AU), which is the average distance between the Earth and the Sun which can be traveled at the speed of light (186,000 miles/second). This allows for the more efficient analysis of distances between celestial bodies, and the stellar bodies outside our solar region. To measure travel distances by space ships, the space community seems to have set on the time – in days or years – that it takes to complete the missions of space exploration.

The best model for understanding and explaining the dimensions of solar system that I have encountered is one which reduces the region to the scale of the typical downtown city block in the Midwest region of the United States. This model, which I found described on the internet, reduces the solar system in size by a factor of a billion. At this scale, the Earth is about the size of a grape. The Moon orbits around the Earth from a distance of about a foot away. The Sun is about the size of an average-sized man; its distance from the Earth is equivalent to an average city block. Jupiter, then, is the size of a grapefruit and it about five blocks away from the Sun. Saturn is about the size of a medium-sized orange, and is ten blocks away from the Sun. Uranus and Neptune would be the size of lemons, and be located some twenty to thirty blocks from the Sun, respectively. This model provides a perspective of the size of the solar system that can be easily grasped and conceptually manipulated to determine an idea of the kinds of practical distances are involved in military space operations. In reality, the idea of "distance" is only useful and valid if one thinks in terms of the distance it takes to go from one point to another in the solar system – given a particular mode of transportation.

As to the development of the structure of the solar system, the most generally-accepted model of the formation of the solar system assumes the progenitor as being the early solar nebulae – the numerous clouds of gas and dust – which had been expelled by the prime explosion, known as the Big Bang. According to the model, the early solar nebula condensed from such a diffuse interstellar cloud. Eventually, it is thought that the cloud collapsed as a result of an initial cooling process. When the cloud could no longer maintain its hydrostatic balance, the tendency of gravity to maintain a uniform level, initiated a condensation and it eventually collapsed. In any case, once self-gravity initiated condensation, the cloud collapse would have been a self-motivating and acceleration. In the course of time, the cloud would have increased its rotational period as it condensed, eventually forming a spinning disk with most of its mass settling and forming in the center. It was this central mass that eventually became the basis for the proto-Sun and the molecules and atoms in the rest of the disk would have been distributed along a central plane of the disk. Finally, some molecules would have condensed into solid particles which, in turn, would have accreted into the planetesimals; the building blocks of the planets.

The catalog of nodes of the solar system (as of this point in time) includes central node, the Sun, and all the objects that orbit around it. This includes planets, moons, asteroids, and comets – and now, a plethora of artificial satellites and other orbiting spacecraft. Scientists place the system's time of origin at about 14.5 billion years ago. There are subregions within the overall region. The so-called inner planet subregion includes Mercury, Venus, Earth, and Mars. The other main subregion is populated by the planets Jupiter, Saturn, Uranus, and Neptune. Between the two main subregions there is the so-called Asteroid Belt. Beyond the orbit of Neptune, there is a vast transition zone into the inter-stellar space that marks the end of the domain of the Sun. It is the place where the dwarf planets and the icy comets, the detritus of the construction of the solar system, maintain an uneasy orbit in the outer reaches it. Within this assemblage of celestial bodies, it is estimated that there are now 8 planets, having 160 known planetary satellites (moons), as well as countless asteroids; some of which have their own satellites. Comets and other icy bodies, and vast areas of highly tenuous gas and dust, which is known as the interplanetary medium (IPM), account for the rest of the total population.

Empirical studies of the "DNA" of the comets, asteroids, planets, and all the other celestial bodies of the solar system tend to reinforce the notion that all of these came from a common ancestor. To begin with, the population of the solar system has a single, common origin. That is to say, it is generally accepted by most (though not all) scientists today that our solar system was created by a single causative effect: the so-called "Big Bang" event. Because of this, one area of commonality among the planets of our solar system, it is hypothesized that they are all composed of the same basic elements. This hypothesis has been supported by empirical studies of meteorites on Earth and by in situ analyses by the probes and rovers on or over the bodies in space. Since an element is said to be any substance that cannot be decomposed into simpler substances by ordinary chemical processes, it can be considered that the elements are the absolute basic building blocks of all matter in space, no matter where we conduct our analyses. Another factor in the presence of commonalities within the solar system has to do with the manner and degree to which the fundamental forces of nature – gravitation, electromagnetism, and the weak and strong nuclear forces seem to operate the same everywhere in the solar system.

On the other hand, these chemical elements have been combined with one another in an almost unlimited number of ways to more complex molecular structures throughout the region. One reason for this variation within the solar system, as geologists are now learning, is that each of the planets has gone through differing periods of development since their respective origins following the Big Bang some 4.5 billion years ago. Fundamentally, these differences are the derivative of spatial and time factors. More specifically, some variations have been the result of internal processes; others have been bombarded and shaped by external forces. One instance of this reconfiguration process is the way that hydrogen and oxygen combine to form the compound – water – throughout the solar system. It is in this area of chemical combinations and manipulations that humans are likely to find some of the factors that account for the differences in the nature of the various bodies in the region. Then, there is the fact that the relative location and movement of all the bodies in the solar system is a function of the Sun's relative gravitational influence on them, as well as the relative influences of gravity among the planets themselves. Another variable that accounts for differences in the anatomies and physiologies which occur within the solar system is the fact that all are affected variably the Sun – both in terms of gravity and heat. Finally, the internal core processes have taken different routes during the development of the various planets.

These common attributes among the bodies of the solar system have enabled geographers and other Earth scientists to apply many of their familiar concepts and technologies towards the study of what might be called "cosmography" of the solar system. Another fact that we are discovering from the perceived commonalities within the solar system, is that our protocols for technology development and the scientific method can be effectively applied throughout the solar system. For instance, one of the technologies that are being developed for human exploration of Mars has been inspired by the magnetosphere that envelops and protects the Earth from the harmful effects of cosmic radiation; the same phenomenon that will impact on astronauts, sometimes with lethal consequences. Other manifestations of technology – our tools for measuring and analyzing structure and process – also seem to be effective in our space explorations. Thus, the technology of optics; our remote sensors of the various parts of the electromagnetic spectrum still work; and our terrestrial alloys and other materials still function as expected. And, most importantly, the laws of physics still are observed throughout the universe: gravity is omnipresent throughout the solar system and the universe. The other forces, including electromagnetism, the strong force and the weak force also appear to operate on the atoms and their sub-atomic particles in a predictive manner.

The human conceptions related to light, heat, and quantum physics also apply throughout the universe, including the solar system. Heat is a form of internal energy that is associated with the random motion of the molecular constituents of both matter and radiation. Temperature is an average of some of the internal energy that is present in a body. Absolute zero temperature is still the lowest possible energy state of any given substance. This is because the behavior of electrically-charged particles interacting with electric and magnetic forces is as expected; charged particles in motion interact with magnetic forces to create fields; and the electric force between a pair of charged particles follows the same laws of physics. Similarly, each particle creates an electric field in the space surrounding it; each particle responds to the force exerted upon it by the electric field in its own position; the way in which changing magnetic fields produce electric fields in space is still the same. By the same token, the optical devices still perform as they do on Earth; the images produced by the microscopes and telescopes still can be relied upon. Optical devices still perform as they do on Earth and spectroscopy continues to be effective in the analysis of these formulations of chemical elements. On an even more fundamental level, atoms still combine to form molecules, whose structure is the stuff of chemistry, along with compounds. In short, the physical laws, indeed, do appear to be universal in their application.

One major difference is in the nearly zero gravity of the interplanetary medium. On Earth, gravity drives everything, from the way life has developed to the manner in which materials interact. Gravity affects much of the fluid behavior that we are used to seeing on our planet. However, in an environment of virtually no gravity, as in the interplanetary medium of space, other factors such as surface tension control the behavior of fluids. For example, surface tension causes drops of liquid to form almost perfect spheres due to the absence of gravity. On Earth, liquid is distorted by the gravity when it is resting on or a surface. The behavior of fluids is the essence of many phenomena in materials science, biotechnology, and combustion science. While the differences that occur between the Earth gravity and the microgravity in space can present engineers and astronauts with some practical problems, they also provide some opportunities for scientists. The microgravity environment of space flight gives scientists a unique opportunity to study the states of matter (gases, liquids, and solids), and the forces and processes that affect them. One outgrowth of this has been the International Space Station, on which a sophisticated laboratory has been constructed to enable astronaut-scientists onboard to conduct a variety of science experiments in a microgravity environment, over periods of time sufficient to carry them out to fruition.

Microgravity also has different effects on materials than the gravity environment we are used to on Earth. The relationships between the structure, properties, and processing of materials are different. Thus, the physical, chemical, electronic, thermal and magnetic characteristics of materials interact differently in the microgravity environment; and the arrangement of the atoms in the material takes different forms. So, there are many differences in the ways in which materials are formed. So the possible methods by which they can be solidified, evaporated, condensed; or dissolved and then separated from a solution are more varied too.

As a practical matter, in the environment of microgravity metals, polymers, semi-conductors, and ceramics materials can be expected to function differently than we might expect, given the fact that we are used to the gravitational environment of Earth. This is especially true with respect to the conduct of any operations in space, but especially military operations, because they are closely intertwined with the application of science materials. As an illustration of this vital relationship, consider the problem of reentry into the Earth's atmosphere which many space vehicles face at the end of their mission. At this point in a round-trip space mission, as the space vehicle reenters the atmosphere, it encounters a blanket of gas which creates drag on the vehicle. This is not all bad, because it provides a natural "braking" effect to slow down the vehicle's velocity during reentry. Unfortunately, the gas molecules also produce severe heating as they flow along the spacecraft's blunt nose. In the early stages of space flight, heat shields were made of ablative (cutting and evaporating) materials that carried away the heat of reentry while they were shedding the very material of the heat shield. Later, the space shuttle program introduced refractory materials – silica tiles and reinforced carbon-carbon materials – that directly withstood the heat of the reentry process. These advances in materials science and technology have been possible, in large part, because of the scientific experimentation that is being done in the microgravity environment aboard the ISS.

In the end, we likely will find that current terrestrial military technology and tactics can be adapted to the geographic environments that lie beyond our Earth's atmosphere. An example of this can be seen with respect to the propulsion technologies which have been developed on Earth, but now are being used effectively in space exploration operations. Thus, "rocket engine" technology, which basically refers to any propulsion system that carries its own integral fuel and oxidizer component, enables these propulsion systems to operate in any medium that lies outside our atmosphere. Incidentally, this quality of self-containment will likely mean that future space explorations into the outer areas of our solar system will require some method for recreating both fuel and oxidizer components. In this sense, it is no different than the imperative for being able to recreate oxygen and water in deep space. In any case, the concept of propulsion, whether by the counter-emission of energy or gases, is now being applied more broadly than the emission of energy from continuous chemical explosions caused by the combustion process. One alternative propulsion technology can be seen in the harnessing of the movement of charged electrons, or ions, in a counter-direction, to propel machines in space.

To illustrate this applicability of the science and technology that has been developed on Earth to the conduct of human space exploration, I refer the reader to an article in the <u>Scientific American</u> (November 2011), written by Peter H. Smith, who is a professor of planetary science at the

University of Arizona. In the article, he discussed the launching of the latest NASA mission to Mars, which actually occurred in 2012. Notable to me is the report in the article that the primary vehicle for conducting that exploration of the surface of Mars during that mission is a small ATV (All Terrain Vehicle), which probably was tested in the Atacama Desert of South America. Since the Mars mission has become operational, the rover ("Curiosity") has landed inside a crater on the planet, and has climbed over the lip of the depression. Outside the crater, the rover vehicle is now taking samples of soil and rocks for some "on-site" laboratory processing. It is gathering these samples with a robotic arm, which appears to be similar to the robotic arms that are being used to conduct deep-sea geological explorations on Earth. The results of the lab work then are being down-loaded to Earth.

Also, as the various space exploration missions have shown, there is much that is familiar in the areas of the solar system that lie beyond the domain of our planet Earth. One great realization that has followed from the initial explorations of the Moon, and from the multi-spectral analysis, is that there are many similarities between the Earth and the Moon, in terms of their composition. In fact, the methodology of looking for comparisons between lunar findings and materials on Earth has also been effectively used in the exploration and analysis of all bodies in our solar system. One of the Moon-analog geographic environments on Earth that has been used to develop hypotheses for study on the Moon, and to train astronauts, has been the vast Gobi Desert of Central Asia. Like the surface of the Moon, this is an apparently cold, waterless place, and like the Moon, it is not a sandy desert, but mostly bare rock.

The Atacama Desert of South America also has been used as an analog model for exploring the Moon. It is an extremely arid region in northern Chile; it is about 700 miles long, from north to south. Like the Moon, it contains extensive pebble accumulations and some sand dunes. Thus, both of these regimes are good places to test lunar vehicles and to train astronauts. Other such analog regions for space exploration include the volcanic regions of Hawaii and Iceland, the geyser regions of Yellowstone National Park in Wyoming, USA, and the submarine fracture zones on Earth, which generally separate ocean-floor ridges and produce much of the underwater volcanism environments on our planet. The tundra and permafrost regions of the Arctic and Sub-Arctic areas of the Earth provide many analogs to what may be found on the Moon and other planets like Mars. In South Africa, ancient basalt rocks, thought to have been formed only a billion years after the formation of the planet Earth, serve as analogs for what may be found on the Moon and other planets of the solar system. Then there is a place on the surface of the Earth that appears to have been custom designed as an analog for the inner planets and the Moon. Its name is the "Craters of the Moon National Monument and Preserve" in Idaho. This is a region that contains the volcanic cones and lava flows; and more than 35 craters which appear to be relatively young, only a few millions of years old. Some of these craters are nearly a half-mile wide and hundreds of feet deep. Here there are some craters with such high lips that, they would give the Curiosity rover challenges like it is experiencing on Mars today.

When it comes to the other bodies of the solar system, the physical geography of Mars has also been found to contain many analog features with that of the Earth. According to a NASA.gov article, entitled Mars Here on Earth, the exploration of the Red Planet has presented surface phenomena that are like many places on Earth. Consider the description of the Martian geography that has become so familiar from the NASA mission reports. These include: erosion features on cliffs and crater walls; gullies; impact basins and volcanoes (Olympus Mons). All of these will immediately be familiar to the geologists and the geographer. Then there is the description of Mars as a planet of extremes; an arid, rocky, cold place, and extensively cratered. These will remind the geographer of places on Earth which fit aspects of that description as well. Death Valley, for instance, has its Ubehebe crater and "Mars Hill," which have many similarities with the geologic features that have been found on Mars. Also, Mono Lake, in California is like the Gusev Crater on Mars. This appears as an ancient basin or depression, which is thought to have been the site of an ancient body of water there. Also, surface features on Mars have been found which remind one of the Channeled Scabland of Washington State, in the U.S., where extensive floods have occurred recently. The resulting surface features that have been wrought by this flooding in Washington is thought to provide clues as to what has produced similar features on Mars, such as the Ares Vallis flood plain where the Mars Pathfinder lander touched down.

Another phenomenon that is found on Earth, and which may lead to the discovery of similar life-forms elsewhere in the solar system, is the Stromatolite. These are ancient, primitive, single-cell organisms that date back on Earth to the about 500 million years ago. They look like mushrooms or pedestals, and they are formed by successive secretions of waste by the single-cell organisms, combined with limestone deposits that are washed up by ocean waves. Today, they are mostly found in the shallow ocean waters of West Australia. The importance of these organisms is that they produce oxygen and, therefore, might explain how our planet has so much oxygen today. It also might point to a possible way to look for the presence of oxygen in other parts of the solar system.

Along with the similarities and the differences that exist between Earth and the other planets of the solar system, there are also these same variations occurring amongst the other planets as well. Thus, the difference among the planets and the "would-be" planets like Pluto, and the largest of the asteroids, seems to be one of relative degree and quantity, rather than of kind. These differences can generally be explained by the fortunes of random place and time. Most commonly, they are manifested in differences in mass and gravitational influences, which must be taken into consideration during mission planning. Also from the perspective of humans who are traveling through, working, and living in space, the differences constitute variables that have to be taken into consideration during mission planning, at best. In the worst case, these variations can be a matter of life and death, literally. One constant, however, is that there is always going to be insufficient oxygen outside the Earth's atmosphere to sustain human life, or to maintain the effectiveness of their air-breathing machines. As a practical matter, this means that astronauts will have to have a portable source of oxygen in their space vehicles and space suits at all times in order to survive, and to be able to function effectively in this space environment. And, it is apparently the case that none of the other planets and bodies in our solar system contains the necessary level of oxygen to enable humans and their air-breathing machines to function effectively either – given our present state of technology.

But there are also other variables among the planets and other objects in the solar system that matter to the human in the interplanetary medium. For one thing, there is the matter of the differing effects of gravity at various points in the solar system that are most significant to human operations in space, or on other planets. So far, in the first two decades of the 21st century, humans have conducted most of their exploration and scientific studies in the region of the solar system that is known as the "Inner Region" which can be understood to mean, within the orbital zone of the main asteroid belt. Our experience, so far, is that gravity is omnipresent, to some degree, throughout not only within our solar system, but the Universe, as well – just not in the degree that we are used to on Earth. But its value is not a constant. Instead, it is a variable that must be constantly factored in during mission planning and operations. Then there are the differences in heat and pressure that have to be taken into account by humans who propose to conduct ongoing operations in the solar system. This reality became painfully clear during the attempts to insert a probe into the lower depths of the atmosphere of Venus.

Some of the variations and similarities in the solar system may be subjective, rather than objective. That is to say, it would seem that, as we venture farther out into the heliosphere, and begin to conduct more intensive operations in our solar system, there will need to be some changes in the ways we perceive and think about our physical and cultural "environment." One of the differences that humans will encounter as they venture out to live and work in throughout the solar system has to do with our concept of time and distance. We, for example, have our own "mental maps" of environment. These will need a lot of work and modification as we venture farther into outer space, where we are learning that the concepts of space, distance, and even gravity – indeed, our very understanding of physics – is forcing a lot of rethinking. For one thing, we are finding that the atom, our iconic "building block" of the Universe, apparently has lost its status in that regard. There is now a subatomic particle called the "quark" that is considered the true basic building block of nature. Interestingly, these and other subatomic particles are said to interact by means of a "strong force" and a "weak force." Also, never mind "warp speed," which requires a great deal of mental work for me to grasp, but I have come to comprehend the ramifications of the speed of light, which is approximately 186,000 miles per second, and which does not vary. Thus, I understand that a bolt of laser energy will take the same amount of time to arrive at its target at a given distance, no matter whether it is emitted from a moving or stationary source point.

In any case, the physicists have continued to use Newton's laws with respect to bodies throughout the solar system, as they have in the past centuries. The concepts of body, mass, inertia, force, momentum, changes in motion and direction, action and reaction on a body; all remain valid in space. Engineers and other applied scientists also have found that the solar system (and beyond) responds as expected to the methods and tools that have proven to be effective on Earth. By the same token, many other terrestrial scientists – including geologists and geographers – are finding many efficacious analogs in space exploration. Geographers have also been able to apply their terrestrial concepts and methodologies in the exploration and spatial understanding of other planets and celestial bodies. As soon as photographs of the Moon and the other planets became available, the geographers must have, almost as a reflex, mentally assigning names to the various features of terrain and imposing some kind of a cartographic discipline of latitude and longitude, just as has been done with respect to our home planet, Earth. So now there is a casual reference to the poles and the equator of the planets, as well as the lines of latitude and longitude when referring to the absolute location of geographic features on the surface of these planets.

On the other hand, it seems that the solar system, as a whole, is like the planet Earth in many respects, and therefore, lends itself to the application of the techniques that geographers have developed for studying planets in general. Consider the structure of the solar system: it can be said to have an atmosphere, a lithosphere, and a hydrosphere, just as the Earth has – albeit in somewhat circumstances, and in a more discrete fashion. The solar system also lies within a larger region in space, the "celestial sphere." The latter is actually a construct that was developed by ancient astronomers to provide a larger context for the heavenly bodies that are seen in the heavens. There is a system of latitudes and longitudes that is superimposed on the celestial sphere, just like the global grid systems on the planets. Traditionally, the celestial sphere coordinate system has served a practical purpose for astronomers as they studied the planets and the stars. Now, in the age of space operations, it provides a navigational reference system for mission engineers and astronauts as well.

The planets of the solar system have all been created from the same basic matter. However, there are now many variations in the structure and composition of the planets, depending on their relative location during their early development, and their spatial relationships with other celestial bodies in the four billion years that have followed. Therefore, we can expect to discover that the planets share many similarities and differences within the solar system as we study them more intensively. It is this variation in their locational distribution, as well as in their structure and constitution, that draw geographers to apply their traditional concepts and methods when studying each of the planets and other bodies in the solar system. In term of structure, the inner planets have the greatest similarities because of their common experiences of origin and early development. We see this in the familiar core-mantle-crust configuration and the recognition of the fact that they all have an atmosphere and a lithosphere, although only the planet Earth is known to have a hydrosphere.

What we are also realizing is that the planets and other bodies of the solar system also are reconstituted clumps of matter, in much the same fashion as the continents on Earth. One major reason for this is that, like the present-day continents on Earth, which have broken apart from the proto-continent, Pangaea, the planets have been dispersed throughout the solar system from the original stellar cloud that bore them. Thus, it can be understood that the "lithosphere" of the solar system is distributed throughout the region, somewhat in the manner of the continents. On Earth, the force that causes the separation of the continents essentially is exerted by the heating and of the mantle material by the core, which powers the process of subduction, extrusion, and other tectonic processes that are at the heart of the "continental drift" process on our planet. Ultimately, it is the energy and gravitational forces emitted by the Sun that exerts the primary control the tectonic process of the planets and other celestial bodies within the Heliosphere, as well. However, the competing gravitational influences of the bodies themselves also can cause internal heating of the body. So, even though the notion of a common, but discrete, solar system lithosphere is still quite crude, I believe that it does help to provide a useful framework for analyzing the various celestial bodies of the solar system.

The principal planets and moons of the solar system share many geometrical attributes as well. For this reason, the familiar concepts and terminologies that are used to describe the sphere that we call Earth can also be applied to these other celestial bodies. That is to say, all of these larger bodies, because they spin on their axis, are essentially spheres and, therefore they can be dealt with geometrically in the same terms. Thus, when the astronauts and scientists of the space exploration programs begin their descriptive analysis of a planet or a large moon, such as Saturn's moon, Titan, there are the familiar references to diameter, mass, gravity and density. Also, polar flattening and bulges on the spheres is noted; these are seen as deviations from the ideal shape of a perfect sphere.

One of the main advantages of the geometrical sphere in the study of the cosmography and "planetography" of the solar system is its familiarity to humans. We have had an opportunity to develop the techniques for conducting measurements of the geometrical sphere over the course of thousands of years on Earth. As is the case with many abstractions of a real phenomenon, many mathematical relationships and metrics, or models, have been developed to facilitate the quantitative analysis of any sphere that is encountered in reality. Among the features of the sphere that are commutative in the solar system is that of the great circle route; the shortest course between any two points on the surface of any planet is still a great circle route, and it always lies on a plane that intersects the planet's center. Also, the spherical shape of all planets can be organized conceptually in terms of hemispheres, poles, and equators. Fortunately, the familiar latitude and longitude coordinate system also can be utilized to determine and describe the position or location of any place or object on all planetary surfaces. And, the length of a degree of arc latitude is still approximately 69 miles, although there will be variances due to any nonconformities of a planet's curvature.

 And so, on December of 1968, when the Apollo 8 astronauts – Frank Borman, Jim Lovell, and Bill Anders – the first humans to walk on the Moon – arrived there, they and the scientists on Earth already had a sense of the familiar. There was a map of the Moon which had been constructed through the utilization of the same principles of cartography and mapmaking which had been developed on Earth. Beyond that, the astronauts and the NASA folks on the ground had "mental maps" which they had developed as the result of sensorial stimuli and logical thinking. It was a case of the perception of reality being quite close to the objective reality. What began with relatively less empirical observation and stronger inferential thinking in the past now has been replaced by more empirical data and computer-assisted logic, to produce what we now know about the solar system. So, in the case of the Moon, the astronauts planted the flag for mankind, walked around a bit, gathered some lunar samples and data – and then returned to Earth. To this geographer, what they also had done was to effectively add to the content of one more node of the solar system. Later manned Apollo missions would add more content to the lunar map when they orbited the Moon and carried out some more surface expeditions.

When the data and imagery from space began to be physically returned by the astronauts, and later streamed by the unmanned space vehicles, the geologists and the geographers began to recognize many familiarities and consonances with their frame of reference on Earth. To appreciate this, just consider the terminology that is being used in the analyses and reports from the NASA space missions during the past fifty years. From these debriefings, a geographer would immediately begin to develop a mental map of the solar system as a single, multi-nodal model, in which the atmosphere, lithosphere, and hydrosphere would be seen in cross-section fashion – like the layers of a golf ball. With the aid of computer-generated animation, I can see a composite "physical geography" (consisting of all the planets and moons of the solar system) which depicts the horizontal variations and the vertical variances as well.

THE MILITARY AEROSPACE REGION

The Military Aerospace Region (MAR) is an example of a geographic functional region; therefore, it is defined by the patterns of interaction which occur among the various nodes that lie within it. A MAR can occur anywhere within the solar system. But, given today's technology, in theory, in practice it is now being manifested on Earth and in "geo-space." The latter refers to the MAR that has developed in the geographic domain which is centered on the zone of artificial satellites that are now orbiting around the Earth. As used here, the MAR is meant to be an all encompassing term, which refers to military aerospace operations in the Earth's atmosphere as well as in outer space. It also serves as a useful paradigm for analyzing space operations that are now beginning to occur in the near-Earth orbit region. And, it also should also serve to as a spatial device for dealing with any area of operations within the solar system. Therefore, I am proposing this model as a possible unified view of military geography in the 21st century, throughout the solar system. It is designed to consider all the traditional geographic variables, as well as the additional ones that are to be found outside the Earth's atmosphere.

The MARs that occur within the atmosphere and in geo-space have much in common in the abstract. The following is a descriptive analysis of these two MARs: both have boundaries which are being dynamically defined by the combination of objectives and technologies; the geometry of both spaces is dominated by the orbit and the curvilinear lines that connect each orbit; within these orbits, there are the "natural" and "man-made" nodes which are interacting in much the same fashion as in any functional region on Earth; and the orbits can be portrayed as vectors which present variations in magnitude and length. In a sense, orbits are like the lines-of-communication that are seen within MARs on the surface of the Earth. One derivative of these shared regional characteristics and attributes is that even though the foreground any analysis of these MARs might involve a particular "man-land" relationship, the geographic concept of the "functional region" will always be either in the background.

As it happens, the majority of military aerospace operations continue to occur within the Earth's atmosphere and, more particularly, within the lowest layer of the standard atmosphere, which is called the troposphere. In a generally-accepted static model of this geographic phenomenon, the upper limit of this subregion is about 10 miles at the equator and approximately 5 miles at the poles. Its lower boundary is composed of both land and water surfaces. The actual definition of this "surface," in terms of military aerospace operations, is somewhat more ambiguous, given the propensity for humans to burrow into the subsurface and into the depths of the oceans to either create military infrastructures or to actually carry out offensive military operations. An example of the latter is the force of nuclear, long-duration submarines which can fire their missiles from below the surface of the ocean.

Prior to the placing of Sputnik I into low-Earth orbit (1957), the practical upper limit of the MAR could be said to have been at approximately 70,000 feet of altitude, which is the operational ceiling for the high-flying U-2 reconnaissance platform. However, in the years following the launching into Earth-orbit of the first artificial satellite, the center of gravity of military aerospace operations has been broadened to include what might be called the geospace. The upper boundary of this geospace MAR is also somewhat ambiguous, although it might be defined as the upper limits of practical military space operations. Therefore, it can be said that the upper limit of military aerospace operations in the geospace domain is itself constantly, a function of the advances in military technologies. The operational ceilings of spacecraft is one measure of upper boundary of the geospace MAR, but it is also defined by the operational range of surface-based ballistic missiles and even the service ceilings of weapons-bearing platforms that launch their missiles from stratospheric altitudes.

So, this new MAR, which has been developing since the middle of the last century, is the "Geo-Space Region." Its boundaries and nodal characteristics derive from the continuing movement of human spacecraft from one point to others within the defined space. These movements – which regional geographers call "commutes" – first began with a series of forays into a very low-orbit in space, around the Earth. These were the exploration and learning sorties, in which humans began to first try to understand the basics of what it takes to operate in space and, more significantly, outside the nurturing cocoon of our atmosphere, and the protective shield of the geo-magnetic sphere. All of the commutes that have been taking place in this region have originated from the surface of the Earth and have been directed toward the orbital zone of the artificial satellites; generally between 200 and 300 miles in altitude.

———

The military operations that have taken place in this MAR during the past three decades have focused on deploying on artificial satellites to perform many of the intelligence, reconnaissance and surveillance (ISR) functions that traditionally had been done by surface-based or airborne systems during the 20th century. Thus, military-dedicated artificial satellites have been launched and placed into low-orbit around the Earth to detect any ICBM launches or nuclear explosions that might occur within the geospace. Others have been deployed in orbit to provide the traditional communications and navigational support that has been by aircraft in the atmosphere, or by stations on the surface of the Earth. However, as soon as these support systems had been deployed in space, there arose a need to provide security for them. So now there have appeared mostly ground-based missile systems on the surface of the Earth to detect and, if necessary, foil any enemy attacks on our artificial satellites. These kinds of offensive military actions have actually only occurred a few times during the history of this MAR, and these were basically either tests, or attempts to demonstrate the capability to shoot down an adversary's satellites.

The Space Shuttle program is an example of how military capabilities can be enhanced from both dedicated military-related research and development, and as a derivative of "civilian" research, development, and operations. This decades-long NASA space operation has developed, tested and implemented technologies and techniques for potential military logistical and maintenance support for orbiting platforms and weapon systems in space. Just consider the vast data base of information and imagery, as well as the experience-base of trajectories and maneuvers that NASA has developed during the course of this program. Indeed, the Space Shuttle experience may prove to be the most fruitful in the development of military space operations because of the multitude of practical problems related to propulsion and payload management that have been solved. The NASA engineers have tackled the very real issues of "cost" during every space mission, for example. This is especially so in the technological and engineering solutions which have been developed in the areas of launching payloads in the most frugal manner, and in reducing "exciting" operations to the routine procedures. Finally, there is the voluminous archive of computer programming code and algorithms that has been developed during the course of the Space Shuttle Program that does not have to be "reinvented" by military space engineers.

At the same time, valuable practical experience is being gained with respect to the precise orientation of platforms in orbit, where there is no horizon against which to measure one's attitude, pitch, or yaw. Also, considerable the practical experience has been gained in the use the gravity attraction from a nearby body to maneuver a spacecraft within an orbit, or to make changes in acceleration in space. Then there is the expanding knowledge and skill-set that has been developed in the area of inertial guidance in space, such as the precise combination of accelerometers, coupled with integrators in that special environment. These have proven to be useful in the guidance and control of not only the platform, but the onboard weapon system as well. This would be a crucial capability in any space environment, but especially within the artificial belt which has become quite densely populated and complex. Since their introduction in the early years of the space age in the middle of the 20th century, these innovations have proved valuable in controlling trajectories of booster rockets and spacecraft in the interplanetary medium, and during planetary operations.

————

One issue that arises when one is trying to draw a boundary, or boundaries, around a functional region, such as a military aerospace region, is that of defining one that is feasible considering the resources available at the time. Thus, following the attack on Pearl Harbor by the Japanese, the "ideal" boundary of the U.S. MAR could be seen as extending over an area that covered the entire Pacific Ocean, including the enemy's homeland islands. However, in January of 1942, the feasible boundary would have had to be drawn somewhere closer to the center of American air power in the eastern region of the Pacific Ocean. The first tentative attempt by the U.S. to extend the feasible boundary westward was the Doolittle raid on Tokyo in April of 1942. But the redrawing of the feasible boundary actually began following the Battle of Midway in June of the same year. Later, a definitive step towards the reconciling of the ideal with the feasible MAR boundary in the Pacific Theater during WWII occurred with the capture of Saipan and Iwo Jima in 1945 by U.S. forces. These two nodes of the region made possible the extension of American air power over the Japanese homeland, culminating in the atomic bomb attacks later that year.

As can be appreciated, the boundary of a MAR is not always presented as a "fine line" of demarcation, rather, it is more often a "fuzzy" zone; not precise at all. Like most physical and cultural boundaries, it is a "frontier" or "zone" within which the core attributes that might distinguish a region tend to blend-in throughout the space that lies on both sides of a hypothetical hard line of demarcation. Such ambiguities often have led to wars during throughout the history of humanity and they might continue to produce such conflicts even in space in the 21st century. Then, there is the fact that boundaries often change or fluctuate along a mean trajectory (like the electromagnetic fields), even when they are manifested by "hard" natural features, such as a river. Indeed, rivers have a history of constant shifting of their courses, much to the consternation of political geographers and cartographers. Even more transitory are the political boundaries that are drawn up by treaties by humans, from time to time. Indeed, history has shown this impermanence even when the boundaries are defended from change by military fortifications.

So, it can be definitively said that boundaries are simply a human construct of the mind; they do not exist in reality, or in objective fact – but their effect on human activities is real. The overall boundaries that are depicted on an aeronautical chart or other projection of a military aerospace region, therefore, are always subjective. One such political-military boundary is the one which delimits the so-called "no fly zone" or the "no strike zone" in a MAR. Another is the line around a "demilitarized zone." On the other hand, the "imaginary" boundary line that describes the area of lethality, which derives from the effective range of an anti-aircraft artillery gun or surface-to-air missile, is effectively "real" to the military aviator who is "working" a nearby target. All of these boundaries may not be explicitly drawn on the chart of a given area of operations, but they can have some very real consequences to military aviators, just the same. In the same vein, the geopolitical boundaries, which reflect the "political" realities of any war, are often expressed in terms of "rules of engagement" within a MAR.

The location of the feasible boundary (as opposed to the ideal boundary) also has to do with technologies of logistical support. In practical terms, the outer limits of this functional region are related to the quantity of fuel that is being carried on-board, or which can be supplemented by in-route refueling – whether on the surface of a body, or in space. Within the Earth's atmosphere, the boundary that is defined by the fuel that is needed for propulsion is normally related to the quantity which can be carried on-board the platform. Within this nominal constraint, there is some flexibility that can be achieved by modifying the nature of the propulsion system, the fuel itself, or by loading fuel in areas of the platform that lie beyond its original design (such as was done with the B-25s of the Doolittle Raid on Tokyo). Another is by reducing the weight of the platform itself, or the payload that is carrying. Then, there is the strategy of "caching," which, in space, can involve stationary "filling stations" on the surface of a planet, moon, or asteroid, or on an orbiting artificial base.

Another factor that determines the location of the boundaries of a MAR has to do with the endurance limits of the crew of the manned platforms. This can refer to the life-support requirements, the biological and psychological needs of the crewmembers or it can relate to their continuing effectiveness in carrying out their assigned tasks over longer periods of time. Because the force of gravity is neutralized in space, biological endurance is less of a limiting factor, even though the length of the atmospheric sortie is only a matter of hours. It is for this reason that space craft are designed for human endurance, in much the same manner as the long-range, nuclear submarines of today. That is, the crew must bring along an adequate onboard supply of oxygen and other life-support resources, as opposed to just the water containers and in-flight food packets from the field kitchen, in atmospheric flight. But even this limiting factor that is attached to space flight has been resolved through the application of technologies, such as those that are being applied to space bases like the International Space Station. In this case, the space station operates more like a naval aircraft carrier. Like the naval platforms, the space platform can be continuously supported by auxiliary supply vessels, whose function is to shuttle between land bases and the carrier, thus extending the limits of endurance.

The lack of gravity in space also can be a positive factor in longer-range sorties in that medium. This is because there is virtually no drag due to the vertical gravitational attraction of nearby planets. In space, the procedure for staying aloft, basically, involves the balancing of several variables, such as the product of masses of the spacecraft and the planet, the distance between the two bodies, and the proper amount of propulsion to make changes in acceleration and to maneuver. Once in orbit, the dynamic tension between centrifugal and centripetal forces will operate to maintain the desired orbit altitude over the planet. (The image that helps me to visualize this is the hammer-throw event in the Olympics). After achieving the desired point in orbit, the endurance factor then becomes mainly a function of the physical and mental capacity of the human crewmembers and, to a lesser extent, to the stresses imposed on the platform and its vital systems. So, there are many positive techniques that work to extend the maximum duration of the sortie in space. These include such measures as the rotation of crews, on the one hand, and a logistical and maintenance strategy that involves the utilization of shuttle platforms which deliver replacement of parts and maintenance crews to do maintenance and enhancing modifications to the hardware systems, as we have seen in the case of the in the case of the International Space Station. In any case, the "endurance" factors that are related to human and artificial systems continue to remain one of the critical determining factors in the drawing of the MAR boundaries as military space operations are extended beyond the geospace domain.

On another level, it can be said that the defining boundary of a MAR is related to the extent of the adversary's competing view that boundary. As I see it, the relationship between opposing views of a MAR's boundaries is much like the dynamic that is present in electronic warfare; that is to say, one of measure and countermeasure. Thus, in the situation where the boundary is in dispute and the objective of each side is to gain control of a space domain, there could be an initial measure taken by one of the adversaries. This is analogous to the initial measure in electronic warfare. So, one side may launch strikes into the domain of an adversary in order to claim control over the perceived region. If the active force is successful, then it can be said that one side has effectively redefined the boundary of the region. Many times, this first gambit is countered by the adversary and the "boundary" of the MAR is then put into play, as it were. An example of this would be the initial strike by Japan on Pearl Harbor, which can be seen as an attempt to establish what Japanese Empire thought one of the boundaries their MAR should be. However, the United States almost immediately began to carry out operations to contest the Japanese definition of their MAR. This is an example of how the boundary of one's "ideal" MAR is often, only temporary, depending on a host of interacting geo-political-military – and technological – factors.

So, there seems to be a natural impulse among humans to "organize" spatial areas and phenomena within them, in many different ways, according to a variety of motives and objectives. This is so, whether the space involves an industrial complex or a system of air defense weapons on the ground, or a contested battle area. Humans seem to instinctively create in the mind some kind of logical order from the seeming disorder that occurs in reality. An example is how we "regionalize" our environments so as to simplify and, therefore, better understand and more efficaciously manipulate them for our purposes. As part of this process, we proceed to categorize spaces and the objects that lie within a defined boundary in order to develop an ordinal census of our region. Then we begin to elicit perceived relationships among the points of phenomena that are observed. In the case of the military aerospace region, we see two logical constructs, the military air force, on the one hand, and the region of targets, on the other. The irony in these endeavors is that while focus our attention on the "big picture," we often fail to see the phenomena within each node. As a result of this problem of scale, to recapture proper perspective between the individual "trees" and the contextual "forest," geographers have learned to operate almost simultaneously at various levels of scale. This process of "zooming" can be very useful in arriving at the optimum level of scale of conceptualization and operation for a given time-space region.

———

The concept of scale also plays an important role in determining what is actually "seen" by eye-enhancing artificial sensors. Thus, we often strive to comprehend the total region at once; to see the "big picture," as it were. At this level of scale, the wider field of view enables the observer to focus more on the network of nodes and linkages within a regional system. It also provides the observer with context as to the spatial relationship of the region within a larger region. On the other hand, a large-scale "blowup" of a smaller portion of area provides much more detail about each node and vector in the region. Counter-intuitively, the map or chart projection that is used to depict this "big picture" is referred to as a "small-scale" map; whereas a "large-scale" map presents phenomena in greater detail. For purposes of being able to comprehend the whole region, regardless of the actual size of the reality that is being observed, a small-scale map is probably best. But if the observer wants to focus on the intrinsic details of the region, the large-scale projection is more useful. Regardless of the scale that is being applied, when it comes to imagery, resolution (or clarity) is important as well. The measure of the latter now is often expressed in terms of the number of pixels that make up an image.

During our analyses of cell-regions within the context of the sortie, we were essentially operating with large-scale map projections, which enabled us to focus on the details of each slice of space-time, in a serial fashion. Now, we are going to "pull back," if you will, and take in the whole region, with small-scale map projections. This will provide us with a "strategic view," as opposed to a "tactical view" of the MAR. From this perspective, instead of focusing on the details of the individual cell-regions, we now try to get a more "global" view of the entire series of cell-regions. In the language of systems analysis, we can now take in the overall system of subsystems, along with their components and their internal and external linkages. Also, we are able to gain an apprehension of the total context of the phenomena population that we wish to observe and analyze, as well as the "intensity distribution" of some particular attribute of the various nodes of the system.

In this way, through the careful observation and analysis of the distribution of objects within a region, many useful patterns emerge to give spatial structure to them. At this point, the concept of "perception" becomes important in the analysis of a region. For one thing, the attributes of the whole region cannot be deduced simply through the analysis of each of the parts in isolation. Another thing that must be considered is that the apprehension of the whole, or any of its parts, is largely in the "eye of the beholder." Therefore, for example, even when there is "hard copy" imagery from optical systems, or other remote sensor systems, the attributes of the region are still often a matter of "imagery interpretation" by trained observers and analysts. This is the essence of the art and science of "targeting" that is done by air intelligence specialists and operations people. This targeting cycle normally begins with the debriefings of air crews and the bomb-damage assessment (BDA) that are done to evaluate the effectiveness of a bombing operation. Another source of input data for subsequent targeting operations is the raw imagery that comes from dedicated reconnaissance platforms, which can be located in the atmosphere or in space. The regional analysis approach is effectively utilized when the imagery interpreters and the intelligence specialists analyze the visual data from a given MAR, and produce a "map" of an area of operations. In this context, the chart depicting the target systems in a MAR is an example of what geographers would call a map of the spatial distribution of some class of objects within a region. In the case where "enemy targets" are the distributed objects. Traditionally, these were usually projected in two dimensions. With the present level of computer technology, however, target systems can now be projected in three-dimensions.

Tactical target systems generally fall into several major categories: (1) ground-based air defense systems, (2) "hard" economic-political systems and (3) "targets of opportunity" that are mobile and can appear at any time, especially during the conduct of air support of friendly ground forces or armed aerospace reconnaissance operations. The second class of targets generally includes factories, electric-power generator systems, communication system nodes, and lines-of-communication targets (such as, harbors, airfields, bridges, etc.). So, it can be seen that the targeting analysis and planning process for these types of target systems is one that requires a detailed understanding a regional system of inter-connected and inter-related set of nodes and linkages, in order to develop an accurate, hierarchical map of the targets to be attacked within a MAR. This also requires a careful analysis of the relative importance and systemic function of the individual target.

Another aspect of regional analysis that can be useful in the analysis of military aerospace operations is that ability to analyze "three-dimensional" geographic regions. This analytical tool has long been used by geographers in the context of agricultural activities, for example. The geographer, Preston James, in his book, Latin America, used the idea of "vertical zonation" to describe and explain the system of horizontal agricultural terraces, which he saw as micro-climate regions. These are most prominently developed in near-Equator environments that also have a mountainous terrain. It seems that the sides of the mountains in these environments offer a series of strata zones which present varied "climate regimes" to the farmer or the herder. This situation provides the opportunity to produce a range of agricultural products that normally are associated with the latitudinal zones nearer to sea level. Generally, it can be said that these vertical zones correlate with the latitudinal zones that one might encounter as one travels from the Equator towards the poles on Earth.

This same concept of "vertical zonation" can be utilized to describe the "operational altitudes" within which the various platforms operate within a three-dimensional MAR. In this case, however, the vertical zones represent a particular environment that affects military operations. In my experience, these can be described as either "threat" or "attack" zones. The first has to do with the fact that modern air defenses also are vertically-zoned in terms of the range capabilities of the surface-based anti-aircraft artillery, the surface-to-air missiles, and the cannon and air-to-air missiles that are carried by other platforms. This means that the threat environment, in terms of air defense systems, is three-dimensional and vertically-zoned as well. Thus, near the deck, where most close air attack operations take place, there are anti- aircraft weapons which can fire at a rapid rate, and which can move their field of fire quickly in response to the movements of the attacking aircraft. At the other extreme of the threat envelope of attacking air forces is the "high-altitude," but relatively-slower attack platform, such as the B-52 bombers in the Vietnam War. These high-altitude bombing operations face their own particular vertically-zoned threat from longer-range surface-to-air missiles.

There is a three-dimensional distribution of the attack and support platforms. These may include platforms that perform: security cap or escort, "flak-suppression and radar-busting, electronic warfare, search and rescue, and command & control functions. Since the beginning of the Space Age, many new vertical zones of platforms have been added, in the ongoing seeking of the "high ground" in the space domain. The most significant addition to the population of platforms has been occurring since 1957. I am referring to the approximately 4,000 artificial satellites into low Earth-orbit within the geospace MAR. Currently, most of these platforms are not dedicated military units in the strictest sense, but military-dedicated artificial satellites currently are in operation in geospace. In any event, all of these platforms can be viewed as nodes in a three-dimensional area of operations. Again, these platforms can be seen as being distributed in a somewhat "static" population within a region, perhaps with the added attribute of density. To do so, however, would require the plotting on a map of "loiter patterns" or orbits since these platforms must always be moving in this regional space. This "static-dynamic" concept of regional distribution of platforms will perhaps prove useful when dealing with any aerospace order of battle in space in the future.

Both the systems and spatial distribution approaches to the study of a region can also prove to be useful when describing a MAR. Generally, each of these often will result in the development of regional maps that are a most efficacious way to communicate "ground truth" in any human endeavor – including military aerospace operations. Such maps are used by geographers and other scientists, as well as by "target analysts" in military aerospace operations. The first approach focuses on the overall pattern of nodes and the linkages that may connect them. It also attempts to describe and explain the processes and interactions that give the system animation. The second type seeks to discern uniformities and variations in a dominant population of objects. It is hoped that such patterns will show the spatial distribution of sets of like objects, conditions, or activities. Since such sets of phenomena seldom occur at the same rate or intensity; variations of these intensities throughout a space are depicted on a map by utilizing contour lines or variations in color, for example. The regions that are depicted by these maps are organized around nodes, and they are therefore called "nodal regions." The second type of map shows regions in which there is uniformity or homogeneity of particular phenomena throughout the space. These are the so-called "uniform" or "homogeneous regions."

LAOS MILITARY AEROSPACE REGION: A CASE STUDY

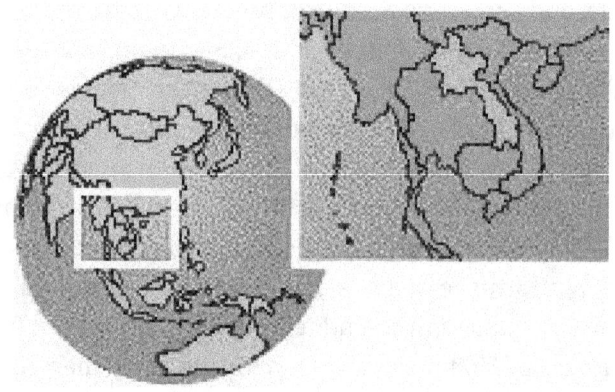

We now take a look at an example of a Military Aerospace Region; in this case, the Laotian theater of the Vietnam War. Like all MARs, it consisted of nodes which were connected by linkages that are called sorties. So, it can be said that one set of nodes in this spatial system were the launch points, the target sites, and the recovery points. The linkages in this system were the sortie flight paths. On another dimension, the MAR consists of natural and cultural phenomena that are distributed over the surface and atmosphere of the area. In one sense, these are the environmental systems that interact with the sortie operation that is being are carried out. One effective way for analyzing these is through use of a geographic methodology called descriptive analysis, which is the traditional way to study the topography, atmosphere, hydrology, and flora and fauna that together define the natural environment of a region. (It is the same methodology that was utilized by the famous expeditions of exploration into new regions, such as the Lewis & Clark expedition). The same analytical approach also has proven useful in the analysis of the military aerospace region. Actually, one could say that there were a variety of environments within the Laos MAR, at any given time. One reason for this is because there were several launch points for the platforms that operated within the Laos MAR; most of them located outside the region. Thus, there were launch sites in Guam, South Vietnam, Thailand, and in the South China Sea that sent platforms in to work over the PDJ and the Ho Chi Minh Trail, as well as the entry points to the Hanoi-Haiphong complex. Each of these launch sites had its own mix of atmospheric variables that affected that initial leg of their respective sorties. Even within Laos itself, and especially during the winter monsoon season, there were various weather regimes operating from hour-to-hour for those platforms that launched from Lima Site 20A at Long Tieng (the MARs headquarters base).

The launch site environment that I was most familiar with was located at Udorn Royal Thai Air Base in northeastern Thailand; located within minutes of the "fence" (the Laos-Thai border). I remember the 0300 mornings when I was scheduled to fly a 12.7 hour sortie over the PDJ. It seems that it was sometime cold enough so I had to wear an insulated flight jacket, especially at our working altitude of about 10,000 feet. In the winter, it would be relatively dry and cold; but during the summer it was hot and humid. The dawn was the best time for the takeoff because later in the day, the Sun would dry out the air on the deck and lessen the lift-capability on the wings. Nevertheless, on most mornings, I would be wringing wet from the perspiration at takeoff, making it difficult to write on my log-book. However, once we got to altitude, the unpressurized cabin would actually feel cold. Therefore, it can be said that the relative humidity was probably the most significant aspect of the natural environment for our aircrew members. But once we arrived on station over the PDJ, the most important element of weather was the layers of cloud cover along our trajectory.

With respect to our C-130 platform, the temperature and relative humidity were perhaps the most significant factors that controlled the decision to launch our command and control aircraft. The reason is that our platform was overloaded at the point of takeoff, due primarily to the weight of the extra fuel that was needed to sustain a 12.7 hour sortie above the PDJ in northern Laos. As far as the pilot was concerned, this meant that our aircraft would require not only enough thrust to overcome the inertia of the weight at takeoff, but also ambient air that was "thick" enough to give the needed lift for a successful takeoff. That is why Cricket launched at first light of dawn, before the Sun would be high enough to burn off the thickening effect of greater relative humidity.

This takeoff-regime analysis can also be applied to the launch profile of any platform in any part of the world, or on the surface of the planets and other larger bodies in the solar system. In all cases, some combination of physical geographic factors will affect the propulsion-lift profile of the platform and, therefore, the load capability of a platform at the takeoff phase. In Laos, the takeoff environment for the CIA and Air America platforms (which included C-46 transports, twin-engine Otters, helicopters, and even a contingent of modified T-28 fighter-bombers) was characterized by an atmosphere full of dust and smoke. The latter was due to the burning of vegetation by the Hmong people, who still utilized slash-and-burn cultivation techniques. Then, depending on the season, there was either fog or low clouds to contend with. Adding to the mix was the short airfields that were usually ensconced within menacing terrain or forest.

All of this meant that, for all manner of platform, the effects of the natural environment were crucial to the launch phase of any sortie. They also heightened the aeronautical takeoff problem in the Earth's atmosphere, which involves such factors as the ratio between the weight of the aircraft and its payload (whether cargo, personnel, or armament) and the thrust that can be produced by the engines (propeller or jet), as well as the "lift" capacity of the wings and other air foils. Therefore, it can be said that, ultimately, the net result of the combination of the takeoff natural environment, on the one hand, and the takeoff capability of the platform, on the other. This was the point in time of the sortie when all the material and propulsion engineering had to come together to combine with the natural forces in order to overcome the gravitational attraction of the Earth.

One negative element of the cultural geography that never materialized during the takeoff phase of our sortie (thankfully) was the threat of an attack on our platform by enemy ground forces at the takeoff site. This likely would have been in the form of AK-47s and 12.7mm machine guns, or even SA-7 shoulder-launched, heat-seeking surface-to-air missiles being fired on our lumbering aircraft at the very takeoff point in the cycle. Indeed, during my jogging sessions along the perimeter fence at Udorn, I often took notice of the fact that the hootches (houses) of some the locals were constructed just outside the fence. I often wondered what would have happened to the takeoff of our aircraft if the enemy had been able to mount a 12.7mm machine gun at the end of the runway, just outside the perimeter. One of the reasons why such an attack never occurred during my tour of duty surely was the force protection security operations which were carried out by Thai military to ensure there was no infiltration by hostile elements into the off-base residential area that grew up just outside the security fence.

In the terminology of NASA space operations, the "payload" that was carried by our aircraft consisted of a "capsule" (about the size transport container) in which, we, the command and control crewmembers performed our duties. This capsule, which was inserted into the cargo bay of the Cricket C-130, was equipped by a suite of electronic and communications systems. Within this approximately 30-foot long airborne command control facility, orbiting the battlefield at an average altitude of about 12,000 feet, the "back-ender" crew managed the air assets from the U.S. Air Force and the U.S. Navy. The airborne battle-staff did its work exclusively through the medium of sound;

somewhat like the situation in a submarine. That is why our platform was equipped with a full range of communications equipment which could operate in many different frequencies. We even had an on-board radio engineer to manage the radio systems. His job was to keep the backenders in continuous contact with all the other nodes in the communications network over the PDJ, and with 7th Air Force headquarters in Vietnam. Thus, we utilized just about every frequency in order to communicate with all the other nodes, on and over the battleground. Communications discipline and encrypting devices were also employed to ensure that we knew with whom we were talking, and what was being said.

With respect to the phenomena that occupied the MAR, it is reasonable to say that one of the most prominent was the population of platforms that operated in the region. (Please note that in this book, I use the term "platform" to refer to all types of aircraft and spacecraft, because this level of abstraction appears to be the most useful way to describe not only airplanes and spacecraft, but missiles, as well). Actually, there were many types of platforms which were used to conduct sorties on the PDJ. One particular group of platforms consisted of WWII-era A-1Es, T-28s, and A-26s, which proved to be highly-effective in the "guerrilla war" environment of that MAR. As a whole, the allied "air force" which operated over the Plainne Des Jarres (PDJ) consisted of a wide assortment of bombers and fighter-bombers; attack aircraft and fighters; helicopters and observer aircraft; as well as transport aircraft and a radar-bearing aircraft; and the C-130 "Cricket" command & control aircraft. There also were platforms that carried out ELINT (Electronic Intelligence) and other specialized duties. One of these was "Cricket" C-130 platform on which I worked as part of the Airborne Command & Control function.

Within the Laos MAR, there also were many orbits that were flown by the platforms. These were orbits which were flown at various altitudes; and some of the orbits were shorter than others, depending on the mission requirements. Some of the orbits were circular; while others were elliptical. On average, however, the orbits were more like pretzel-like ovals. The dimensions of the orbits could encompass an area with diameters of as much as 600 miles. Within this area of operations, platforms operated within air masses that could be contain multiple layers of dense clouds in the summer months, and a chronic patchwork of cloud cover during the winter months. Within this varied airspace environment, the platforms operated in an atmospheric envelope, at working altitudes that ranged from the surface to 12,000 to 20,000 feet. However, the B-52 bombers "worked" at altitudes of about 30,000 feet. The orbits themselves can be seen to be a series of cell-regions. These can be analyzed in time-sequence fashion, especially as they travel from one geographic environment to the next.

———

The orbit phase of our Cricket sortie over the PDJ resembled the NASA space missions in many respects. Thus, the descriptions of the transit phase of NASA space missions bring back memories of the operations onboard the Cricket during our 12.7 hour sortie. Like the NASA missions, the essence of the "payload" of the Cricket mission was the communications equipment that we used to maintain "situational awareness" and to carry out our command and control mission. Unlike the NASA missions, however, the weather environment was the most important aspect of our operational environment. In space, weather (as we know it) is not a factor, but the electromagnetic radiation environment is more acute overall. However, electromagnetic radiation was a major component of the electronic war that went on between the U.S. Air Force and the enemy air defenses in the northeast regions of Laos, and along the Ho Chi Minh Trail.

The "cultural" environment in Laos was extremely crucial to military aerospace operations there. By the time I came onto the scene in 1971, the NVA and their Warsaw Pact advisers had constructed one of the most comprehensive and sophisticated air defense systems in the world to protect their logistics operations along the HCM Trail. At the northern hub of the line-of-communication (LOC), the air defenses consisted of a network of radar sites for detecting our attacking aircraft, and for subsequently guiding the fire of the surface-to-air missiles and the anti-aircraft artillery that complemented they controlled. All of these air defense systems were connected by land-line communications to central control bases that choreographed their fire so as to maximize the lethal effect on the intruders. On the other side of this electronic warfare, there was the U.S. ground and air-based radio beacon systems which served as navigational aids to the attacking aircraft. Later, there would be deployed the Wild Weasel aircraft that served a radar site hunter-killer function to either degrade or destroy the radar fire-control sites on the ground. These specialized platforms essentially searched for the tell-tale electromagnetic "signature" beam of the radar-controlled air defense sites. They then would fire their air-to-surface missiles which would simply "ride" the electronic path back to the original emitter on the ground. The electronic war was essentially a matter of initial measure and consequential countermeasure, in a kind of duel in which radiation beams were used to gain the upper hand, whether that meant a successful attack or an effective counterfoil.

––––––––

As a geographer, I recognized that within the PDJ MAR, there was a geographic object that was created by the myriad of voice communications occurring there. These created a virtual "web" that would rival in intensity a portion of the computer-based internet of today. There were literally hundreds, if not thousands, of transmission and receiver points that were "spinning" at any hour of the day. Our own command and control aircraft capsule was equipped with the radio technology that enabled us to communicate on every phase of the radio spectrum. Thus, we had UHF and VHF capabilities which enabled us to talk with the "gomers" (indigenous ground troops) on the ground. We also had an array of HF and other receiver/transmitter points along the spectrum in order to communicate with the various allied aircraft that operated over the battle zone. Within these regions of the radio phase of the electromagnetic spectrum, there were a multitude of frequency points that I imagined to be discrete "nodes" in a region. Thus, it can be said that the MAR over northern Laos could be seen as a functional region in which the central node was the Airborne Command and Control Aircraft (Cricket) and the subsidiary nodes were the transmitter/receiver points on the ground and in the air.

In a way, in monitoring and moving along this network of radio communications, I was behaving in very much the way in which modern internet "surfers" monitor the various websites on the worldwide web computer network today. Another similarity between the two internets is seen in the imperative to maintain security over our communications, and in the way we also sought to maintain "situational awareness" of the battlefield. And, like the strategy and tactics that are used to prevent "hacking" on the computer internet, we utilized similar tactics to counter any "hacking" of our voice communications. Among these was the constant channel-hopping to gain some modicum of secured communications. Finally, we used encryption to keep our communications secure, as is done on the computer internet.

Thus, it can be said that we on Cricket operated the radio portion of the spectrum like a piano keyboard; using the channel-switching technique in our communications with friendly ground liaisons which we call Forward Air Guides (FAGs). The FAGs were invariably Hmong tribesmen and Laotians, as well as a few Thai "volunteers," who were trained in radio communications and essential English by the CIA. They would typically initiate contact with Cricket on a radio channel that we constantly monitored. Most of the communications, other than the periodic "ops normal" signal, would be a report of enemy "TIF" (troops in the field) or "TIC" (troops in contact). The latter meant there was an active assault or artillery attack against a particular forward base (Lima Site). Meanwhile, the intelligence operatives on Cricket would forward the necessary information about ground situation to "users" that included the onboard Battle staff commander, 7th AF headquarters, or CIA headquarters at Long Tieng. In most cases, however, the intelligence was passed on to a Raven Forward Air Controller (FAC) who, in turn, would direct the attack aircraft against the targets. Simultaneously, the second intelligence specialist onboard Cricket would pass the pertinent information to a fellow crew-member; an airborne air traffic controller next to him. The controller would then assume "air traffic" control of any air support resources that might be coming on station. Thus, it can readily be appreciated that this web of communications would be operating at a great level of intensity during these operations. This also was an example of combat air operations interacting with the geographic elements within the radio phase of the electromagnetic spectrum, otherwise known as the radiation environment.

———

The radiation environment over the PDJ also included other phases of the electromagnetic spectrum. These were manifested in three kinds of battlefield situations. One of these involved the air-to-air combat between two aircraft in which one or both utilized radar or infrared radiation to detect, track, and attack the other. Then, there was the radar battle that existed between the enemy fire-control systems for their SAMs and AAA, on the one hand, and the friendly Wild Weasel aircraft which were equipped with radar and infrared sensors for both detecting hostile signal emissions, and directing missiles against the ground air defense weapons that were controlled by radar emitters. Toward the end of my tour, in 1972, the shoulder-launched SA-7 SAMs that relied on the infrared phase of the spectrum for weapon guidance began to appear in northern Laos, especially along the perimeter of the Ho Chi Minh Trail. On the other side of the battlefield equation, the infrared portion of the spectrum also was integral to the effectiveness of the U.S. Air Force A-10 operations against the trucks and tanks that plied the North Vietnamese lifeline to the Viet Cong forces in South Vietnam.

Another web, this one being constructed by the combat support sorties that were flown over the PDJ, was constructed by the "strands" of their trajectories. These represented many support missions that were occurring simultaneously to enable and facilitate the success of the attack sorties. Among such support sorties, there were included air reconnaissance platforms, some operating in the visible light portion of the electromagnetic spectrum; while others were equipped with sensors that focused on radar and infrared emissions. The latter usually was emitted by objects whose "signature" indicated that they were trucks, tanks, or artillery pieces. Also, functioning much like the P-51 escort platforms of WWII, there were the F-4 Phantoms that watched over the rest of the force, ready to respond to any forays by Mig interceptors from North Vietnam. Under their umbrella of protection, other platforms performed their air attack, bombing, electronic warfare, and transport missions, to name but a few. Other attack platforms, such as the A-7 bombers, launched from naval carriers in the South China Sea, while other platforms worked numerous close air-support operations in concert with the indigenous Laotian T-28 converted trainers, which were based at Long Tieng. Then, there were the B-52 heavy bombers that flew "arc light" bomber attacks over the Laos MAR from time-to-time at an altitude of about 30,000 feet. And, when they were not working combat air patrol missions over the PDJ, the multi-mission F-4 Phantoms would also perform air reconnaissance, fighter-interceptor, and fighter-bomber operations. Further adding to the strands of sorties, was a fleet of smaller transports, medivac helicopters, SAR (Search and Rescue) flights, and sundry other support platforms which carried out their duties over the PDJ. At the same time, the Twin Otter aircraft of the CIA also would be flying repeated transport sorties in support of the Lima Sites.

———

Our own C-130 platform was modified to carry extra supplies of fuel, so that the day-long mission could be carried out without the need to refuel after takeoff. Instead of normal cargo that was carried by a Hercules transport aircraft our platform carried a thirty-foot long capsule within its cargo bay; it was outfitted with a row of airliner-type seats along one side of the cabin. On the other bulkhead, plastic displays and communications equipment faced these seats. This was the airborne command center (call sign Cricket), in which the two battle-staff commanders, the pair of air traffic controllers, and the two airborne intelligence specialists managed the air operations space, and directed the incoming allied air assets to their respective targets. It was the intelligence specialists on-board who coordinated requests from friendly ground forces for tactical air support. The other member of the "backenders" crew was the radio specialist who enabled us to remain in constant contact with 7th Air Force Command headquarters, as well as all the other nodes of the network.

The air intelligence specialists' charge was to maintain contact with the CIA Case Officers who functioned as "command advisors" for ground operations on the battlefield. Other communication nodes included the Laotian (Hmong) General Vang Pao, who had direct command of his indigenous guerrilla infantry forces, and the on-site bi-lingual Forward Air Guides who might be Laotian or Thai "volunteers." At the same time, the Air Controllers onboard Cricket performed their air control function over our combat air space. Another part of my job, as an airborne intelligence officer, was to keep track of the geographic environment within our area of operations. This involved a continuing analysis of this environment, with special emphasis on those features that were most likely to affect the various daily air operations over the PDJ, and the environs of the Ho Chi Minh Trail. All of this would be recorded on my mission log and would provide material for my post-mission report to 7th AF headquarters.

The post-mission reports also would be incorporated into the pre-mission briefings by other intelligence officers. Along with these and other inputs, such as the daily meteorologist reports the intelligence officers were able to develop daily intelligence pre-mission briefing for the crews, just prior to takeoff for the next 12.7 hour mission. Once airborne, the MAR charts would be constantly updated with any "real time" intelligence, as necessary. I would note here that all of this plotting of phenomena on the charts was done with pens and grease pencils, on paper and on the plastic medium that covered the wall charts. Today, I often contemplate what it would have been like to have the 21st century computer and telemetry technology in Laos; to have the ability to "stream" imagery and data to the 7th Air Force Headquarters in South Vietnam for processing and feedback, almost in real-time fashion. Even better, what it would have been like if the intelligence could have been automatically processed by micro-chip programs and relayed to the ultimate "users" – the warfighters on the ground and in the air.

Most of the recoveries of the local sorties; those flown by the Raven FACs, the CIA and the Hmong T-28 pilots, occurred within Laos, mostly at Long Tieng (LS20A). The Allied out-of-country sorties returned to their home bases in South Vietnam, Thailand, or far away as Guam. These landings were generally uneventful, as long as there were no major maintenance or damage issues. There was one set of long sorties, however, whose recoveries were always dangerous, if not hazardous. These were the U.S. Navy sorties which recovered at one of the naval carriers out at sea. Even when there was good weather, the carriers always presented moving runways, on which landings required the same precision that the space station rendezvous do today in space. Indeed, the carrier pilots with whom I spoke described the experience, even under the best conditions, to be like throwing a ball into a small, bobbing basket on a windy day. Also, the carrier had to be oriented properly with respect to the winds, in order to provide the headwind that the air platform would need to decrease its speed, as necessary, on its approach to landing. The location of the naval carrier on the sea also had to be carefully coordinated and timed, very much like the air refueling platforms, and the International Space Station.

———

An important part of the pre-mission briefing dealt with the natural and cultural environment that would be encountered during the course of the sortie. My part of the briefing was an analysis of the military geography with which we would be interacting during the course of the sortie. Both the natural and cultural elements of the overall environment in which the air operations would take place were analyzed, both individually and as a holistic environment. (I prefer to use the term "natural," rather than "physical" so as to distinguish these from the "man-made" artifacts, which are also physical. So, when I refer to natural geographic features which affect military aerospace operations, I am speaking mainly of the atmosphere, lithosphere, and hydrosphere). The following is a discussion of the natural and cultural geography of the Laos MAR, especially the manner in which it affected military aerospace operations there.

The spatial analysis of the MAR in northern Laos begins with two basic metrics: the two-dimensional surface and the three-dimensional atmosphere in that space. The overall three-dimensional area of the MAR is bounded by the 18-degree (N) and 23-degree (N) lines of latitude; and by the 102-degree (E) and 105-degree (E) lines of longitude. The third dimension reaches a height of about 35,000 feet, which was the altitude at which the B-52 bombers operated in that MAR, and is also the approximate effective ceiling of the Soviet SA-2 surface-to-air missiles. Perhaps a more practical measure of the dimensions of this MAR would be the time it took for the various platforms to arrive at their operations or target points following their launch from their naval carrier or a land base. From my empirical observations, I estimate that the carrier-based platforms were able to arrive over the target site, deliver their ordnance, and return to base within two hours of flight time. The same time frame was established for the B-52 sorties over Southeast Asia, even though their home base was located on Guam, an island territory situated near the Mariana Islands in the South Pacific. The sortie of the C-130 on which I worked, arrived on station over the PDJ within half an hour of takeoff from Udorn Royal Thai Air Base; it was located some 75 miles south of the "fence" (Mekong River) which separated Thailand and Laos. In other words, the air time from base to target was like a daily commute for all American air assets. However, like most commutes, the normal flying time to the area of work could vary, according to such factors as bad weather or the appearance enemy aircraft along the sortie flight path.

———

As far as military aerospace operations were concerned, the significant natural geographic features included a mountainous region to the north and the east – the Annamese Mountain range. This was a roughly north-south axis along which the logistical system, the Ho Chi Minh Trail, was laid out. This cordillera also formed a political divide in Indochina between the Vietnam to the east, and Laos, Thailand and Cambodia to the west. It is a rather continuous range for about 700 miles, which also roughly describes the north-south length of the PDJ MAR. Although its highest peak (Linh Peak) is only about 9,000 feet, the cordillera is nevertheless extremely rugged in its local relief and, therefore effectively channels human movement through the few substantial passes, especially in the northern end of the range. The most important of these passes, in terms of military geography, was the Mu Gia Pass in the northern end of the system. It was a vital access route for moving war materiel from northern Vietnam to the Viet Cong in South Vietnam. Historically, this pass had long been part of a main overland civilian transportation network that linked northern and southern Vietnam, as well as northeast Laos and Cambodia. Aside from these mountains, there also was the extreme relief that was to found along the foothills of the Luang Prabang Range in the north and west of Laos. West of these mountainous regions, the other significant feature of the topography is the so-called Plaine des Jarres (PDJ) – so named for the thousands of very large burial jars which are distributed throughout that area. This relatively flat terrain was the site of much of the ground combat which occurred in northern Laos, and which was mainly fought by the proxies of North Vietnamese Army (Pathet Lao) versus the proxies of the United States Central Intelligence Agency (Hmong irregulars) from about 1955 to 1975.

Another major feature of the natural geography that greatly interacted with U.S. military aerospace operations was the triple-canopy rainforest that covered the Ho Chi Trail. The first time I saw a triple-canopy rainforest, it was from the ground up, in the Philippines, during my training in jungle-survival at the Pacific Jungle-Survival School on "Huk Mountain," near Clark Air Base. I remember looking up through the thick thatch-roof of vegetation, during the middle of the day, and not being able to see the Sun, except in small areas where humans had cleared the trees. Later, during my tour in Laos, I would find that the thick canopy of vegetation also would have made me effectively invisible from any aerial visual observation. More importantly, I would have been protected from air attacks, even in broad daylight. By the same token, the nights were absolutely dark on the jungle floor.

Thus, as I would discover later, this same triple-canopy type of tropical rainforest would provide the North Vietnamese with a natural camouflage and cover from American bombs throughout the course of the Vietnam War. It was not surprising, therefore, that when the American phase of the Vietnam War began in about the early 1960s, the erstwhile economic trade network evolved into the sophisticated military transportation-communication system which was dubbed the Ho Chi Minh Trail by the U.S. Military. By the time I had arrived on the scene, in 1971, this "trail" had been transformed by the North Vietnamese into an advanced road and highway system, which rivaled that of any relatively advanced "developing nation." Millions of tons of supplies and armaments were sent through this logistical spatial system by the North Vietnamese in support of Viet Cong guerrilla and NVA operations in South Vietnam.

Generally speaking, the rugged terrain and the patterns of vegetation cover served to effectively channel several types of human movement within the MAR. Thus, the few natural passes through the Annamite Range determine the patterns of human ground movement throughout the region of the cordillera. The resultant predictable patterns of ground movement then did much to dictate the patterns of movement by the aircraft that performed interdiction operations over the region. Of course, all of this unwanted attention from U.S. air combat platforms then resulted in the North Vietnamese drive to build up their air defenses in order to shoot down the enemy aircraft. Therefore, it can be said that the natural topography and vegetation, as well as the weather regimes caused by the elevations, were basic factors in the construction of a particular "cultural" military aerospace region there. For their part, the NVA saw the natural geography of the cordillera as a highway system which could be fortified and camouflaged against enemy air strikes. The U.S., on the other hand, viewed the natural geography of the Ho Chi Minh Trail as a targeting problem to be solved with appropriate weapons and tactics. So, this is how a configuration of natural terrain and vegetation that developed along the western side of the mountain range – largely in response to weather patterns – would come to be called the "Ho Chi Minh Trail" during the Vietnam War.

To the west of this natural-cultural region lies the "Plainne des Jarres" (PDJ); so-called by the French during their colonial domination of the region because of the huge stone jars that dot this region. The PDJ essentially is a plateau; it is characterized by relatively flat terrain, overall, but there are significant differences in local relief within it. As far as vegetation is concerned, the PDJ can be described as a grass-land that is interspersed by trees, the latter occurring mainly on the rims of the plain and along streams that crisscross the region. This place has a climate that is dominated by the monsoon rainy season (May through October) and a relatively dry season (November through April). One result of this monsoon climatic regime is that the region is generally covered by clouds during the rainy season. There are also some highlands along the western and southern edges of the PDJ which are characterized by a series of intermontane valleys and small basins. Perhaps the largest and most prominent of these valleys was the one called Long Tieng. It has long been the center of Hmong culture in the southern part of the PDJ and, during the period of the CIA clandestine military operations in Laos, it also served as the headquarters for both the agency and the Hmong guerrilla army which was commanded by General Vang Pao. Approximately 100 miles northeast of Long Tieng, there was another political-military headquarters, this one was the center for the Communist Pathet Lao forces that opposed our allies on the PDJ. In most ways, the military conflict that was occurring on the northern plateau of Laos was separate from the one taking place along the Ho Chi Minh Trail. So, it can be seen that the PDJ and Ho Chi Minh Trail were two subregions of the overall Laos MAR.

Unlike the HCM Trail subregion of the Laos MAR, the PDJ had no terrain features that would channel human movements (except on a very local scale). Instead, the plateau looked more like an abstract plane, at least for purposes of military aerospace operations. Indeed, the most significant natural geographic feature that affected military operations there was not the terrain or vegetation, but the weather. In the winter rainy season, the cloud cover would significantly hamper military air operations and provide the enemy ground forces with the opportunity to conduct offensive operations on the PDJ. Using a chessboard metaphor, this was their turn to move their pieces (tanks, artillery and infantry) southward from their home bastion. Their gambit would be directed against the network of forward CIA-Hmong Lima sites that had been established during the previous dry season (under the cover of overwhelming air superiority). So, the battle for the PDJ can be seen as an alternating series of chess moves that were dictated by the cycles of the Monsoon seasons.

———

Toward the end of 1972, a cutting-edge technology was introduced to the Laos MAR by the U.S. Air Force; one which was to be particularly affected by a particular feature of the natural geography of the PDJ. I refer specifically to the myriad of limestone "sinkholes" that pervade the soil structure of this region. These so-called "karst" soil regimes were created by long-term percolation of water and carbonic acid which worked to eat away the limestone rock and form large sink-holes in the surface ground have. Over time, some of these sinkholes would be covered over by a thin thatch of vegetation and soil. This, in effect, became a pseudo-surface on the PDJ. The U.S. Air Force, in this case, discovered the military geographic consequences of this natural phenomenon when it introduced the F-111 fighter-bomber to the air war in Laos. That is to say, it was learned, the hard way, that the Terrain Avoidance Radar technology, which enabled these aircraft to fly and fight very close to the deck, and at very high speeds, contained a fatal flaw.

Thus, in order to perform these dangerous maneuvers, the aircraft were equipped with a dynamic radar system that provided real-time indications of the terrain below the aircraft. This technology enabled the air attack aircraft to hug the terrain (below the view of the ground-based radar), to essentially "outrun" air defense guns and missiles. However, it was a risky tactic because it was completely automated. In fact, the reaction time that was required to maneuver was too short for manual actions; it required the quicker response of a computer-controlled altitude-correction system. In a sense, TFR was simply an extension of existing auto-pilot technology; and it presented the same dangers to the pilot. So, the sensor part of the system had to provide "real-time" information to the reactive parts of the system in order to fly at the desired altitude, "just above the deck." Unfortunately, when the TFR scanned the limestone sinkholes over northern Laos, it relayed faulty data to the analog computer on board the platform. The upshot of the matter was that the "automatic pilot" actually caused the aircraft to drop down to the "real" surface, hundreds of feet below the apparent deck. As a consequence, at least three F-111s were lost in this first test of the TFR system. This was a case where the sensor subsystem relayed spurious data about the relevant geography back to the processor subsystem. It is also a metaphor for the classic dilemma that is faced by all pilots, but especially military pilots: do you trust your eyeballs, or your flight instruments? Ultimately, the technology of the sensor, the processor, and the actuators advanced to the point where the TFR could detect the "terrain" in a much more sophisticated and precise manner.

But, as we have seen, there is another kind of geography, the cultural geography, which affects a sortie. On the ground, it is manifested in the man-made layer of "cultural" features that is interwoven with the natural features, to present an overall military geographic environment to the sortie. In a less-industrialized region like the PDJ and its environs, these would include roads and bridges; buildings and dams. These are considered to be targets for interdiction, as well as the disruption of communications and power-production. However, their very value as targets of air attack also tend to make them hazardous to military aerospace operations in their vicinity, as well as the approaches to the targets, threatened by the air defense systems which were deployed to defend such "high-value" targets. This part of the "environment" consisted of enemy air defenses, including possible air interceptors (the enemy air defense order of battle). Utilizing photo-reconnaissance and other imagery analysis, the air crews were briefed on the location of the air defense resources, including the radar or infrared guidance systems that were identified.

Adding to the complexity of the overall cultural geography that affected military air operations over Laos was the "political geography" of that region. One of its main features was the fundamental "rule of engagement" that prohibited military air attacks, or even flyovers, in certain areas of Laos; but especially in the northern and eastern parts of the country. The reason for these restrictions stemmed from the fact that Laos was officially neutral with respect to the Vietnam War between North Vietnam and the United States. As a result of the Geneva Accords of 1954, Laos had been divided into three regions of political hegemony: those of the right-wing Royalists; the Communist Pathet Lao; and the Neutralists, respectively. As all political geographers understand, the "real-politik" situation on the ground can generate much different political maps than the theoretical construction does. Thus, the overlay of reality on the nominal map of northern Laos depicted a political geography that included "no-fly zones" and "no attack zones" which would not be found on any conventional map of that region. Other no-fly zones, which were not created by formal agreement, included the highway that was being constructed by the Chinese to link that country with Thailand and Burma. All of these were part of the cultural geography of Laos during the Vietnam War. The point to be made here is that the political geography in a MAR can be just as important as natural geographic features in their effect on military aerospace operations.

Simply stated, the objective of the CIA on the PDJ was to cause the North Vietnamese Army as much trouble as possible in their attempts to achieve their objectives there. The key to this mission of interference and obstinacy against the NVA and their Pathet Lao clients were the Hmong striker troops. This CIA "client army" was commanded by the Hmong General Vang Pao, who proved to be a brave and highly-effective commander. Part of the reason for his success was his ability to assimilate the modern weapons and tactics of the air-mobile guerrilla war that had been developed by the Green Berets in South Vietnam with the Montgnard tribesmen there. Thus the Hmong battalions were trained and equipped by the CIA to conduct what we would might be called "air mobile" attacks against the NVA and Pathet Lao forces that confronted them on the PDJ. The CIA and Vang Pao even managed to form a highly-effective squadron of T-28s, known as the "Chapakawas" (phonetic spelling). These were converted trainer aircraft which were fitted with guns, rockets, and bombs; these reconstituted fighter-bomber platforms were mostly flown by indigenous Laotians. The home base for these CIA-Hmong forces was Long Tieng (aka Lima Site 20A), which was located between the capital of Vientianne and the southern edge of the PDJ. This also was home base for an assortment of light transports, ranging from helicopters to light fixed-wing transports. The latter consisted of light, "bush pilot" type aircraft with names like Otter, Porter, and Beaver. This fleet also was known as Air America; it was a client airline of the CIA in Laos.

Long Tieng also was the central node in a functional system that included a network of other nodes, called "lima sites." From this base on the southern edge of the PDJ, the CIA and Vang Pao commanded the air mobile operations, while the Lima Sites served as forward bases whose mission was to both serve as staging bases for offensive missions by the Hmong irregulars; and to "hold the ground" that had been won, against the NVA and Pathet Lao enemy forces. Daily operations included long-range reconnaissance and helicopter-borne infantry missions against the NVA and their Pathet Lao proxies throughout northern Laos. The Hmong irregulars also utilized helicopters and Air America transports to respond to enemy attacks on the Lima Sites. These were a network of fortified forward bases (called Lima Sites) which very much resembled the redoubts that the Green Berets established among the Montgnard tribesmen in South Vietnam. Indeed, many of the "contract" operators in Laos were former Green Beret types. And, some of the Lima Sites in Laos also were manned by Thai "volunteers." That is to say, the Thais generally served as garrison troops on the Lima Sites that also served as artillery bases. The idea was to use these Lima Sites both as launching sites for offensive air-mobile operations, and as defensive redoubts and reconnaissance positions. The 105mm and 155mm artillery fire-support bases served both offensive and defensive functions.

The Communist air defense order of battle in Laos and in all the MARs of the Vietnam War was mostly ground-based, because we had virtual air supremacy, or at least situational superiority, over most of the areas of operations in that conflict. This state of affairs also enabled allied aircraft to operate in what was called a "permissive environment" with respect to Mig interceptors. However, the threat environment from enemy air defenses increased as one neared the Ho Chi Minh Trail or the Mu Gia Pass to the east. By the time I arrived in-country in 1971, the enemy air defenses had undergone major upgrades, at least around the Haiphong-Hanoi complex and the Ho Chi Minh Trail. With the support of the Warsaw Pact, the air defenses along the Ho Chi Minh Trail and northeast Laos had become sophisticated, integrated systems.

The NVA had also incorporated the Soviet Doctrine of War which had evolved during WWII in all the MARs. That is: build a mass of relatively low-cost and low-tech, but reliable and easily-maintained weapons to overwhelm the enemy. This doctrine was implemented following the German invasion the Soviet homeland in 1942, when the Russians built and deployed a huge number of relatively low-tech aircraft, tanks, and artillery to combat the German invasion of their motherland. It was felt that masses of cheaper, relatively low-tech weapon systems would be sufficient to counter more advanced, but fewer, German weapon systems. And it proved to be the case. The same thinking drove the weapons and tactics of the NVA air defense systems, especially within the region which comprehended the capital, Hanoi and the main port, Haiphong, and along the Ho Chi Minh Trail in Laos. However, in the Vietnam War situation, the communists also applied some of the most advanced technologies of the Warsaw Pact when they felt they needed an extra edge. In this war, they relied more on such anti-aircraft weapons as the 12.7mm heavy machine gun, the 14.5mm gun, the 23mm ZSU-quad gun, and even the shoulder-mounted, infrared-guided, surface-to-air missile, the SA-7. All of these air defense weapons shared the same virtue; they were light enough to be carried by individual soldiers or on pack animals. Or, they were often dismantled for portage and then reassembled on site.

The NVA air defense systems also can be seen as manifesting the vertical zonation that was discussed earlier. Recall that the farmers in the tropics utilized the strata of micro-climates along the side of a mountain to cultivate crops that normally would be found in higher latitudes. In an a somewhat analogous fashion, the North Vietnamese first created a layer of AAA weapons which included the 12.7mm machine gun and 23mm cannon. It was designed to form a veritable "curtain" of fire at heights below 2,000 feet. Most of these were guns were optically-directed and, they were not coordinated by any overall communication system. Their main role was to provide "point defense," against low-flying aircraft. The next stratum of air defense ordnance was produced by larger AAA weapons, such as the 23mm and 37mm guns. They effectively raised the air defense ceiling to approximately 20,000 feet. The highest stratum reached up to about 30,000 feet and was formed by 57mm and 75mm anti-aircraft artillery. Moreover, as the air war over North Vietnam continued, the composition of the strata became more and more sophisticated, and deadly for attacking aircraft and aircrews. The efficacy of NVA air defenses against attacking allied bombers, especially over the Haiphong-Hanoi complex, improved dramatically when the Warsaw Pact countries installed a state-of-the-art integrated air defense system there during the later years of the 1960s. One of the most significant additions to the air defense armament population in that region was the introduction of the radar-controlled surface-to-air missiles, such as the SA-2. Ultimately, other SAMs were integrated into this upgraded system, including the SA-4 and the SA-6 which were designed to operate at lower heights than the SA-2. Also, many of the larger anti-aircraft-artillery sites were placed under radar control and integrated into the overall command and control system of the air defenses.

In terms of military geography, what had happened was that the zonal strata of ordnance had been made more intense at every level. Now, the incoming attack aircraft had to deal with not only "barrage" fire from undirected AAA, but also with ordnance that was more accurate and coordinated by radar and communications subsystem. The upshot of the matter is that the North Vietnamese eventually built what would turn out to be one of the most advanced and formidable air defense systems in the world at that time. This is also when electromagnetic radiation became a "geographic" factor in terms of the military aerospace sortie in Southeast Asia. Now, the mix of hindrances and hazards to the sortie mission included, not only the clouds, fog, thunderstorms, lightning, and the other elements of natural geography – but also, the elements of cultural geography, which included the enemy air interceptor, as well as the anti-aircraft artillery, surface-to-air missiles, and the electromagnetic radar and infrared fire-control systems that gave them precision.

This major change in the air defense threat profile that was presented to attacking aircraft then prompted changes to the weapons and tactics which were utilized by the U.S. Air Force to strike the relevant ground targets. Probably the most significant of the countermeasures against the advanced NVA air defense system was to develop and deploy the so-called "Wild Weasel" aircraft system. It was based on the F-105 fighter-bomber platform. Like many other aircraft that had been originally designed for duty in WWII, Korea, and the Cold War, this platform had been redesigned and renovated for a new, specific mission in the Vietnam War. In this case, to counter the threat posed by the NVA fire-control radar sites. The essence of the new F-105 platform was its payload: a transmitter/receiver system that actively searched for the tell-tale radar signals that were emitted by the fire-control radar sites on the ground. The other component of the payload package was an air-to-ground missile that was specifically designed to detect and follow the ground radar beam back to its source, and to destroy the site.

The Wild Weasel system greatly enhanced the survivability of the other allied attack aircraft as they worked to lay down ordnance on their targets in North Vietnam. But to the military geographer, the greater significance of this development is that the electromagnetic spectrum was becoming another facet of the natural geography that interacts with military aerospace operations. I have continued to study the electromagnetic battlefield in the decades that have followed my tour of duty in the Vietnam War, and during this period of time I have seen how electromagnetic energy has become a paramount factor in modern warfare. This form of energy is now used in almost every area of military aerospace operations. Even during the period of the Vietnam War, electromagnetic energy was being utilized for intelligence, surveillance, and reconnaissance; and to conduct communications and even to direct ordnance as part of the new generation of "smart" weapons. In addition to the Wild Weasel program, there was a move to provide many other U.S. platforms, like the B-52 and F-4 weapon systems, with on-board "chaff" and other electronic countermeasures in order to be able to fly and fight in the highest-threat environment.

The same dynamic occurred along the Ho Chi Minh Trail. Thus, in Laos and in North Vietnam, the NVA simply deployed relatively vast numbers of AAA and SAMs to counter the relatively high-tech military air systems of the U.S. air forces. Their adoption of the Soviet doctrine of warfare along the Ho Chi Minh Trail also was expressed in terms of the principle of redundancy. That is, within the thick vegetative cover of the corridor region, the NVA had seemingly created layers of anti-aircraft artillery and surface-to-air missile systems. These air defense weapons included a of 75mm and 57mm anti-aircraft artillery, a medium-level 37mm, and a lower-level tier of 23mm AAA guns. At the same time, the triple-canopy of vegetation was a physical-geographic complement to the triple-canopy air defense systems that the NVA deployed along the Ho Chi Minh Trail. In the later 1960s and early 1970s, the NVA enhanced their AAA threat with the addition of SAMs (Surface-to-Air Missiles) along the HCM trail that wound southward from North Vietnam, along eastern Laos, and into South Vietnam.

Expressed in the terminology of military geography, the natural and cultural geography of the HCM Trail system affected all U.S. air operations, whose overall mission was to interdict the flow of supplies, arms, and men through this line of communications. This was manifested in the way that the NVA utilized the clouds, haze, vegetation and other natural obfuscations to deny the U.S. attack platforms a clear shot against the tanks, trucks, bicycles, and foot traffic along the trail. They then armed this transit system with air defenses to make the interdiction even more difficult. In a way, it reminded me of the Hadrian Wall defense that the Romans constructed to keep out the incursions of the Celts in Britain. Again, consistent with the Soviet doctrine of redundancy of weapon systems, the NVA constructed a multi-layered system of transportation routes which, like the triple-canopy of the forest, included subterranean as well as surface routes, and "pit stops" and caches all along the course of this logistical system. The latter included subsurface bivouacs and refueling stations throughout the PDJ to maintain the flow of men and supplies from North Vietnam to South Vietnam.

To summarize, it was the combination of triple-canopy rainforest, monsoon rains, clouds and fog, as well as smoke and cinders produced by the indigenous slash and burn agricultural practices, that presented the main natural geographic challenges to military aerospace operations in northern Laos. As for the cultural geographic factors that affected such operations, it appears that these can be grouped into three main subcategories: the geographic manifestations of the political realities that existed during the campaign in northern Laos, especially in the way in which these political entities interacted with enemy air defense systems; the smoke and haze that was emitted as a result of the practice of slash-and-burn agriculture by the indigenous farmers in the region; and, of course, the NVA air defense system.

In this particular geographic environment, the combat aircraft had to operate within constricted spatial channels which subjected them to a set of micro geographic elements. These included the weather factors that affected visibility over the target and impacted on the ultimate phase of the sortie. In terms of the delivery of ordnance on ground targets, the clouds, fog, or smoke obviously created difficulty with accuracy. On the other hand, these geographic features also created cover for the enemy ground positions, as well as the air defenses that protected them. The same dynamic occurred with respect to vegetation and terrain features around the targets.

So, we have seen how the conceptual tool of the geographic region can be used effectively to conduct an analysis of a military area of operations, from many different perspectives, to arrive at a better overall understanding of the spatial nature of human operations in that area – in this case, the relevant human operations were the military aerospace operations. Next, we turn to a closer look at the geographic environmental systems as they relate specifically to these operations.

THE ENVIRONMENTAL SYSTEMS OF THE SOLAR REGION

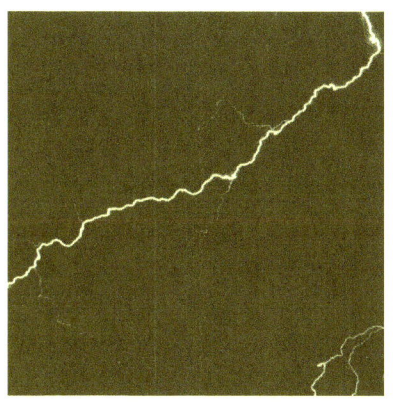

As we have seen, a major focus of military geography is the study of the effects of natural and cultural environmental phenomena on military aerospace operations. This has been the case within the Earth's atmosphere and now, in outer space as well. The following is an analysis of the "natural geography" that interacts with military aerospace operations throughout the region that we call the solar system. The precise dynamics of the interaction between military operations and the physical geography is complex, and can be said to be "situational," in both a temporal and spatial sense. That is to say, there is no single "environment" or interaction dynamic that occurs between this human activity and the natural environment. Rather, there are various subsets of the total environment. These are the systemic processes that transverse the dimensions of the solar system.

Within the Earth's atmosphere, for instance, there is the geographic phenomenon known as the "weather mass." To the meteorologist, this refers to a large body of air which has presents internally uniform conditions of atmospheric pressure, humidity, and temperature at a given altitude. In space, the atmosphere is replaced by the interplanetary medium and the "weather mass" refers mainly to the field of charged particles called the solar wind. Like the weather systems of the Earth's atmosphere, the fields are systems in which the charged particles radiate and interact around large bodies, such as the planets, to form other types of fields. And, like the weather systems that affect military atmospheric operations, the fields of charged particles affect military space operations. So, one can imagine that in future military space operations there will be pre-mission briefings by "astro-meteorologists" to military astronauts.

Here I will refer to these systems as geographic fields as well. Indeed, this phenomenon also appears to be very much like the uniform region of the geographer. So now, following a half-century of space missions into the solar system, we have discovered several kinds of fields of charged particles throughout the heliosphere. The largest of these is the solar wind that is emitted by the Sun in the first place. As this field of charged particles propagates outward through the heliosphere, it encounters many "islands" of matter called celestial bodies. It also encounters magnetic fields, similar to the geomagnetic sphere that surrounds our planet. These also can be seen as "particle fields," within which there are as many processes and movements as are found in the Earth's oceans. They also can be described as nearly uniform regions of electric charges and ions. Therefore, it appears that these "fields" also can be studied as another form of geographic regions, which contain matter and energy, and within which electromagnetic processes occur. In summary, atoms, in their many configurations, can be described by many names.

Radiation Environment

The term radiation usually refers to the mechanism by which energy is emitted from a source and propagated through a surrounding medium. Examples of radiant energy with which we are familiar include light (electromagnetic radiation) and sound (acoustic radiation). Another form of radiation, radioactivity, is a property that is exhibited by certain types of emitting energy by subatomic particles that usually occurs spontaneously, but it can also be induced by artificial means. In both cases, it is an attribute of individual atomic nuclei. The nuclear process involves the decomposition of an individual nucleus into a more stable configuration. When the decay process occurs, certain particles of electromagnetic energy also are emitted. Radioactive decay, on the other hand, is a property of several naturally occurring elements and of some artificially produced isotopes of the elements. All of these various types of radiation occur throughout the solar system, and are therefore part of the physical geography that affects and interacts with military space operations.

Very often, the radiation environment that confronts military aerospace operations is in the form of an electromagnetic field. This phenomenon can be seen as presenting an internally uniform combination of energy and matter. The fields are produced by charges that form around a force that is produced by a magnet, electric current, or a changing electric field. Magnetic force refers to an attraction or repulsion that arises between electrically-charged particles because of their motion. The magnetic field itself is produced by the dynamic effect that is exerted by two moving electrical charges. There is magnetic resonance when the absorption or emission of electromagnetic radiation by electrons or atomic nuclei produces a magnetic field. In the case of the individual planetary magnetic fields, the magnetic field around the planet results from the interaction of the planet's own magnetic field (which is largely generated by the dynamics of its core) with the solar wind. The nature of the interaction is a function the size and composition of the internal magnetic core of the planet and its interaction with the materials that directly surround it.

———

The most-used working model for the planetary magnetosphere in the solar system is the geomagnetosphere that occurs around the Earth. It is the magnetic field which is projected from the center of the Earth, diffuses toward the poles, and envelops the globe. This force field serves to protect the atmosphere and the surface of our planet from most of the harmful cosmic and solar radiation that arrives to interact with it. Without this vital magnetic "blanket," our atmosphere would not remain intact, and life on Earth could not exist. But even the residual geomagnetic storms which do occur can do great damage to power grids on our planet. Within the Earth's magnetosphere, there are two doughnut-shaped zones that are centered over the Equator. These zones are occupied by significant numbers of energized protons and electrons that are trapped within the Earth's magnetic field. The inner zone of this field extends outward from the atmosphere to an area that lies from approximately 600 miles to 3,000 miles above the surface of the Earth. The outer belt extends from about 7,000 to 13,000 miles above the Earth's surface. Together, these zones are called the Van Allen Radiation Belts; named after the scientist who first discovered them. Of the two, the inner Van Allen belt contains a relatively dense flux of high-energy protons. These are thought to be the outcome of the decay of neutrons, which are produced by the interaction of energized cosmic rays of galactic origin with the Earth's atmosphere.

I find it useful to think of the Earth's magnetosphere as a kind of "windshield," except that the wind in this case is a solar wind; one which consists of electrically-charged particles which flow in the same way that the wind would in a wind-tunnel. This "windshield" serves as an obstacle to the flux of charged particles (or plasma) that is related to the solar wind; these are deflected around the Earth by the bow shock that occurs as a result when the wind interacts with magnetosphere. However, some plasma particles from the solar wind can pass through the magnetopause. When these interact with charged particles from the Earth's ionosphere, the result is the occurrence of a dynamic process, which is referred to as "reconnection." The upshot of all this interaction is that there is a release of energy which, sometimes, causes a reconfiguration of the magnetosphere which is called a "geomagnetic sub-storm." Sometimes there is also a precipitation of energized particles into the ionosphere, thus giving rise to "auroras," which are luminous displays in the Earth's upper atmosphere. These are manifested mainly in the higher latitudes of both hemispheres: the aurora borealis in the north; aurora australis in the southern hemisphere.

Within these magnetospheres, there is a region in which the combination of magnetic phenomena and the high conductivity results in the process of ionization. Ions are essentially neutral atoms or molecules that have been converted to electrically-charged atoms or molecules. The magnetosphere appears as an elongated balloon which streams out to form a tail that stretches downstream, away from the Sun. The upper boundary of a magnetosphere is always on the sunward side. Its size is generally determined by the counter force that is generated from the center of a planet, and it is at this point where it first encounters the incoming solar winds. The latter is the boundary point where the pressure of the incoming solar radiation is equal to the pressure of the outgoing magnetic field; it is called the magnetopause and its thickness varies from planet to planet. On the leeward (the side facing away from the Sun) the force of the solar wind is less, thus allowing the geomagnetic field to extend up to as much as several astronomical units outward.

———

When we analyze the magnetic fields that occur throughout the other planets and some moons of the solar system, we find that Mercury has a surprisingly Earth-like magnetic field, especially considering its small size and slow rotation. The current dynamo theories that describe the phenomena usually require a thoroughly-molten core and quite rapid planetary spin, but neither of these occurs on Mercury. At this point in the studies of the solar system, it can be said that a general theory which can describe the mechanics of the planetary magnetic fields is still a work in progress. Consider that even more rapidly spinning bodies, such as the Moon and Mars, lack magnetic fields. Then there is the case of Mars and Venus, which are thought to be completely solid, as evidenced by the seeming lack of an internally-generated magnetic field. Mars is no longer generating a dynamo, but it may still have remnant magnetized material which was part its ancient crust.

Then, it is possible that Venus may lack a magnetic field because, although its core is fluid, it does not circulate; or simply because the core is solid and hence is incapable of supporting a dynamo. So far in the space exploration of Venus, no evidence of a metallic core, such as occurs on Earth, has been found on that planet. Thus, Venus does not appear to have a significant magnetic field that can be attributed to an interior dynamo. It is thought that this condition may be due to slowness of rotation of the planet or that the core is solid and, therefore, is not able to support a dynamo. Because Venus lacks a detectable magnetic field, its bowshock occurs just a few thousand miles above the surface of the planet, held off only by the ionosphere. The latter, however, is not sufficient to prevent gradual loss of water from the planet.

The magnetosphere of Jupiter has been described as a "cavity" that is created as a result of the interaction of the solar wind with the planet's magnetic field. The field is relatively huge, compared to that of the other planets; it extends up to about 5 million miles in the direction of the Sun and outward almost to the orbit of Saturn in the opposite direction. This makes Jupiter's magnetosphere the largest and most powerful in the solar system. In terms of volume, it is the largest known continuous structure in the solar system, after the heliosphere. The existence of this magnetosphere was first inferred from observations of radio emissions in the 1950s; and it was directly observed by the Pioneer spacecraft in 1973. It is now thought that Jupiter's magnetic field is generated by electrical currents which flow in the planet's core, which is made up of metallic hydrogen. This magnetic field also forms a torus (a doughnut-shaped cloud of gas and plasma) around the planet, and it causes the torus to rotate with the same angular velocity and direction as the planet. The torus, in turn, loads the magnetic field with plasma, thus forming a pancake-shaped structure. Interestingly, the magnetosphere of Jupiter is shaped more by Io's plasma, rather than by the solar wind.

The magnetosphere of Saturn was observed by the Pioneer mission in 1979. It too refers to the cavity in the flow of the solar wind which is created by the planet's internally-generated magnetic field. Here again, the shape of the magnetosphere being the result of the behavior of charged particles from the interplanetary magnetic fields. Its inner magnetosphere, like those of the Earth, traps a stable population of highly-charged particles which travel along the lines of the whole magnetic field. This field is filled with plasma originating from both the planet and all its moons, but the main source of this plasma is the small moon, Enceladus, which ejects as much as 1000 kilograms/second of water vapor from geysers on its south pole. Adding to the complexity of Saturn's magnetosphere is the fact that its moons, Titan and Hyperion sometimes cross its magnetosphere, and thus trap more charged particles which also collide with atoms in Titan's atmosphere. This is an example of the behavior of the gas giant planets that shows that they are like a solar sub-system. Although Saturn's magnetic field is not as large as that of Jupiter, it is still 578 times more powerful than that of Earth. The behavior of the electrically-charged particles in the spaces between rings is also influenced more by Saturn's magnetic field than by the solar wind.

———

The largest magnetosphere by far, however, is that of the Sun. It is generated by the convection of heated material from the center of the star, which cools as it rises to the surface, called the corona. Normally the interaction between the energy that is emitted from this magnetosphere is emitted in a somewhat constant fashion and then propagates outward throughout the heliosphere. Occasionally, however, solar flares that contain great amounts of energy explode from smaller portions of the magnetosphere and a powerful solar storm is generated. In both cases, a field of highly energetic particles is emitted, but in the case of the solar flares a much greater magnitude of energy and particles is produced. The movement of the magnetic field through the solar system is known as the solar wind. It is essentially a flux of particles, consisting mostly of protons and electrons that combine with nuclei of heavier elements in smaller numbers. These are accelerated by the high temperatures of the solar corona at velocities great enough to escape the Sun's gravitational attraction. When the solar wind encounters a planet's magnetic field, a shock wave is produced. The portion of the solar wind that does not interact with the planets continues to travel to the edges of the heliosphere, and eventually diffuses into galactic space.

As is the case in the Earth's atmosphere, we can say that there are generally-prevailing solar winds being emitted from the Sun's surface. However, there are occasional storms or eruptions on the surface that produce great gusts of energy and charged particles. These are the solar flares, which are characterized by a sudden and intense brightening of a small portion of the Sun's chromosphere. The flare develops in just a few minutes and can last as long as several hours. Most of the energy consists of ultraviolet radiation, intense X-rays, cosmic rays, and some less energetic particles. Most flares are almost invisible in ordinary light because the energy release takes place in the transparent atmosphere and only the photosphere, which contains relatively little energy, can be seen in visible light. Solar flares vary greatly in size, from giant events that shower the planets with particles, to bright flashes that are barely visible. It is these occasional solar storms that pose the greatest hazard to the artificial systems in space; especially those that are dependent on electrical circuits of one type or another, including computer hardware systems. The sudden gales of charged particles can destroy or damage electronics that control navigation and communications, for example, and they can even cause artificial satellites to be thrown out of their orbit and, potentially, fall down to Earth.

———

One thing that appears to be common among the planetary magnetospheres is that the solar wind interacts with the magnetic fields when it bombards a planet at supersonic speeds, and then forms a bow-shock on the planet's sunward side. This bow shock is made up of a stream of charged particles that are emitted and propagated in all directions from the Sun, and throughout the solar system. These are ejected from the upper atmosphere of the Sun, and eventually form the heliosphere (a vast bubble in the interstellar medium which surrounds the solar system). In one sense, this heliosphere acts very much like the Earth's troposphere. That is, these solar winds are responsible for "weather" events, such as lightning and dust storms, especially on planets and moons that lack magnetospheres.

From the perspective of the geographer, the planetary magnetic fields, the magnetic cores that produce them and the solar winds, can be seen as forming a unique spatial system. When we view the magnetic fields that occur around the Sun, and some of the planets and moons, what emerges is a pattern of local fields that interact with a major flux (the solar wind). Also, the spatial distribution the magnetic fields within the solar system, appears as a region of moving and dynamic "clusters," of electromagnetic energy. Therefore, the locations and patterns of movement of these fields can be analyzed by using many of the existing propagation and distribution models that have been developed in the science of geography. An example of an internal analysis of a field is a NASA program, which is utilizing its Cluster and Mars Express Missions to compare the loss of oxygen from the respective atmospheres of Earth and Mars, when the stream of solar wind arrives at each of the planets. The differences in the loss rate will help to understand the relative efficacy of Earth's magnetic field in deflecting the solar wind and protecting our atmosphere (nasa.gov).

Aside from the magnetic fields, it seems that the most important and pervasive of these systemic environments consists of electromagnetic radiation, in all its wavelengths and frequency profiles, as it occurs spatially throughout the solar system. Its effect on bodies in the solar system, including spacecraft, is ubiquitous in the sense that they are all exposed to it. However, the type, amount, and intensity of the radiation can be highly variable in many instances. Thus, the amount of radiation that reaches the surface of a body is primarily dependent on the distance of that body from the source of the radiation, the Sun. Another variable derives from the combination of radiation types that is emitted from the Sun and the stars. Then there is the situation in military space operations that is likely to become more common in the 21st century; which is the radiation that is emitted by the electronics of the spacecraft systems and, possibly, from nuclear explosions.

———

The Use of the Electromagnetic Spectrum in Military Operations

As far back as World War I, we see the beginnings of the utilization of the electromagnetic spectrum as a military tool or weapon by the air forces of the time. It began with the radio portion of the spectrum, as the various wavelengths in this phase of the electromagnetic spectrum were used in navigation and communication. By the start of World War II, the utilization of the EM spectrum remained primarily within the radio portion. However, the radio signals were now also being utilized in the form of "radar." Then, during the Vietnam War era, the modern amplification and stimulation of emitted radiation began; it involved the use of electrical power to amplify and shape the emissions from antennas. And today, the infrared and microwave portions of the EM spectrum also are being used as a military tool; which can be used to detect, track, and guide weapons onto targets.

Now, since the turn of the 21st century, military aerospace operations have ascended beyond the altitude of the protection of the geomagnetic sphere. As a result, the hazards of the electromagnetic spectrum have become more significant to military space operations. Another factor that tends to increase the threat of radiation in military space operations has to do with the duration and culmination of the radiation on a particular object. In the context of human space operations, injury from radiation is caused by time of exposure to ionizing radiation, as well as the power of the radiation. Within the overall threat profile, the radiation at any given place and time occurs in various forms, and also differs in the degree of damage that occurs. One such threat is derived from gamma rays, because of the theoretical degradation effect of such rays on nuclear weapons and other space systems. The response to this perceived threat has been to "shield" the components of such systems with lead and other matter.

There are other variables involved. For one thing, the damage is dependent on the type of ionizing radiation that is involved, as well as the portion of the body that is exposed to the radiation. Other factors include the duration and the cumulative dose that is administered. As far as humans are concerned, the tissues and organs that are most affected include the lining of the gastrointestinal tract, the bone marrow, and the skin itself. The resulting trauma involves the destruction of cells that are so vital to the maintenance of tissue structure and function. The symptoms that result from the intensive radiation of the gastrointestinal tract or the bone marrow constitute a condition that is called radiation sickness. The symptoms of radiation sickness vary in their severity. But, as a practical matter, they very likely may cause a mission emergency. Even moderate doses of radiation can severely depress the immune system, thus leaving the body open to bacterial infections. In lower doses, an irradiated person is generally able to survive, but the nausea, vomiting, and malaise may still occur. These symptoms might be of lower intensity, so that the exposed person is still able to continue to carry out his or her duties in spite of measurable depression of the bone marrow. In any case, this is a situation that must be considered in mission design and planning.

Even on Earth, under the protective layer of the geomagnetic field, and since the time of World War II, there has been an increasing hazard to humans from the increasing use of x-rays and other parts of the electromagnetic spectrum in many occupations and other mundane settings. In the medical setting, for example, there are many new devices and treatments that increasingly expose humans to a variety of electromagnetic radiation, from the X-ray through the infrared, and into the electro-optical domain of the laser beam. At the same time, plutonium and uranium are being used in increasing quantities and ways to treat all sorts of cancers and other human diseases. Then there is the nuclear waste that derives from nuclear reactors, such as those that provide the heat to produce the water steam that turns the electric turbines. Ultimately, there is the fact that nuclear uranium waste retains about one-half of its radioactive strength, even after more than 20,000 years. So, whether military operations are carried out within the atmosphere of the Earth, or beyond it, the immediate and the total load of radioactive and electromagnetic radiation on the human body and on man's machines (especially the components that depend on electricity), face an increasingly significant environment of radiation that will affect the success of military operations.

The machines that are deployed in outer space are especially vulnerable to the effects of the various forms of radiation. Solar cells, integrated circuits and sensors can all be damaged by radiation. Even within the Earth's "safe zone," radiation storms can damage electronic components of space-related systems. The microminiaturization and digitization of logic circuits has only heightened the vulnerability of satellites to damage by radiation. Finally, it can be said that the threat environment created by radiation is much more complex than that which is posed by the "constant" fluxes of radiation in the solar system. For one thing, there are special regions of radiation throughout the solar system that will pose their own profile of radiation threat to both manned and unmanned spacecraft and systems there.

———

Consider that NASA's Curiosity Radiation Assessment Detector (RAD) collected radiation data for about 3.5 days, during August of 2012. Its mission was to quantify and measure the effects of the various wavelengths of electromagnetic radiation on man and machine on Mars, especially from inside Mars's Gale Crater. One reason for this effort was to determine what the effects of radiation would be on future manned missions to the Red Planet. This planet was chosen for such a study because it has no global magnetic field like the one found around Earth. Also, the tenuous atmosphere on Mars permits significant amounts of ionizing radiation to reach the surface of the planet. Previous measurements taken by another spacecraft, the Mars Radiation Environment Experiment, were done to specifically measure the dangers to future human astronauts and colonizers on Mars. This earlier orbiting spacecraft found that radiation levels in orbit above Mars were 2.5 times higher than they are at the International Space Station. NASA has calculated that a three-year exposure to such levels as are found on Mars would be close to the safety limits that are currently established. Local levels on the planet's surface would probably be somewhat lower and they also might vary as to specific location, depending on altitude and the presence of local magnetic fields. Consequently, even before reaching a planet such as Mars, astronauts would be subject to radiation during the nine-month journey.

———

Even before reaching the transit region in outer space, manned and unmanned spacecraft will encounter another radiation event that could negatively affect the humans and their machines. This is the so-called Van Allen radiation belt that is comprised of two torus-shaped layers of energetic particles that surround the planet Earth; it is held in place by the geomagnetic field and extends from 600 to 40,000 miles above the surface. Within this geographic field, two distinct belts of energetic electrons form an outer belt; whereas a combination of protons and electrons make up the inner belt. The latter is distinguished by the fact that it contains lesser amounts of alpha particles. Both belts pose a hazard to satellites, which must protect their sensitive components with adequate shielding if their orbit spends significant time in the radiation belts.

Another radiation field that poses a threat-environment is the artificial radiation that has resulted from high-altitude nuclear explosion testing that was done during the 1960s, by the United States and the Soviet Union. One in particular, the Starfish Prime test, produced a radiation belt that has the greatest intensity and duration of any of the other similar artificial radiation belts. Starfish Prime was a high-altitude nuclear test that was conducted by the U.S. in 1962. The explosion occurred at 250 miles above a point that was located 19 miles southwest of Johnston Island in the Pacific Ocean. It was only one of five such tests that were conducted by the United States in outer space during that era; an exoatmospheric test, as it was called. The subsequent electromagnetic pulse (EMP) which occurred was larger than had been expected, and it caused electrical damage in Hawaii, some 900 miles away. The Starfish Prime belt has already been known to cause damage to artificial satellites orbiting the Earth, such as the United Kingdom Ariel 1 and Telstar 1. It also has caused damage to the Soviet satellite Cosmos V. Ultimately, all of these satellites were completely unable to continue to operate, all failing within a few months of the Starfish detonations.

———

The Chemical Environment

Just as the electromagnetic spectrum of radiation can be seen as a virtual region, the Table of Elements can be treated as a virtual region of the basic chemical building blocks of the Universe. In military space operations, the dictums that "knowledge is power" and "forewarned is forearmed" are especially relevant with respect to the relevant chemical environment. Even before humans decided to leave the familiar chemical environment of our own planet and to engage in military operations in the solar system, we had realized that the atmosphere is not simply an inert body of air. Rather it is more like a container of atoms of elements that are constantly forming and reforming molecules of chemicals. If we consider the solar system as a region of elements, we can apply the concepts and methodologies of the geographic region to gain spatial insights into the matter. Therefore, we can say that the solar system is like an ocean in terms of the spatial distribution of the chemical elements. That is, it is a spatially variated distribution of the elements, which can be mapped in order to discern patterns that may be relevant to human activities in space.

This kind of geographic analysis has been done with respect to the Earth's atmosphere for at least three centuries, and especially since the beginning of the Space Age. The imperative for this advance in our knowledge of our atmosphere became even more relevant during the 1970s, when the recognition of the hazards of "smog" increased. This contamination of the chemical environment of our atmosphere had grown with the advent of the Industrial Revolution, the internal combustion automobile, and by the development of the megalopolis in North America and Europe, primarily, but in other local areas in the developing world as well. Eventually, it became obvious that when Sunshine irradiated the exhaust from all combustion engines, it resulted in the creation of a concoction of interacting compounds, beginning with the highly reactive ozone. During this same time period, scientists began to also draw attention to nitrates that jet-propelled aircraft emit in the stratosphere, along with sulfates and water vapor. It also was suspected that the chemical aerosols could stimulate the formation of water droplets, thus altering the cloud cover pattern, perhaps resulting in little understood effects on climate. Then there was the contention by some scientists that a single nitrate molecule, through constant cycles of reaction, could destroy many molecules of ozone. This has taken on greater significance, since it is now well-accepted that the tenuous layer of stratospheric ozone is all that obstructs harmful ultraviolet rays from reaching the surface of the Earth.

Moreover, now that space operations are occurring with increasing frequency and intensity outside of our relatively well-understood atmospheric environment on Earth, it has become imperative that we develop methods and technologies to constantly gauge the chemical environment within any operating environment in the solar system, including the interplanetary medium and the atmosphere of other planets, as well as their surfaces. Even the operations of the NASA space shuttles already have been found to have an effect on the chemically fragile portion of the atmosphere through which they pass, as a result of the emission of chlorine from the spacecraft. The chemical composition of the plasmas and zones of particles that occur throughout the solar system also can have an effect on military space operations.

Throughout the developing military geospace domain there is the underlying reality of an operating environment in which there are chemical elements. Therefore, wherever military air or space operations happen to take place in the solar system, they still will confront the same reality of atoms having atomic numbers and chemical attributes, each corresponding to one of the chemical elements. Most elements exist as isotopes, which have differing numbers of neutrons. All isotopes of an element exhibit the same chemical behavior, although isotopes can be separated on the basis of differences in atomic mass (i.e., nearly the sum total of the quantity of matter contained in the masses of the protons, neutrons, and electrons that make up the atom). It is this atomic mass that is unique to each element and, therefore, can be utilized as a kind of "signature" in the conduct of space reconnaissance or in the development of precision ordnance in space.

It also is the case that communication of information occurs as the result of chemical analyses that are done of both organic and inorganic matter anywhere in the solar system. That is, the scientist elucidates the various properties and attributes of an element or compound, and then is able to catalog them for further processing and analysis. In fact, whether the analysis is being done remotely or in situ, what is happening is that the chemical element is transmitting information to a receiver. With this in mind, consider the Handbook of Chemistry and Physics (W.M. Haynes, Editor-in-Chief, 91st edition, 2010-2011), in which a summary of properties of physical and polymers is given. I suggest here that these properties can also be utilized to detect, track, and attack targets throughout the solar system. In all these cases, the methods of analysis used in chemical environments on Earth continue to be effective throughout the solar region. These involve the measurement of some physical property of the target, such as density, refractive index, absorption or polarization of light, electromotive force, and magnetic susceptibility. All these properties can be detected and measured with the appropriate technological devices. With the continuing advances in the technology of space exploration, we now have the capability for doing these types of chemical analysis remotely, which will be the more likely scenario in military space operations. Thus, with the utilization of the remote sensing systems that have been developed for scientific studies in space, we can maintain constant awareness of the relevant chemical environment throughout the course of the sortie. The same technologies also should prove useful for detecting and attacking targets.

However, in order gain specific, relevant information about a given chemical environment on the surface of a celestial body, one must also be able to do in situ analysis of a sample of the rock and soil. One of the most efficacious strategies that have been developed by NASA involves around an orbiting science laboratory which deploys a "robot geologist" (a rover) to do the actual in situ soil and rock analyses. The key to this strategy is a telemetry subsystem, which allows for the nearly real-time relaying of the data that is gathered by the rover. The data can be relayed to laboratories that are located in space or on Earth for analysis and scientific application studies. Thus, the newest of these rovers – Curiosity – will be equipped with the newest generation of onboard "chemical element reader" to measure the chemical ingredients in Martian rocks and soil. It also will have onboard the so-called "Alpha Particle X-Ray Spectrometer (APXS) instrument, which utilizes the power of the alpha particles and X-Rays to bombard a target. Such a procedure then causes the target to give off its own characteristic alpha particles and X-Ray radiation. These emissions are taken-in and processed by the X-Ray detector, which reveals which elements and what quantities of each are contained in the samples of rock and soil. Meanwhile, the rovers that are being sent to Mars are being equipped with an onboard mass spectrometer; it will be able to separate elements and compounds by mass for identification and measurement. The gas chromatograph deals with soil and rock samples that are heated until they vaporize, and then separates them into the resulting gases for analysis. The laser spectrometer will measure the abundance of various isotopes of carbon, hydrogen, and oxygen in atmospheric gases such as methane, water vapor, and carbon dioxide. In terms of military space operations, the fact that wavelength of an element can be detected and measured by the spectrograph means that it can be used to guide and control the detonators of ordnance.

Another aspect of chemistry in military space operations is in the area of materials science. This is defined by dramatic improvements in the way that we manufacture and manipulate the various objects of space technology, including the so-called nanotechnology. Now, in the current "silicon and composites age," materials matter more than ever. Therefore, in the laboratories of NASA and of their partners in the private sector and academia, scientists are working hard to create foundations for tomorrow's products of the space age. These include ultra-smooth coatings that repel everything from spacecraft surfaces to self-regulating materials; that alter their properties in order to create custom materials that can be utilized in space. Then there are the so-called "smart" materials that can be programmed to reflect electromagnetic energy in a "camouflage" form to outwit enemy beams; they can be used to improve everything in space operations, including the development of space suits that are custom-fitted for all sorts of specific missions. Finally, there is the so-called Moore's Law, the central tenet of the Silicon Age, which describes a principle of materials science that says: every 18 months, we are going to find a way to cram twice as many components onto a finite chip.

———

The Gravitational Environment

Gravity itself is one of the four major forces of nature; the least powerful, but the most ubiquitous and relentless these phenomena of physics. It plays no part in determining the internal properties of ordinary matter (which is left to the so-called strong and weak forces), but because of its universal application, it controls the movements of the bodies in the solar system and the rest of the Universe. Since the beginning of the Space Age, and with the sending of spacecraft into the solar system, a great deal of empirical data has been accumulated. As a result, we now know a great deal more about the gravity of the celestial bodies and about the nature of gravitation itself. For one thing, we have learned that gravitational fields have many common internal characteristics, regardless of their location.

So, based on these empirical observations and several centuries of inferential logic, a considerable body of knowledge and understanding of gravitation has been accumulated. The earliest theories about gravity were derived from attempts to explain the relative locations and movements of the planets in the sky. This is a scientific quest that any geographer today would immediately recognize. Thus, just as a geographer would view a map of the distribution of cities, towns, and villages across a two-dimensional plane on the surface of the Earth, the astronomers, mathematicians, and physicists have been making observations of the bodies that are distributed throughout the "celestial sphere" that the ancient Chinese astronomers saw in the skies. These observations and the subsequent attempts to explain them have led to the development of many models and theories, from which some "laws" have been constructed. These laws can also be understood to be hypotheses to be tested against reality in the continued refinement of our understanding of why the spatial arrangement of the Universe is the way it appears to be.

The first iterations of models were based on the assumption that the spatial arrangement of the bodies was done by a supernatural force of some kind or other. In the tradition of western culture, this force was a deity who acted as an architect of the elegant structure that was seen in the skies. This geocentric model was replaced several centuries later, and with considerable consternation and struggle, by a heliocentric model. The latter model was based on a combination telescopic observations and logical inferences, and these scientific efforts resulted in the development of spatial models of the solar system. These applications of the scientific method also have produced theories and derivative hypotheses that could be tested by further empirical observations. Parallel to the work that was being done to rigorously describe what was being observed, other scientists and thinkers were looking for that "holy grail" of geographic science: the causative variable or set of variables that determined the spatial behavior of the celestial bodies. One answer came in the form of Sir Isaac Newton's laws of motion and his law of universal gravitation. The latter is especially relevant because it explains the relative location and movement of not just the planets, but those of the comets and even the smallest particle in the Universe as well. Albert Einstein thought that Newton's gravitational force was really a by-product of what he saw as the essence of gravity; that is, as a fabric of space-time. In other words, according to this model, the gravitational field was seen as being like a trampoline in which bodies made depressions in the fabric according to their mass. So, in this view of gravity, the attraction between two bodies was still a function of their combined mass and their distance from each other, at any point in time and space. There was an additional variable that is called the centrifugal force, which is a fictitious force that applies only to the movement of a body on a circular path; the body is kept in a particular "orbit" according to the relative tension between the centrifugal and centripetal forces on the orbiting object.

Since the beginning of the space age, we have been able to make more precise empirical studies of the gravitational fields, including those emanating from the various planets and bodies of the solar system. The upshot of this is that the quantitative facets of this geographic phenomenon have been tested and confirmed during the course of the Space Age. For example, the Apollo astronauts used a gravimeter and an accelerometer during the missions to the Moon; these are two of the main instruments that are used for measuring variations in the gravitational field of celestial objects. Later space missions would measure the gravitational attraction fields of the other planets as well. The more recent space missions have had the benefit of gravity maps that have been constructed by specialized missions such as the European Space Agency's efforts to create a highly detailed map of the Earth's geode, a global model that "illustrates the subtle variations in the gravity field" of the oceans (spaceflightnow.com 3/17/2009). The methodology is to deploy a triad of accelerometers on a spacecraft which will compute the difference in the pull of gravity throughout the mission, "every second for about 20 months." The output of such gravity mapping missions is a complete map of a planet's gravity field.

As a result of all this study and rigorous thinking, it is commonly accepted that the value of gravity of the planet's surface is a function of a combination of factors. These include: the gravitational attraction of the planet; the centrifugal force caused by the planet's rotation; and tidal vibrations that are caused by nearby bodies. The practical gravitational environment, as far as military space operations are concerned, resembles the trampoline that was described earlier. It is also accepted that this "trampoline," in its natural state, is populated by bodies with differing masses, each causing "depressions" in the trampoline according to their absolute and relative mass value. Therefore, as a practical matter, this means that the conduct of military space operations will need to be cognizant of the gravity "topography" within the specific area of operations.

With respect to the distribution of gravitational influence among the bodies of the solar system, these exercises in gravity mapping have provided empirical support for the hypothesis that there is some relationship between the size/mass of a body and its planetary gravity. They also have enabled scientists to develop a map of planetary gravitational values. Thus, using the planet Earth as the reference point, the larger values can be assigned to Venus and Mars, as well as the Moon. In the case of these close neighbors of our planet, the more dominant factor seems to be the distance of the body from Earth. On the other hand, the values that are assigned to Jupiter and Saturn seem to show how size/mass can trump relative distance. Thus, Jupiter and Saturn seem to exert a greater gravitational pull upon Earth than any of the other planets. They also cause tremendous effects on smaller bodies which orbit within their planetary domains. These include their tributary moons, as well as any comets and asteroids that might venture into their neighborhood.

In orbital mechanics and aerospace engineering, a gravitational "slingshot," or gravity-assist maneuver refers to the use of the relative movement and gravity of a planet or other celestial body to alter the path and speed of one's spacecraft. Gravity assist can be used to accelerate (both positively and negatively) and thereby redirect the path of the spacecraft. This so-called "assist" is provided by the motion of the gravitating body as it pulls the spacecraft. The technique was first proposed as a mid-course maneuver in 1961, and has been used by interplanetary probes such as the Mariner 10 and the Voyager probes. Due to the reversibility of orbits, gravitational "slingshots" can also be used to decelerate a spacecraft. In terms of military space operations, these maneuvers are used to gain a better angular position in relation to the opponent. In a contest of maneuvers, as in a "dog fight" situation, these maneuvers can be used by both adversaries gain an offensive advantage, or to evade an attacker's gambit. They can be used by both of the opponents to strive for an offensive position, or in the case of disengagement maneuvers, to help facilitate an escape.

It may be reasonable to say that the total effect of gravity on a spacecraft during a sortie or mission is such that space operations might be called "gravitational" flight. Newton assumed the existence of such a ubiquitous force occurring between all massive bodies, one that does not require bodily contact, and therefore a force that acts at a distance. The net effect of gravitational forces is ever-present, occurring even in the relatively "low" gravity domain of the interplanetary medium, and increasing in intensity and complexity within the gravitational fields of planets, moons, and some larger non-planetary objects, such as the more massive asteroids. The actual effect on a spacecraft also varies as to the phase of the trajectory or orbit in which it is operating at any given space-time event.

Current scientific thought tells us that the gravitational force attraction on an object is caused by the presence of an adjacent second object. So, as soon as our spacecraft escapes the dominant attraction of the Earth, the gravitational formula changes upon entering the micro-gravity environment of the interplanetary medium, and again as we approach the gravity domain of another body. Thus, for example, weight is a consequence of the universal law of gravitation, that is, any two objects (spacecraft and earth, for example), because of their respective masses, attract each other with a force that is directly proportional to the product of their masses, and is inversely proportional to the square of the distance between them. In practical terms, that means that more massive objects weigh more in the same location; but the farther an object is from the earth, the less its weight. This is the essence of the initial calculus with respect to operations at the launch point, and in space itself. And, once in space, the smaller mass and radius of the Moon, compared with those of the Earth, combine to make a given object on the moon's surface weigh one-sixth of what it would weigh back on Earth. Furthermore, because of all the mass that is present in the Universe, each point in space has a property which is called its gravitational field, or the acceleration of gravity, at the point where the object is located.

The Earth's gravitational attraction was one of the major impediments to spaceflight, until the middle of the 20th century, when sufficient rocket-power was developed to enable the spacecraft (or its mass) to achieve initial liftoff, and to generate enough velocity to escape the bounds of Earth's gravitational field (at a height of about 60 miles) and, finally, to achieve adequate velocity to enter into the desired orbit for the mission. However, the earth's gravitational pull on the rising spacecraft subsides only gradually. At an altitude of 100 miles, it is still 95% that of gravity's effect at the surface of the earth. Then, at approximately 1,680 miles above the earth, its effect is reduced to about 50% of the original effect. Finally, the gravitational pull of the earth becomes negligible at distances beyond several million miles. The spacecraft is now in the microgravity environment of the interplanetary medium.

Then, the gravitational influence of the Moon gradually becomes predominant with respect to the space vehicle. At this point, the pull of gravity is only 1/6th of the Earth's gravitational pull. As far as space operations are concerned, the gravitational pull of the Earth becomes negligible only at distances of several million miles, except when the Moon and lunar gravity (one-sixth of the Earth) becomes predominant. So now, a much different set of variables come into play. The calculus now comes down to the total lift (propulsion) force of the rocket engines being able to exceed a force twice the weight of the spacecraft hardware at takeoff. Only then is the space vehicle able to rise at an initial acceleration equal to the standard gravitational (g force) force of 37 feet per second per second. Then, as the propellant mass of the rocket is consumed, the vehicle continues to lighten and the rate of acceleration also increases. However, as the spacecraft encounters other massive bodies (natural and cultural) the influence of gravity varies as to the distance and the mass of the other bodies. Thus, when spacecraft approach within the gravitational influence of Mars, or the bodies of the Asteroid Belt, or especially the Jovarian bodies, the gravitational equation will be different.

Flight into Earth orbit is achieved by launching a rocket vertically from Earth's surface and then tilting its trajectory so that flight is parallel to the surface, at the same time that the spacefaring portion of the vehicle reaches orbital velocity at the desired altitude. Once in space, the effects of gravity on aerospace operations change almost completely. For one thing, the techniques for dealing with the usual "friction" now had to be relearned in space; all of the lessons that our species had learned about motion and "weight" had to be relearned in the new gravitational profile that was presented differently during the various phases of the sortie. The powered ascent phase itself was divided into two operational events: a vertical-rise segment for surface clearance and the orbital insertion segment. Each of these had their own gravity profile to deal with. Once in space, the gravity profile is characterized by the relative absence of gravity within the spacecraft. An illustration of practical effects of this can be seen in the reports by astronauts that even a simple and "mundane" task, such as the squeezing of toothpaste from its container, requires a new set of muscle movements and hand-eye coordination than is required on the surface of the Earth.

What all this means to military aerospace operations is that the design of the aerospace machines, whether within the atmosphere or outside it, must take into consideration the changing gravitational environment during the entire course of the sortie. As a spacecraft, such as an artificial satellite moves through a planet's gravitational field, it experiences forces because of irregularities in that field, resulting perturbations on the orbit of spacecraft. Therefore, the military planner now has to come to terms with the different effects of gravity in space when proposing to conduct military aerospace operations in that milieu. The navigator now has to deal with a gravity profile that is qualitatively different than on Earth. And, the weapons officer in space, too, must consider a dynamic gravity profile in space and during planetary operations in solving the weapons delivery problem. Even more critical is the effect that prolonged exposure to zero gravity has on the human physiology. Astronauts who have had to spend just a few weeks in such an environment have reported the significant loss of muscle mass and strength. There have also been reports of lessening of bone density equivalent to that of the human aging process on Earth. These effects on human physiology can increase the possibility of bone breaks and the other such problems that are associated with brittle bones in aged humans. These could prove disastrous to a human crew in a prolonged space mission.

A special effect of gravity on space operations has to do with an "object" within the spacecraft; more particularly, a human body. So, once a spacecraft escapes the gravitational pull of the Earth, the gravity environment changes. The lack of gravity causes a paradigm-shift in the effects of gravity on the human astronaut. Within the Earth's atmosphere, gravity is manifested in inward "pushing" against the human organism, which affects all parts of the anatomy and physiology, including the skin and circulatory system, and most significantly, the blood supply to the brain. In space, the effect of gravity on the human body is reversed: now the low-gravity condition causes and outward "pull" on the organs of the body, tending to a complete destruction of its "integrity." To counter these effects a special space suit is needed to hold together the physical integrity of the systems of the human body.

Another gravity-related effect on military operations is called perturbation. This refers to a deviation in the motion of a celestial object that is caused either by the gravitational force of a passing object or by a collision with it. With respect to the artificial spacecraft bodies, the effects of perturbation may be likened to the effects of air "turbulence" in intra-atmospheric flight. One situation, in particular, where these effects are likely to significantly affect military space operations would be within the Jovian domain. Here, the effects of perturbation on smaller bodies is heightened and more complex because of the effect of Jupiter's on the nearby planets and on the asteroids. In general, throughout the interplanetary medium, the subtle effects of the gravitational forces of the planets can cause unwanted disturbances to the trajectory of a spacecraft.

The nature of gravitation has also proven useful in space operations. For example, the "gravity-assist" maneuver is being successfully utilized to provide a "boost" to the velocities of spacecraft in a variety of situations. On one occasion, this technique has been employed by NASA at least once to rescue an Earth-orbiting communications satellite when the launch vehicle failed to achieve its intended mission orbit. Other robotic spacecraft have used this technique to achieve their targeted orbits, higher up in the Sun's gravity well. Voyager 2, for example, was launched in August of 1977 and did a flyby of Jupiter with a "gravity-boost" by Jupiter on its way to the domain of the Jovarian planets. Its twin, Voyager 1 launched the following month and did the same thing, and actually reaching Jupiter before Voyager 2 did. Voyager 2 then obtained a gravity assist from Saturn, another from Uranus, and finally climbing all the way to Neptune and beyond.

One interesting gravity-related feature that is found in space is the Lagrange point. These are geographic objects in space that are utilized in a variety of ways in space operations. These are located in unique points in space, and they are being used by spacecraft that are sent there maintain a stay in an absolute place, or hover, if you will. They can do so because of the dynamics of competing gravitational attraction and the centripetal forces. The existence of these spatial points in space was postulated by a French-Italian mathematician named Joseph-Louis Lagrange in the 18th century. He believed that in a two-body system, such as the Earth and the Sun, there would be certain nearby points where a third object could be positioned, and which would remain in place relative to the other two objects. In the case where the Earth and the Sun are the first two bodies, there is a Lagrange point 1 (L1) which is located about 4 times farther from Earth than the Moon. All together, there are five Earth-Sun Lagrange points.

NASA has used the Lagrange points in many of its space missions. Typically, a spacecraft is sent from Earth into space, where it will continue moving along the pre-planned trajectory, unless it is affected by the perturbation caused by the gravity of another celestial object, such as one of the planets or moons. Even if the spacecraft were to attempt to stop in the middle of space, the gravity of the objects around it would eventually pull it in some direction. Another change in the vector of the spacecraft would also occur at certain points of the two bodies, combined with the third body's (our spacecraft) centripetal force, which would keep the third body in a constant location relative to the other bodies. The theory of the Lagrange points has been proven correct by repeated space missions, and these points of dynamic tension are now being used by NASA as virtual "parking lots" for spacecraft in the solar system.

At one Lagrange point in particular, known as the L1 point, and which is located about 900,000 miles from Earth and toward the Sun, one finds the space telescope known as SOHO (Solar and Heliospheric Observatory). Its mission basically is to study the effects of the Sun on the Earth. Because this spacecraft is not in Earth-orbit, it commands an uninterrupted view of the Sun at all times, and is able to collect information about the solar wind an hour before it reaches the Earth. L2 is located at the same distance from Earth as is SOHO, but in an opposite location. Also found here is the WMAP (Wilkinson Microwave-Anisotropy Probe), which was launched in 2010. It has been able to take the most detailed pictures of our Universe to date. By collecting microwave radiation from over 13 billion years away, the age of the Universe and its components has been determined more precisely than ever before. It also has provided substantial evidence for the validity of the Big Bang theory. L3 is located directly behind the Sun. No missions have been planned for that point because of the difficulties involved in communications with a spacecraft there. L4 and L5 are positioned on the Earth's orbit, about 92 million miles in front of and behind our planet as it travels around the Sun (nasa.gov).

In terms of military space operations, the Lagrange points in the solar system might be seen as potential intermediate space bases, which can be utilized in the same way that islands in the Pacific are used even today. Historically, this air-power strategy was employed with the American forces in WWII. They utilized the islands of the Pacific Ocean, as well as their aircraft carriers, as "stepping stone" air bases against the Japanese home islands, which were situated beyond the range of any continent-based aircraft, except for some that were located on mainland China. In space, we face the same situation with respect to the regions of deep-space, which are situated beyond the "feasible" reach of Earth-based spacecraft. However, through the use of the Lagrange points as "stepping stone" bases, the effective range of military space power can be extended outward. Another technique which might be used in conjunction with the Lagrange points is the "gravity assist" technique, which utilizes the gravitational "drag" from another body to "slingshot" the spacecraft farther along its trajectory, and thus extending its range in a cost-effective manner.

THE SPHERICAL LAYERS OF THE SOLAR SYSTEM

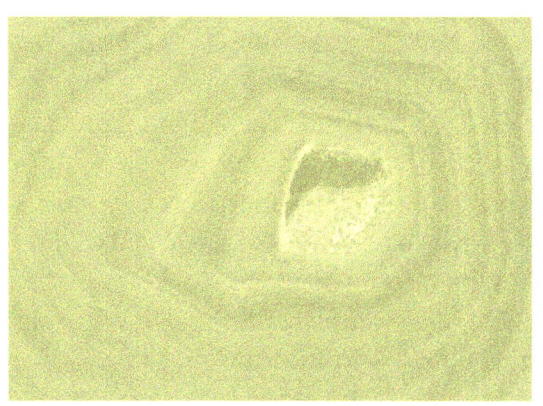

The general term "atmosphere" refers to gaseous envelopes that surround many of the planets (including Earth), and some of the larger satellites of the outer planets, of the solar system. One reason for studying the atmospheres is that they contain a great deal of information about the celestial body's origin and historical development. In terms of planetary exploratory operations and possible future military space operations, the various planetary atmospheres present environmental challenges and opportunities, just as the Earth's atmosphere does at the present time. These vary in terms of chemical composition and structure, but there is a general commonality that enables one to view them as discrete instances of an overall atmosphere of the solar system. The following is a comparative analysis of the known atmospheric regions of the solar system.

The original source of the gases that occur in the atmospheres is thought to have been a nebula that was produced by the Big Bang event; the components of which were subsequently propagated throughout the heliosphere. This was the prime cloud of material and energy from which the solar system formed. The cloud was rich in volatiles and particles, and must have been the ultimate source of the atoms in the present planetary atmospheres. At greater distances from the central point of the nebula, the constituent material in the gas cloud tended to settle along an extensive plane around the Sun. Then, as the material cooled, great clumps of rock and other materials grew and accreted to form many bodies of various sizes throughout the solar system. Ultimately, through the force of gravitational attraction, some of these original bodies grew large enough so that, if the gases around them were cool enough, they could accumulate an atmosphere from the volatiles of the gas cloud. The atmospheres were ultimately derived from the lighter elements which remained in gaseous form, while the heavier elements settled down toward the core.

Another source of materials that makeup a planetary atmosphere related to the process of differentiation within the planet, as a result of which volatile materials were expelled from the planet's interior by volcanic activity and via other thermal escape valves. The lightest gases, such as hydrogen and helium, would be quickly lost in space, but those that were held by the gravitational attraction of the planet would leave a mix of elements and particles that are peculiar to the particular nature of these processes. On Earth, some water vapor would have been condensed to form liquid water, and it seems to have contributed to the ocean which was created approximately 4 billion years ago. Additional water has been contributed to the total inventory of water on Earth by impacting comets, which have been found to be composed of about 20 percent water. It was in the oceans that free oxygen first appeared, with the arrival of photo-synthesizing life around 3.5 billion years ago.

So, why do some of the planets and other larger bodies have significant atmospheres, while others do not? Even though the chemical composition of the original atmospheres has changed significantly during the billions of years since its origin, the total inventory of elements on which it is based has not. However, there also are processes that continue to contribute to the particular "abundance" of each planetary atmosphere. One of these is represented by the continuous impacts on the planets by asteroids, comets, and streams of micrometeoroids. In general, it appears that the planetary atmospheres will continue to be increasingly important to the conduct of military space operations in the solar system, especially as the tempo of launches and landings continues to rise.

It is the so-called "secondary atmosphere" that is presented to military space operations today; it is the current upshot of all the historical processes that have created it. Thus, on each planet and some moons there will likely be a more or less unique set of atmospheric pressures, temperatures, and chemical compositions. These must be evaluated in terms of their particular effect on the material makeup of the spacecraft, including erosion caused by heat or cold degradation. Also, the effect of the various chemical molecules on fittings, linkages, and other joints will have to be considered in the mission planning. By the same token, the "geography" of atmosphere on the various planets and moons must also be factored in when planning for ordnance fusing systems that rely on changes in atmospheric pressure. Within Earth's atmosphere, lightning strikes have proven to be a major hazard during launching operations. However, a more acute atmospheric hazard to space operations occurs during the re-entry phase of a mission: friction and the resulting heat that is generated by the kinetic energy of the re-entry vehicle. It will be interesting to see how the atmospheric conditions on other planets will affect entry and launch operations there.

Light's effect on the atmosphere is itself varied: It can cause ionization (the removal of electrons from an atom); dissociation (which involves the destruction of a molecule); scattering (which causes changes in a photon's direction); and absorption (which occurs when a photon's energy is absorbed). All of these effects are significant to future military space operations. Thus, for example, ordnance that involves explosives often includes a fusing system that is based on perceived electromagnetic energy radiation. Therefore, depending on the particular dynamic that is occurring in a given planetary atmosphere, the fusing mechanism will likely need to be "custom" calibrated to operate in a particular radiation environment. In general terms, it appears reasonable to assume that weapon systems that work well in the Earth's atmosphere or the interplanetary medium may not perform in the same manner on other planetary atmospheres.

Atmospheric pressures and densities are another set of geographic phenomena which military space operations must confront. A useful base-line template for the analysis of differences among the planetary atmospheres is the so-called standard atmospheric model. It postulates a given vertical distribution of temperature, pressure, and humidity, which is considered to be a global average of these parameters. Like most models that are formulated to describe Earth phenomena, this one makes a set of assumptions that may or may not be realistic on the other bodies of our planetary system. One of these assumptions has to do with the lower gravity environment on some of the other bodies of the solar system. Thus, when applying the template to other bodies in the solar system, it is important to factor in the following differences: (1) there is no acceleration of air in a vertical direction on some of other bodies; and (2) the assumptions that air is dry and that acceleration of gravity does not change with height may not be true on some of the other bodies in our planetary system. On the other hand, even though the real atmosphere may vary considerably from the assumed standard atmosphere, the latter is still useful for such purposes as pressure sensor calibrations and spacecraft designs, among others.

Here is what we already know about the Earth's atmosphere, which commonly refers to the lowest layers of air that encircle the planet in outward concentric envelopes; like the skins of an onion. For practical purposes, the atmosphere is estimated to be about 450 miles deep (from the surface of the Earth to the reaches of space), but there is no real defined boundary between the atmosphere and space; the domain of the atmosphere simply gradually fades into space. This so-called "ocean of air" is characterized by several layers that are defined by their relative distance from the surface of the Earth. In the language of the geographer, within this overall "region of air" there are several subregions of air that occur at various heights above the Earth's surface. (These layered regions can be described by the geographic concept of vertical zonation). They range from the turbulent troposphere which lies just above the surface of the Earth, to the rarified exosphere region where the atmosphere gradually merges into what historically was once called the black nothingness of space. Of course, as a result of significant space exploration which began in the latter half of the 20th century, we now see "space" from a more enlightened perspective. Nevertheless, the atmosphere of Earth is the planetary atmosphere that we humans know best. We have learned about its structure and its chemical composition as a result of several thousand years of in situ, and more lately, remote analysis. So it is natural that we have abstracted every facet of this geographic feature, down to a set of scientific notations and formulas, to construct an overall model. The essential variables in such a model include temperature, moisture content, atmospheric pressure, and chemical composition. Now this model is being used as a template for developing a more general theory of planetary atmospheres within the solar system. Hypotheses drawn from this conceptual model are therefore being tested in order to develop the knowledge of the various atmospheric regimes of the solar system.

An external analysis of the Earth's atmosphere shows that it can be described as a container of gases, at least at one level of abstraction. Within this gaseous envelope, however, there also are molecules of water in gaseous, liquid and solid form; molecules of chemical elements; dust and other particles; and even organic molecules are added to the mix. There is a veritable sub-table of elements occurring there. When some of the solids combine with water vapor, they often produce precipitation, in all its forms. One of the many states of water that is most significant to military atmospheric operations is manifested by the clouds, which pose a differing set of challenges to military aviation, depending on the type of cloud and its vertical zonation. Then, just to add some even more dynamism to the atmosphere, there are many "geographic regions," including electrical fields and magnetic fields that then combine into electromagnetic fields. Like any geographic region, these fields have magnitudes and densities of phenomena within them. They also contain flowing "streams" of charged electrons and radiations that affect aerospace operations in many ways; some are beneficial, others not. At or near the surface of the Earth (during takeoff and landing) there are other geographic objects that cause problems for military aviation. Examples of those are related to soils and biological matter that reside in the frontier zone of land and air. These include such things as birds and dust, as well as general "FOD" (Foreign Objects Detection) that can foul or even knock an aircraft engine out of commission.

The gases and particles of the atmosphere rise in a stepwise fashion as horizontal fields, extending from the surface of Earth to altitudes beyond 400 miles, then gradually dissipating, until they eventually merge with the solar wind's charged particles. Some of the most relevant of the gases and particles are those that are spewed out by the volcanoes that occur throughout the surface of the planet. Others are carried by winds from source points, such as the Sahara, the Gobi, and other great deserts. Since the turn of the 20th century, a considerable contribution the particle and chemical mix has been made by human industry and warfare. Most of the atmospheric gases are concentrated within the zone which is called the troposphere; it extends from the surface of the Earth to a height of from 6 to 9 miles, depending on the latitude and season. The movement of the gases is generally a function of convection, which is a process of turbulence caused by the relative heating of the Sun. The troposphere is where virtually all water vapor exists, and where virtually all "weather" occurs. At the top end of the troposphere, there is the so-called tropopause, where temperatures fall to -112 degrees Fahrenheit. All this is important both to the operation of aircraft engines as well as the human component of the military aviation system. Both components breathe oxygen and require a certain density of this gas for optimum performance. Rocket engines, by the way, carry their own internal supply of oxygen with which to oxidize the combustion of rocket fuel.

Neither the whole of Earth's atmosphere, nor any part of it, is a uniform mix of gas, water, dirt, or other elements. Instead, what is presented to the military aviator and the scientist is a dynamic and complex mixture of molecular systems mixtures within each region of air space. Nevertheless, there is a kind of gross zonal variation between the horizontal series of layers that emanate from the surface of the Earth, toward space and the solar system. Each of these layers is distinct in terms of its composition and dynamics, and within these horizontal zones there are vertical "envelopes" of atmosphere which further define the anatomy of the atmosphere. Stirring these kettles of the atmosphere are the winds, which are energized by the Sun and by the reradiation from hot surfaces of the Earth itself. It should be noted here that, although the atmosphere itself is important as an intrinsic area of study by scientists such as meteorologists, it is most important to the military aviator in terms of how it interacts with military aerospace operations. That is, the military aviator views the atmosphere as a complex system of forces and phenomena that dynamically pose enablers and hindrances to the central tasks that relate to "flying and fighting." Moreover, it consists of a particular combination of gases and particles that apparently is unique in the solar system.

Because of the primary importance of this gas within the atmosphere, oxygen deserves to be treated with special attention. If I had to make a blanket statement about the relationship between oxygen and military aerospace operations, it would be that: In the proper dosage and administered at the appropriate time, oxygen is beneficial to that human endeavor. However, the presence of too much or too little oxygen, at any given space-time, also can be detrimental to the efficacious conduct of such operations. Within a certain envelope of our atmosphere, nature provides just the right dosage of this gas to maintain human life and to facilitate the "controlled explosions" that provide the propulsion that is needed to fly and fight. Thus, at altitudes above 12,000 feet or so, humans must either carry oxygen in portable containers, or they must construct larger "cocoons" of atmosphere, in order to fly and fight. By the same token, the controlled explosions that provide jet or rocket thrust must have their own "oxidizer" system, along with fuel, to function effectively in the atmosphere or outside it. However, these "portable atmospheres" can also prove fatal, under the wrong conditions. Already, there have been instances of fires and explosions in the astronaut's capsules when oxygen is exposed to a spark or other such catalyst. Even outside the capsule, an improper mix of fuel and oxidizer can produce an explosion in the propulsion system of the rocket, which almost always results in catastrophe for the space mission.

————

Above the Earth's atmosphere, there are many systems of charged particles or plasmas that are interacting with the solar wind. The outer layer is the geomagnetosphere, which is the primary agent of interaction with the solar wind. This is a region of steady magnetic forces that are produced by sources that occur above and below the planet's surface. The geomagnetosphere also includes two oval-shaped belts, or zones that are centered over the Equator. These contain significant quantities of energetic protons and electrons that are trapped within this magnetic field. The inner belt extends from about 600 to 3,000 miles above the Earth's surface; the outer belt, from about 7,500 to 13,000 miles in altitude. Just below the magnetosphere of the Earth is the region of charged particles which is called the ionosphere. Within this region of the atmosphere there are a sufficient number of ions which are caused by mainly solar radiation; these affect the propagation of radio waves within its area of influence. The lower boundary of ionosphere lies at approximately 30 miles above the surface of the Earth, but the effect on radio waves is most pronounced at heights above 50 miles. The ionosphere is useful in the conduct of military aerospace operations, due mainly to the fact that radar and radio radiation is refracted by it. In practice, this effect extends the reach of radar transmitting sites out to as much as 2,200 miles. Ionization in the ionosphere region is chiefly affected by solar radiation at X-ray and ultraviolet wavelengths. The particles which are ionized are mainly molecules of nitrogen and molecular and atomic oxygen.

Like all geographic regions, atmospheres have boundaries, which are an important subject of scientific inquiry by geographers. One of the things that have been learned from these studies over the years is that boundaries are rarely absolute lines of demarcation, either in the nature or in the cultural geography. Instead, boundaries tend to be transitional zones in which combinations of variables gradually change in terms of their proportional profile. In the case of the atmosphere, the lower boundary is nominally the surface of the Earth – including the dry land and the bodies of water. In practice, however, the lower boundary must be described in the plural; and it too is a frontier transition zone, rather than a line of demarcation. It is a short-hand conceptualization, which is of no real utility when one is studying the relationship between the boundary and human activities, except in very small-scale, immediate terms.

Some Earth scientists have suggested that the atmosphere of our planet actually should be seen as originating from some point below the surface of the crust. The argument for such a "greater atmosphere," is centered on the reality that there is no sharp line that separates the surface from the relatively dry air above it, by any measure of relevant properties. Thus, for example, oxygen occurs at sufficient density and the atmospheric pressure to sustain human life and the efficient functioning of air-breathing machines even hundreds of feet below ground. Beneath the surface of the Earth, the lower boundary of the atmospheric region is quite variated. It comprehends the transition zone between the domain of air and the domain of the lithosphere. The upper boundary of the atmosphere is equally ambiguous and transitional, extending outwards for hundreds of miles above the surface. However, at altitudes of approximately 100 miles, the conditions that are present in the lowest subregion, the troposphere, have already been attenuated by the time an aviator reaches the stratosphere. Based on the data that has been received from rocket probes; and especially the drag that has been encountered by artificial satellites at altitudes of several thousand miles, have demonstrated that the terrestrial atmosphere extends to great distances.

I view "cultural" boundaries being acknowledged in space, in much the same way as coastal boundaries that have been demarcated by international law on Earth. That is, they are not totally arbitrarily defined; rather, they have tended to be based on some technical criterion, such as the effective range of an ICBM, or some other ballistic device. In the case of international coastal sovereignty on Earth, the coastal boundary is sometimes defined as the distance from shore in which "national control" of the coastal space can be enforced. On the other hand, physical boundaries often are determined by some measure of "intensity" of a particular attribute. Within the realm of electromagnetic radiation, the intensity may refer to a shift in frequency, which is sometimes referred to as the "Doppler effect." That is, when an apparent shift in the frequency from light waves or chemical spectra occurs. Therefore, in the case of upper and lower boundaries of fields of energized particles or chemical elements in space, it might prove useful to apply the "Doppler effect" principle in the definition of these physical boundaries.

————

Historically, and until the latter decades of the 20th century, the troposphere has been the primary "workshop" of the military aviator on Earth. What we call "air" is the medium that provides the lift to the airplane, in much the same way that the water of the oceans and seas gives "lift" (buoyancy) to naval vessels. The winds of the troposphere are also like the currents of the oceans; they can be a positive force with respect to velocity (and fuel consumption), or they can be a negative, frictional force – depending on the direction one wishes to go. And, most clouds and weather systems occur within this region of the atmosphere. Clouds can serve as helpful camouflage or as a hindrance to visibility in the conduct of military atmospheric operations, again depending on the warfighter's perspective. Another significant aspect of the clouds and weather that occurs within the troposphere is related to the turbulence caused by thunderstorms, a disturbance which can reach the heights of the tropopause. As far as the military aviator is concerned, it is the clouds that are most significant to military aerospace operations above the surface of the Earth. Clouds transport water vapor, sensible heat, and the momentum of the Earth's rotation to the upper portions of the troposphere. They also portend the turbulence and other atmospheric disturbances that can impact the ability of military aircraft to reach the target. In some ways, clouds are like the terrain and the vegetation that occur on the surface of the Earth. That is to say, they can be seen as distinct formations which have mass and structure; in other words, there can be said to be a kind of "geography" of clouds. Therefore, just like mountains or forests, clouds are usually perceived as being obstacles to the visibility of an aviator. On the other hand, they can also provide "stealth" cover against attacks from enemy air interceptor aircraft and ground-based anti-aircraft systems as well. Some types of clouds harbor extreme electrical energy which can bring down an aircraft, or simply render its avionics unusable, or disrupt any electrical-based systems, especially computers. The same is true for the guided missiles that are part of the military aircraft's arsenal.

The Earth's clouds also occur only within the troposphere. At the highest levels, there are thin and wispy cirrus clouds, thin and sheet-like cirrostratus clouds, and smaller, puffy cirrocumulus clouds. At mid-levels, there a gray and sheet-like altostratus clouds, and the gray and puffy altocumulus clouds. And at the lower levels, there are the gray and sheet-like stratus and nimbostratus clouds. Here are also found the dark and lumpy Stratocumulus clouds. However, the most significant to military atmospheric operations are the giant Cumulonimbus clouds which produce thunderstorms, and top up to 50,000 feet above ground. These objects are made up of great clusters of tiny water droplets or ice crystals in the air. Clouds may float more than 33,000 in altitude or they may drift so low that they touch the surface as mist or fog. There is always a certain amount of water vapor in the air, and water vapor often rises and cools, and since cool air cannot hold as much water as warm air, as they rise, the vapor particles start to form droplets (condense) around bits of dust, pollen or salt. As more water vapor condenses, the droplets grow in size and clouds begin to form. Finally, the droplets become so heavy that they clump together and fall to the Earth. If the temperature is high enough, they come down as rain; otherwise they land as hail or snow.

———————

There is a global system of winds that course through the Earth's atmosphere, like rivers of water on the land; these are the so-called "jet streams." These wind systems result from the kinetic energy that is contained in the jet streams, partly derived as a result of heat transport that occurs from the equator to the poles. In their trajectories aloft, across the surface, these rivers of wind are directed in their course by the spinning movement of the planet below the atmosphere. This phenomenon manifests itself in the "apparent" deflection of the path of an aircraft within a given coordinate system (such as the latitude-longitude coordinate system). Actually, the flight path is not deflected; rather the Earth moves beneath it. The jet streams refer to any of several high-speed air currents that flow, more often, in an eastward direction within the stratosphere. They are a feature of upper-air circulations; continuous bands of relatively strong winds that can reach speeds of more than 100 feet per second; and these wind fields they can grow stronger with an increase in height.

Jet Streams are located above areas in the terrestrial atmosphere where there are particularly strong temperature gradients (slopes); especially where frontal zones meet. In those situations the atmospheric pressure gradients tend to cause the resulting winds to increase in speeds, as the height increases. The result is a discernable, relatively thin, flow of air bands that present many of the characteristics of water currents in an ocean. They tend to occur relative to adjacent air fronts, such as the Polar Front in the northern hemisphere of the Earth. In this particular case, the front is considered to be the boundary between the polar and mid-latitude regions of air. In winter, this boundary tends to extends between 30 degrees latitude (N) and 60 degrees latitude (N). Winter fronts also are characterized by stronger temperature contrasts than are the summer fronts. For that reason, jet streams that are located more towards the equator in the winter, and tend to be more intense at that time of season. These jet streams are generally located east of major continents and flow at speeds of up to 230 feet per second. The normal pattern of these jet streams is for two or three current systems to occur in each of the terrestrial hemispheres. There is a so-called Polar

Front, which occurs in the mid-latitudes. This is generally considered a region where differing weather system fronts meet to form all sorts of storm phenomena. It also is the jet stream system that was most relevant to the aviators of the strategic nuclear bombers during the Cold War, which flew an "over the pole" trajectory in their mission planning. During the northern hemisphere summer, another set of jet stream systems appears in southeast and southwest Asia, as well as over tropical Africa. These affect the nature and magnitude of the summer monsoons in those regions.

So, the jet streams can be described generally as relatively narrow bands of air that tend to flow in an easterly direction in the stratosphere, like ocean currents. The stratosphere itself is located at altitudes of about 6-8 miles above the Earth's surface. These jet streams flow at speeds of about 300 miles per hour, over thousands of miles, at altitudes between 30,000 and 160,000 feet above the Earth. During my years of service in the Strategic Aerospace Command, I often heard B-52 pilots and crewmembers discuss the effects of the jet streams on their platform's vector. The sum of the matter that I was able to garner from these conversations was that these jet streams could add momentum to an aircraft that flies along its stream, but it also creates a negative force against aircraft that are flying "upstream," as it were. That is, the aircraft's "true velocity" is actually somewhat greater than what their internal instruments indicate. Thus, early aircraft flew at speeds of about 100 knots (nautical miles) and the air that supported them was blown over the ground by wind speeds of up to 40 knots. It was therefore necessary to factor in the speeds of the relevant winds to determine a platform's "true velocity" with respect to the surface.

———

We know less about the atmospheres of the other planets and moons of the solar system than that of our own planet. However, it would seem that what we have learned about the climates and weather regimes on Earth can be applied to the understanding of the somewhat analogous phenomena throughout the solar system. Thus, all the other planetary atmospheres can be compared to Earth's atmosphere as a point of origin for analysis. Consider that some of the major aspects of our atmosphere include atmospheric pressure, temperature, and compositional properties. These also can be found on other planets and moons. Furthermore, the variances that occur seem to be a matter of degree, rather than of kind. Because of this essential commonality, the Earth's atmosphere can be used as a working model for the other atmospheres in the solar system. On another level, the individual planetary atmospheres within the solar system can be seen as subsets of an overall solar system atmosphere. From this perspective, one can analyze the various characteristics and properties of the atmospheres as common systemic phenomena. Also, such a conceptual viewpoint makes it easier to compare and contrast the various atmospheres, as though they were regions within one heliosphere. In the end, it can be seen that the nature of all planetary atmospheres is determined by the interplay of the same physical forces and chemical processes – all of which are energized by the Sun and, to a lesser extent, by the energy emitted from the internal regions of the individual planet.

So, we can see that the phenomenon that we call the "wind" on Earth has its counterparts on other planets and moons of the solar system. Based on our lengthy experience with our own planet, we understand that the winds and other movements of air are basically a function of the rotation of a planet on its axis. These global movements of air then are made more complex due to other factors, including the differences in adjacent atmospheric pressure and the macro-scale topographical variations on a planet. On a micro-scale, the topographical patterns add another level of local complexity in wind patterns, as does the relative location between land surfaces and large bodies of water. The upshot of all this complex activity is manifested by the various systems of air movements that travel in many directions throughout the surface of a planetary sphere. Winds are also agents of transportation, of dust and other particles along the various strata of the atmosphere. And, the winds also play a major role in determining the nature and degree of climatic and weather regimes in the troposphere.

Outside the planetary atmospheres, an expanded model of winds in the space age now includes the solar winds. As we are discovering through our space explorations, the solar winds that flow throughout the interplanetary medium are one of the major geographic forces in the solar system. They also have an effect on the physiology of planetary atmospheres, especially those that do not have a robust magnetic field to deflect and attenuate the ultimate force of the solar wind on the planet, such as the planets Earth and Venus. On other planets, such as Mars, these raw solar winds also cause many of the classic weathering effects of the winds on Earth. Solar winds are an emission of charged particles from the Sun that propagates in all directions outward toward the heliosheath, which lies at a distance of about 20 AU. As they are propagated throughout the solar system, these charged particles are attenuated by the distance from their origin. But this propagation is not uniform; the flux of charged particles sometimes encounters magnetic fields from some planets which deflect its flow. So, the solar wind can be seen as an analog to the local atmospheric winds on Earth and other bodies in the solar system. One important thing that has been learned from space exploration is that flow the solar wind is made more complex and turbulent as it encounters planetary magnetospheres and other fields.

———

Planetary winds are generally a function of temperature. However, the effective temperature within an atmosphere derives from many complex variables. As in the case of any input-processing-output system, the atmosphere receives inputs of energy from various sources. The most important of these is the Sun, but there are inputs of reradiated heat from the surface of the planet and from the inner regions of the planet, as well. Overall, however, it can be said that variations in temperature within planetary atmospheres generally are primarily a function of the rate of the incidence of sunlight on a particular part of the atmosphere. Within the atmosphere, there can be local spatial variations due to macro factors, such as the inclination of a planet's axis, and to its rate of rotation with respect to the Sun. Micro factors can include the composition and structure of the atmosphere and the nature of the topographical relief below it.

On Earth, the most significant differences in the rate of insolation and temperature occur because of the juxtaposition of larger bodies of land and water. In some cases, the process is seasonal, as in the case of the lands of southwest and southeast Asia which lie adjacent to the Indian Ocean and the Pacific Ocean. This causes the so-called "monsoon effect" wherein winds flow from the region of high pressure to the adjacent region of low pressure. Sometimes the high pressure center is over the land; other times it's over the body of water. Thus, the direction and strength of particular wind regime depends on the relative air pressure of adjacent land masses and large bodies of water (especially an ocean, a sea, or a large inland lake). The same kind of phenomenon may eventually be found on the other planets and moons of the solar system, although the bodies of "water" may turn out to be methane or some other volatile. In any case, such "monsoon" conditions, if they exist, still require much more empirical study.

Planetary winds, like water, tend to flow in a vertical and horizontal manner along gradients in the atmosphere. These gradients occur, primarily, because of macro and micro differences in atmospheric pressure within a region of air. The upshot of this is that wind speeds also can differ due differences in the "slope" of the gradient between a points of high pressure and points of low pressure. Near a planet's surface, winds tend to flow around regions of relatively low pressure, in various directions. Thus, we now are discovering the there are equatorial belts on other planets within which northern winds flow in well-defined patterns of direction. That is, in conformance with the Coriolis Effect, they flow in a counterclockwise direction in the northern hemisphere, and in a clockwise direction in the southern hemisphere. Higher up in the atmosphere, at about 6 to 8 miles above a planet's surface, the pressure systems are organized into sequences of high-pressure ridges and low-pressure troughs, but the "local" manifestations of atmospheric pressure are more complex due to differences in planetary heating and topography.

One of the most succinct, yet comprehensive theories that attempt to describe and explain the phenomenon of wind within a planet's atmosphere is the so-called Hadley model of wind cells, which is named after a scientist named George Hadley. It attempts to rationalize a global system of wind patterns that occur, generally, between two reference points: the Equator and the Poles of the Earth. According to this theory, the Earth consists of a single wind system in each of the northern and southern hemispheres. Essentially, it describes a pattern of wind movement in which warm air rises at the Equator and descends at each of the poles of the Earth. In other words, the theory describes a convector belt that essentially goes northward and southward, respectively, from the Equator to the poles. Within this conveyer belt, however, there is considerable jostling and twisting from the north-south directions because of other global phenomena. One of these is the so-called Coriolis Effect which explains why north-south winds are deflected by the turning of the Earth on its axis.

———

The various elements of weather that occur on Earth also serve as working models for the analysis of similar phenomena throughout the solar system. Consider the phenomenon we call lightning on our planet. One way to characterize electricity is to say that it is generated in storms. It is often described as a sudden flow of huge amounts of electricity between two clouds; between a cloud and the ground; or between two parts of a single cloud. These flows of electricity occur when a region of a cloud acquires an excess positive or negative electrical charge, which is sufficient in magnitude to break down the resistance of air. These charges are contained in droplets of water or particles of ice within the cloud. Since larger particles within the cloud fall faster than the surrounding lighter ones; this is thought to be the mechanism by which hail is produced. The lightning process basically begins with the discharge event and continues as charged particles travel from the upper to the lower regions of the cloud. The result of all this is the formation of a huge electric field.

The storms that are usually associated with lightning are meteorological phenomena that are a function of the degree and duration of solar insolation that energizes the ground beneath them. But, because this insolation load is unevenly distributed over a planet's surface, it gives rise to differences in adjacent portions of the air with temperature differentials above the ground as well. These differentials then produce differences in air masses aloft, which then cause winds to move from one place to the other; higher pressure towards lower pressure. In the troposphere the more intense solar radiation over humid surfaces produces abundant clouds. Some of these clouds, especially in the tropics, but in other regions as well, can produce violent storms: with all of their attendant thunder and lightning, rain and hail, winds and turbulence. These storms have a significant impact on the conduct of military air operations. They produce phenomena such as wind shear, air turbulence, lightning, which affect not only the takeoff and landing stages of an air combat mission, but also navigation, acquisition of target, and the efficacious application of ordnance on the target.

———

The phenomena that we refer to as tornadoes or cyclones on Earth are another manifestation of the effect of heat that is contained within the atmosphere. These are localized and especially violent storms which pose a direct and lethal threat to military aerospace operations; at every phase of the sortie. This phenomenon is defined as a violent, rotating column of air which usually extends from a thunderstorm, whether it touches the surface below or remains aloft in the atmosphere. A tornado can be described as a small-diameter column of violently rotating air. More precisely, it is produced within a convective cloud and frequently comes in contact with the surface of the Earth. This phenomenon occurs in every continent except Antarctica. It occurs most commonly in the mid-latitudes (between 20 degrees North and South latitude). This violent phenomenon is frequently associated with thunderstorms that develop in regions where cold polar air confronts warm tropical air. The most violent tornadoes are known to reach wind speeds of 250 mph or more, and the path of destruction and death can be greater than a mile wide and more than 50 miles in length. Most tornadoes develop during the spring and summer months, but they have been known to occur on every day of the year. They are most likely to pop up in mid- afternoon and early evening hours of the day. Although tornadoes occur in many parts of the world, they occur mostly in the United States, in the region that lay east of the Rocky Mountains. They normally occur during the spring and summer months. The thunderstorms that cause tornadoes develop in warm, moist air in advance of easterly cold fronts. These thunderstorms often produce large hail, strong winds, and the tornadoes. However, in point of fact, no place is safe from tornadoes; in the late 1980s, a tornado swept through Yellowstone National Park, Wyoming. Its path of destruction left a scar of destruction up and down one of Yellowstone's mountains.

As yet, there have been no reports of tornadoes occurring anywhere else in the solar system. However, there are data to indicate that there are huge wind storms occurring on many of the planets, most prominently Mars and Jupiter. Scientists are only beginning to understand these phenomena, but it can be said that they are huge, powerful, and long-lasting by several degrees of magnitude, compared to those on the Earth.

Another element of planetary atmospheres has to do with the amount and type of moisture within it. On Earth, the precipitation within the troposphere occurs whenever droplets of water in the atmosphere coalesce and develop masses that are too great to remain aloft. They then fall to the surface of the earth as rain, snow, hail, sleet. The form it ultimately takes depends on the manner in which the droplets were formed during their formation, as well as the temperatures that are incurred during formation. Precipitation may take the form of rain, snow, sleet, or hail. Rain is the most common form of precipitation; it consists of tiny droplets of liquid water which group around these droplets to form even larger droplets. Finally, when these droplets become too heavy to remain suspended, they fall as rain. Rain may fall in various time regimes: in the form of brief afternoon showers, as steady 3-day rainfalls, as a brief tropical deluge, or as a fine haze or mist or drizzle which never falls vertically. Rain also can come down in the form of a steady three-day rainfall, or as a deluge during a tropical rainstorm, or as a fine mist or drizzle. As snow, it occurs when water vapor in the atmosphere goes through a "slice" of sufficiently cold air and forms ice crystals around nuclei of the droplet of water.

To achieve this degree of scientific understanding of atmospheric moisture on other planets and moons in the solar system still requires considerable work. Thus, the meteorological models and techniques that have been developed to study the Earth's atmosphere are now being utilized to study the atmosphere and weather on the other planets and moons of the solar system. To explain how this works, I will refer to the in situ techniques and technologies that were employed to monitor activities along the Ho Chi Minh Trail of Vietnam. Thus, for example, sensors to measure such things as heat emissions and chemical signatures on the ground, beneath the dense canopies of tropical forests, are also being utilized to study the atmospheres of other planets. It is hard to say where such a remote sensing and processing technique originated, since the Apollo missions and the Igloo White Operation in Laos were essentially contemporaneous; during the early 1970s. In any case, both of these experiences have provided many cross-cultural innovations in technique and technology that have advanced the studies of the atmospheres on Earth and throughout the solar system.

As an example, NASA has been applying the techniques of remote implantation to place nuclear-powered geophysical laboratories at several sites on the Moon, ever since the Apollo missions. These more sophisticated sensor-units have included seismometers, magnetometers, plasma and pressure gauges, instruments to measure heat flow, and laser reflectors. Also, the lunar laboratories now are supported by orbiting satellites to support their geodetic studies. Obviously, these remote sensors are much more sophisticated and powerful than those that were implanted by the U.S. Air Force along the Ho Chi Minh Trail of the 1970s, but they employ the same principle. One technological advance is the development of longer-lasting battery units that enable the sensors to remain operational for longer periods of time. Also, more recent advances in micro-processing technology have provided the sensors with the capability to capture much more raw data during a given unit of time. And, this data can be processed into useful information even prior to transmission to Earth bases for more comprehensive analysis.

Over the course of several space missions and as a result of continuing studies by ground-based telescopes, we have learned that the Moon has very little atmosphere, mainly due to its small size and mass. Its surface gravity is only one-sixth that of the Earth. Therefore, because of the absence of an effective atmospheric shield to protect the surface from the repeated bombardment of countless asteroids, comets, and other particles of all sizes, the Moon's surface is covered by craters of varying diameters and depths. One outcome of this phenomenon is that the Moon has become a valuable model for studying the geography or locational analysis of the crater features on other bodies of the solar system. For example, the distribution of the craters and their morphology provides a tool for inferring the historical geography of crater impacts on Earth, and perhaps gain some insight into the frequencies of occurrence on a planet.

More recently, the Spanish government is providing a weather monitoring station to be carried by the Mars Science Laboratory rover (Curiosity). This in situ unit is designed to measure and provide daily and seasonal reports on atmospheric pressure, humidity, ultraviolet radiation at the Martian surface, wind speed and direction, air temperature, and ground temperature around the rover's position. The pair of small booms on the rover mast will record the horizontal and vertical components of wind speed to characterize air flow near the Martian surface from breezes, dust devils, and dust storms. Meanwhile, a sensor inside the rover's electronic box is being exposed to the atmosphere through a small opening and will measure changes in pressure caused by different meteorological events such as dust devils, atmospheric tides, and cold and warm fronts. A suite of infrared sensors on one of the booms will measure the intensity of infrared emitted by the ground, which will provide an estimate of ground temperatures. A sensor on the second boom will track atmospheric humidity. Both booms will carry sensors for measuring air temperature. At the same time, an array detectors on the rover deck that are sensitive to specific frequencies of sunlight will measure ultraviolet radiation at the surface and correlate it with changes in other environmental variables.

On Mercury, there are only traces of atmospheric components, including atomic helium, sulfur, and hydrogen. There are also transient flows of charged particles from the Sun. It has no equivalent to the ionosphere of the Earth. However, there are solar winds and magnetic storms. Therefore, if we were to plot a value of intensity for the atmosphere at this particular location on a three-dimensional grid, it would probably be the lowest in the solar system. The first significant telescopic data about Mercury in the wake of the Mariner mission in the 1990s discovered sodium in the atmosphere. Later, more refined and powerful Earth-based telescopic studies showed variations in the atmospheric components, from place-to-place and over time. Also found, was evidence of condensed material, perhaps ice, in the permanently-shadowed craters that lay near the poles. The only "weather" on that planet is the result of charged particles that have been emitted by the Sun. These are propagated outward through the solar system and interact with the substantial planetary atmospheres. However, these solar winds remain over Mercury for only a few hours at a time before escaping into space. Any gases that accumulate during the planet's night are quickly dissipated by the brilliant morning sunlight.

Venus has a hot, carbon dioxide-rich atmosphere. We know this from Earth-base observations and from several space missions that have been conducted by NASA and the Soviets. Based on the mission reports, it is evident that the planet Venus lacks a strong magnetic field, and its extreme heat is generated in the lower atmosphere or surface. By utilizing instruments such as an on-board telescope, orbiting spacecraft have been able to acquire detailed measurements of the planet's atmosphere and clouds, which previously had been impenetrable by Earth-based telescopes. By 1990, the Magellan mission to Venus arrived near the planet and was inserted into a near-polar elliptical orbit. Then, in 2012, the Venus Express mission detected an unexpected cold region in the planet's atmosphere that was considered to be cold enough to enable carbon dioxide to freeze out as ice or snow (esa.org). We also now know that Venus, like Earth, has a substantial atmosphere, and that it is 90 times denser than Earth. The heavier gravity of both planets makes it possible to retain these dense atmospheres. Unlike Earth, however, Venus has maintained a large percentage of carbon dioxide in its atmosphere, probably, at least in great part, because of the lack of any biological activity to wash it out. Consider also that: the global climate of Venus is strongly driven by the most powerful greenhouse effect that is to be found in the solar system; winds are important in the transfer of heat (energy); and water vapor is a critical ingredient in its weather patterns.

In terms of military space operations, perhaps the most significant aspect of the Venetian atmosphere is that it is covered by a constant veil of clouds, which is itself evidence of a substantial atmosphere. In fact, it has the most massive atmosphere among the so-called terrestrial planets, which include Mercury, Venus, Earth and Mars. Conditions on Venus are extremely harsh, in terms of atmospheric pressure and temperatures, as well as the chemical environment, which includes sulfuric acid. Under these conditions, common materials such as aluminum and glass approach or reach their melting point; and Titanium and magnesium can combust and organic compounds dissolve in the supercritical carbon dioxide environment. Probes can even be disabled by chemical corrosion, short circuits can be caused by condensation and little-understood mid-air electrical shocks can also pose a hazard to man-made space systems. Overall, the combined effect of pressure and heat, as well the presence of corrosive chemicals such as sulfuric acid, can also negatively affect military space systems within the atmosphere of Venus.

As the Russians have discovered from painful first-hand experience, the gaseous envelope around Venus is composed primarily of carbon dioxide (96%) and molecular oxygen (3.5%). It extends downward, from the edges of space to an altitude of about 60 miles above the planet's surface. Temperatures within the atmosphere vary with distance from the surface, dropping to approximately -148 degrees (F) at a height of 60 miles. Another significant type of feature that has been identified there is the vortex. The most notable of these is the so-called Bizarre Vortex, which is located in region of the South Pole, and which seems to change shape every day. There also is a permanent system of vortices and clouds that are whirling about the poles at great speeds. The pattern of cloud formation near the equator of Venus indicates that the upper layers of the atmosphere circulate around the planet once every four days; this suggests the presence of winds that reach speeds of more than 300 miles per hour. These winds exist at high altitudes, but the atmosphere near the surface is relatively calm. The atmospheric circulation is faster on Venus than on the rocky planets, with the cloud-level atmosphere spinning about 60 times faster than it does near the surface of the planet.

Telescopic observations of Venus present a brilliant yellow-white, essentially featureless surface. Its obscured appearance results from the surface of the planet being hidden from sight by a continuous and permanent cover of clouds. Because Venus' orbit lies within the orbit of Earth, the planet exhibits phases like those of the Moon when viewed from the Earth. Later telescopic observations, both from the surface of the Earth and in space have yielded some more data about the physical geography of Venus, many of which are of interest to the military geographer. In this context, the atmosphere of this planet presents some intriguing characteristics. Thus, despite the high temperatures that are found on the surface and the atmosphere, the gaseous envelope that surrounds Venus is found to contain "clouds." In these clouds, there is carbon dioxide, sunlight, and even water. These, notably, are the prerequisites for photosynthesis and, therefore, oxygen. There is also evidence of nitrogen molecules in these clouds. The cloud pressures are approximately the same as on the surface of the Earth. Together, these properties of the clouds present the most Earthlike extraterrestrial environment known to date.

Mars presents a different atmospheric profile that those of Earth and Venus. Based on observations and chemical analysis by the Viking landers, it is evident that Mars has a very thin atmosphere. It exerts less than 1% of Earth's atmospheric pressure at the surface. The evidence derived from observations so far also indicates that there are only small amounts of water present in the atmosphere today. The typical temperature in the lower atmosphere is about -100 degrees (F), but there are internal variations, as well as seasonal differences. Thus, in the summer, the temperature in the atmosphere which lies just above the surface can peak at about 62 degrees (F). Above this lower layer, the temperature decreases with elevation, at a rate of approximately 2.7 degrees (F) per mile of altitude. Therefore, Mars would show up as a low point on the contour map of the solar region, as being only slightly higher than Mercury.

From spectral observations, we know that the Martian atmosphere is composed mostly of carbon dioxide, 95% by weight. This is an extremely dense concentration, especially when one considers that it is nine times that of Earth's atmospheric content, even though the Earth has a much more massive atmosphere, overall. Mars experiences cycles with respect to carbon dioxide that are similar to that of water in Earth's atmosphere. Thus, carbon dioxide falls like snow at the winter poles and sublimes into gas in the spring. Scientists have calculated that approximately 8 trillion tons of carbon dioxide leaves and enters the planet's atmosphere on a seasonal basis. Photo-chemical reactions also produce trace amounts of molecular oxygen, carbon monoxide, nitric acid, and ozone. We also know that the lower levels of the atmosphere supplies gas to the planet's ionosphere. At the same time, constituents at the top layer are lost to space. Both of these processes affect the isotopic composition of the gases that remain. Water vapor also appears to be present in the atmosphere; it is uniformly mixed up into altitudes of between 6 to 9 miles.

Jupiter has a well-developed system of clouds which are separated both vertically and horizontally, in the same manner as on Earth. From the data gathered by surface-based telescopes and NASA spacecraft, we have learned some interesting things about this giant planet. As an example, there is a stable underlying pattern of latitudinal currents which have remained relatively constant for several decades. On the other hand, the atmosphere can also be quite dynamic. Both of the Voyager spacecraft have revealed a variety of cloud forms which can be accompanied by cyclonic and anticyclonic storm systems. Thus, it can be seen that the main characteristic of the atmosphere of Jupiter is that of constant motion, despite the long-duration of some phenomena.

In terms of size and structure, the atmosphere of Jupiter is about 40,000 miles deep, and presents many series of clouds that are layered like those in our atmosphere. A space probe that actually descended into Jupiter's atmosphere has sent back data that depicts visible clouds that are formed of aerosols, mostly composed of ammonia, with very little water vapor. As the depth increases, cloud particles turn into ammonia crystals and ammonium hydrosulfide droplets, followed by a layer of cumulus clouds that consist of water ice crystals, followed by cumulus clouds that are made up of water ice crystals and droplets. These clouds are naturally colorless, but they do take on faint hues as the result of interaction with phosphorous and sulfur dust particles. There also is a temperature gradient present in the atmosphere; at 43 miles above the surface a temperature of 1,970 degrees Fahrenheit was detected. The layer that lies above 50 miles presents no clear boundary, and the gas simply becomes hotter as one descends. Temperature and pressure continue to increase as one as one descends to the core.

Because Jupiter is a gas giant, it is thought that it primarily composed of hydrogen and helium; the later comprising about a quarter of its mass. The set of data that has been captured so far indicates that the interior of the planet consists primarily of liquid and metallic hydrogen, which resembles the composition of the interior of the Sun. Helium, carbon, nitrogen, and hydrogen were detected. This is thought to like the primordial composition of the planets of the solar system. Spectral analyses also showed evidence of methane. At the present time, the discussion of the Jovian atmosphere is primarily one of exploration and abstract scientific study. However, it is quite conceivable that, sometime in the future, the gases of the other planets of the solar system will be seen as a source of chemical resources that might be applicable to military space operations.

Overall, conventional scientific modeling of the planet Jupiter posits the planet as a set of neat layers, with a dominant outer gaseous envelope surrounding a rocky core which consists of heavier elements. However, new computer simulations, based on more recent data, suggest that Jupiter's rocky core has been liquefying, melting, and mixing with the rest of the material in its interior. The core region is now thought to be surrounded by dense metallic hydrogen, which extends outward to about 80% of the radius of the planet. Rain-like droplets of helium and neon appear to be precipitated downward through the layers, thus depleting the quantity of these elements in the upper atmosphere. And, like the Earth, Jupiter projects a magnetic field which is the strongest and most complex magnetic field of any planet in the solar system. This is a fact that will prove to be significant to military operations near and on the planet Jupiter in the future; that is to say, the magnetic field will affect the effectiveness of any electromagnetic weapons and technology that might be applied.

The atmosphere of another gas giant, the planet Saturn, consists of hydrogen (96%); helium (3%); and small amounts of methane, ammonia, ethane, as well as aerosols of water ice, ammonia, and hydrogen-sulfide. The yellow-gold bands that have been observed in the planet's atmosphere are thought to be the result of extremely fast winds that can reach up to 1,000 miles per hour around the equator. Rising heat from the planet's interior may also contribute to the presentation of these bands. The formation of Saturn's rings is only one of the mysteries of the planet. They appear solid, but they actually are made up of particles – mostly dirty ice – which range in size from small grains to big boulders. The Cassini mission captured images of a March, 2011 storm on Saturn, the largest storm that has ever been seen from a close distance there.

Another member of the Jovian system is Uranus. Like the other gas giant planets, Uranus has a ring system, but it is less complex than the other. As of 1978, nine distinct rings had been discovered. Voyager 2 (1986) observed two more rings, and the Hubble Space Telescope captured images of two additional outer rings during the period from 2003-2005. These rings are thought to be composed of water ice and some additional dark radiation-processed organics. The majority of Uranus's rings are opaque and closely-spaced; there is some dust, but the majority of the content consists of larger bodies. It is thought that the rings are relatively young, perhaps not more than 500 million years old, and that they may have originated from the fragmentation of several moons that once existed around the planet. Uranus's atmosphere is similar in composition to that of Neptune. It is also similar to that of Jupiter and Saturn in some ways, such as the fact that they are composed primarily of hydrogen and helium, as well as traces of hydrocarbons and nitrogen.

Neptune contains a higher proportion of "ices" in the form of water, ammonia, and methane than Jupiter and Saturn. For this reason, both Uranus and Neptune are place in a separate category, the "ice giants." The atmosphere of Neptune is notable for its active and visible weather patterns. Voyager 2 (1989) did a flyby of the planet's southern hemisphere and discovered a "Great Dark Spot" which is comparable to the Great Red Spot on Jupiter. These weather patterns appear to be driven by the strongest sustained winds on any planet in the solar system; wind speeds of as high as 1,300 miles per hour have been recorded. Because of Neptune's great distance from the Sun, its outer atmosphere is one of the coldest places in the solar system, with temperatures approaching -218 degrees Centigrade. Temperatures at its center are approximately 5,000 degrees Centigrade.

I would note, in summary, that as we have done a comparative analysis of the individual planetary atmospheres, I suggest here that each of these regimes can also be seen as being individual "islands" on the interplanetary medium within which they are distributed spatially. This view, I contend, can be quite useful for planning and conducting military operations on a medium that has many similarities to the oceans of the Earth.

———

Climate and Weather Regimes in the Solar System

So, what might be called "climate" on a particular planet is actually referring to a system of meteorological patterns that occur over a significant length of time in a particular area. It should be noted here that the "significant length of time" on the other planets takes on a much different meaning when the planetary periods are measured in terms of weeks and months, rather than days, as is done on Earth. The composition and physiology of a particular meteorological system is itself a function of several global factors, such as the body's distance from the Sun, the net incidence of sunlight on it, the reflectivity with respect to the sunlight, among others. Another global factor, which is only present on some of the planets, is the magnetosphere, which modifies the solar energy before it can penetrate to the level of an atmosphere. The diurnal rotation of a planet and its angle of inclination will also affect the time and duration, as well as the intensity, of sunlight that strikes particular hemispheres and local regions on the surface of a planet. Scientists have also discovered, over the six decades or so of space and planetary exploration that climates are defined in terms of latitude, in the same way as on Earth. Thus, the equatorial latitudes generally receive more insolation than the polar latitudes do. By the same token, the surface heat tends to rise along the equators and descend at the poles. In other words, the surface heat budget and exchange systems on other planets are similar, globally speaking, as on Earth.

Table 1

PLANETARY DISTANCE FROM THE SUN AND ITS TEMPERATURE

Planet	Distance from Sun (AU)	Mean Surface Temperature (F)
Mercury	0.4	-297 to 800
Venus	0.7	896
Earth	1.0	57
Mars	1.5	- 82
Jupiter	5.9	-202
Saturn	9.5	-202
Uranus	19.2	-328
Neptune	30.1	-328

What this table seems to suggests, is that the spatial pattern of planetary distance-temperature zones does reflect some kind of correlation between temperature and the distance from the Sun; similar to the relationship between the latitudinal distance from the Equator and temperature that is seen on the surface of the Earth. Distance from Sun, the star's luminosity (its ability to absorb thermal radiation), the eccentricity of a planet's orbit, and the axial tilt are all factors that determine a planet's rate of absorption at either the atmosphere or on its hard surface. Nonetheless, it can be said that the Sun is the primary heat engine in the solar system. At its surface, temperatures reach approximately 9,900 (F). However, there is a secondary set of planetary heat engines located in the interior of celestial objects that result from the early formation of the celestial objects. This heat is caused by gravity, nuclear fission, and decaying radioactive materials. The amount of internal heating depends on mass; the more massive the planet, the more internal heat it has. The internal heating process within the planets also powers tectonic and volcanic activities. Earth has the most internal heating because it is the most massive of the terrestrial planets. But there are other forces of physics that are operating within the terrestrial atmospheres.

In any case, "heat" is commonly measured in terms of "temperature." The term "heat" refers to energy that is transferred from one physical body to another as a result of relative difference in temperature; the transfer occurs from hotter body to colder body. More generally, heat transfer refers to all phenomena and the mechanisms by which they convey energy and entropy (disorder) from one location to another throughout the Universe. These mechanisms include convection, thermal radiation, and conduction. Conduction refers to the transfer of energy and entropy between adjacent molecules; convection refers to the movement of a heated fluid, such as air; and radiation is the transfer of energy in the form of electromagnetic radiation from its emission at a heated surface to its absorption on another surface. Therefore, the transfer of heat, whether in the heating of an enclosed space, or a pan of water, or in a natural condition, such as a thunderstorm, usually involves all of these processes.

If atmospheric pressure can be seen to exert a kind of "squeeze" on a given unit of air, it is thus keeping molecules of oxygen closer together, according to the amount of atmospheric pressure that is present. On Earth, we are used to viewing atmospheric pressure in terms of its effects on the weather, both on near the surface and in the atmosphere. We understand that it is greater at sea level, and less so at the higher altitudes of the atmosphere. That is why it is more difficult to breathe at higher altitudes (generally 12,000 feet or more). The gradual lessening of the "squeeze" effect on the molecules of air also reduces the volatility of the electrons of the atoms and, therefore, "cools" things down, so to speak, and it therefore it tends to get gradually colder at higher altitudes. Therefore, two of the main characteristics of outer space are the cold, and the absence of oxygen. It is the total heat load in the atmosphere, which is derived from the combination of radiation from the Sun and the core of the planet, and which is reradiated the surface domain, which provides the heat in the atmosphere.

On Earth, this heat is variably distributed within the planetary atmosphere. The spatially adjacent differences in the distribution of heat energy produces a mechanism whereby the atmosphere attempts to regain some kind of equilibrium in this respect. This mechanism generally is the movement between the regions of pressure, from those of high pressure to those of low pressure. The agent that is used to accomplish the sought-after equilibrium is called wind. It is this cauldron of gases; of oxygen-content and temperature-differences, in which all military aerospace operations are carried out in the atmosphere of Earth. But, there are also local atmospheric conditions which acutely affect the conduct of these operations. On the other planets of the solar system, however, this atmospheric model may not prove to be as useful, in its present construction. Therefore, each of the atmospheres on the planets and their moons must be studied further to test the validity of the Earth model. The net temperature of the atmosphere, as well as the surface of any celestial body in the solar region is the result of an input-output process. The input variables were discussed earlier. The outputs, or losses, of heat are what bring the ultimate planetary temperature(s) to a net sum value. There is, for example, planetary tidal dissipation; it is related to gravitational stresses, which tends to bring the net value down. This suggests an interfacial nuclear source of internal heat production, which is a counterforce to the heat produced by the Sun. Radioactivity processes are another source of heat. Radioactive disintegration occurs when unstable atomic nuclei dissipate excess energy by spontaneously ejecting an alpha particle.

The gas giants have much greater internal heating than do the terrestrial planets. Jupiter has the most internal heating with core temperatures of 36,000 (K). For the outer planets, internal heating does more to power the weather and the wind than sunlight, as is the case with the territorial planets. The internal heating within gas giants raises temperatures higher than one might expect from the amount of solar energy that is received. In the case with Jupiter, it is approximately 40,000 (K) warmer than the nominal temperature. Indeed, it has been said that Jupiter operates almost like the Sun within its regional domain. This results from a combination of greater distance from the source of solar heating and the planet's internal source of heat, which is marginally greater than the heat from the Sun's rays. As a result of this, we often refer to Jupiter's domain as the Jovian subsystem. It also includes Saturn, Uranus, and Neptune, as well as their satellites. All objects that lie within this Jovarian domain are greatly affected, in terms of their heat budgets, by the individual effect of Jupiter, and the combined effects of Jupiter and the other so-called gas giant planets.

———

Within the planets, there are many factors that affect the variations in internal heating. The presence of a metal core and a less dense rocky mantle can differentiate temperature within a planet, as can differences in the composition of material in different parts of it. The implication is that the core of a planet may have completely melted at some time in its history. And, as a result, the heavy materials would sink to the bottom, and the lighter materials would have floated to the top of a planet. The presence or absence of an atmosphere also is a tremendous determinant of surface temperature. A thick atmosphere makes a planet much warmer than a thin one, and a very thick atmosphere serves as a regulator, making temperatures fairly even between night and day, as they soak up and hold heat, or reflect it back to space. Then there is the volcanic activity, which has a significant effect on the temperature of the planet's atmosphere and surface temperatures. All of these factors are a major focus for study on Venus, which is being intensively studied by Earth-based space-based telescopes and other sensors, as well as those that are being integrated into space probes. It is from these sources that we know that the surface pressure on Venus is 93 times higher than on the surface of Earth, and that its atmosphere is almost completely made up of carbon dioxide.

The normal albedo of a celestial body also is a factor in determining the heat within its atmosphere and on its surface. The albedo is a measure of the fraction of total incident light that is reflected back to the original source of the light. By way of reference, the normal albedo of snow is nearly 1.0; that of charcoal is about 0.04. Generally, the higher the albedo number, the brighter a body in the skies appears to an observer. It also may be noted here, that albedo values, or more precisely, relative albedo values could potentially be used for military space reconnaissance and space attack operations in the future; in a manner similar to the use of infrared reradiated energy is used at the present time.

One useful way to address the relationship between albedo values and surface temperatures on the planets of the solar system is to think of the differences we may have experienced in relative surface heat when engaging in outdoor activities at a lake site in the summer, in the Midwest Region of the United States. Just by feeling and touching, we noted that the concrete sidewalk that led down to the lake was extremely hot when compared to the grassy areas on either side of it.

Then, when we stuck our toes into the water, we immediately felt the relative coolness of the water surface. What we experienced was a manifestation of the respective albedo values of the sidewalk, the grassy surface, and the water surface. We also might have taken note of the fact that on cloudy days, the heat of these surfaces all decreased, to greater or lesser degrees. In the winter, we noted that the air just above the surfaces, regardless of their composition, got colder when there was snow and ice on the ground, even though the Sun was shining brightly. The point here is that the same dynamics are in play between solar radiations, surface reradiations, and cloud cover on the planets, moons, and asteroids as well.

Table 2

PLANETARY ALBEDOS AND TEMPERATURES

Planet	Albedo	Surface	Temperature (F)	"Water/Ice?"
Mercury	0.11	Thin Rocky Crust	-297 to 800	None
Venus	0.65	Thick Clouds	896	None
Earth	0.37	Rock/Liquid	57	Varied
Mars	0.25	Rocky	-82	Polar?
Jupiter	0.25	Gaseous	-202	None
Saturn	0.47	Gaseous	-202	Some
Uranus	0.51	Gaseous	-328	Some
Neptune	0.41	Gas/Rock	-328	Some

Sometimes the military intelligence with respect to an area of operations or, more specifically, to a target system is derived not so much from the albedo numbers themselves, but by what they suggest. That is to say, a table of quantitative values itself can provide some intuitive information about a planet, even before they are plotted onto a map or chart. This kind of inferential analysis does not even require precise measurements; rather, it is the very process of abstraction from the precise quantitative values that can often provide important insights into the spatial "ground truth" in an area of operations. An example of this was the way in which the intelligence specialists of Igloo White at Nakon Phanon, Thailand drew inferences about potential truck parks simply by analyzing the patterns of acoustic data that was captured by sensors that were implanted along the Ho Chi Minh Trail. Astronomers too utilize the "wobble" (or blinking) factor of the various celestial bodies to draw inferences about some of their properties.

So, upon analysis of Table 2, we can appreciate that the albedo of a planet shows some direct correlation with the temperature of the body. The ordinal depiction of the data also seems to show some relationship between the planetary albedo with the structure and composition of the body. Thus, we see that the terrestrial bodies tend toward a much lower albedo score than do the gas giants. There are two anomalies in this relationship, however; these are the planets Venus and Earth. The explanation for this seems to the fact that both of these terrestrial planets have a relatively thick and layered cloud cover. Thus, the temperature on the surface of Venus is approximately 896 degrees (F), which is probably due, in great part, to the greenhouse effect that is created by an atmosphere that is composed mainly of carbon dioxide. On the other hand, Venus's relative closeness to the Sun is offset somewhat by additional cooling, which is possibly due to efficient eddy conduction and subsequent transfers of heat by means of eddies in turbulent flow.

Mercury lacks a significant atmosphere. Therefore, there is no cloud cover to modify its albedo reading. The intensity of sunlight on Mercury's surface ranges between 4.59 and 10.61 times the solar constant, so one may conclude that the rate of reflectivity of incident sunlight is unexpectedly low, because of the composition of its surface and the lack of a significant internal source of heat. The time of its rotation, plus its nearness to the Sun, might also explain the wide variations in the rate of incident sunlight as well. Thus, for Mercury, the Sun was the major source of heat at its early stages of development, but the planet then rapidly shrank in size and brightness, and it finally became an insignificant factor within a few millions of years. However, radioactive materials began to decay to the point of non-existence within a few tens of millions of years. Heating from collisions with other bodies is now believed to have been a continuing factor. Finally, the heat which was received during this early term of the planet's development and was stored has leaked to the surface, and radiated away.

Even though we commonly say that the planet Mars, for example, is "cold and dry," this does not give much information as to the local temperatures on the surface of the Red Planet. The same can be said about all of the other planets, especially after we get to know them up close with our space probes and rover vehicles. Fortunately, the modern scientists have at their disposal a broad array of tools to accomplish the job of gaining useful information about planetary temperature patterns. Thus, there is in place a network of scientific observation stations, which are located both on the surface of the Earth and in space which are providing the ability to conduct the empirical studies from many different scalar levels, angles and perspectives, and over longer periods of time. The Earth-based observatories have the advantage of being able to acquire serial data at a particular time of a terrestrial day and, thereby, develop a composite picture of a body's periodic changes with respect to heat levels on or just above its surface. Adding to the power and resolution of the observational sensors at these observatories is the computer technology, which has enabled the scientist-observer to develop an ongoing data base of imagery for analysis. Furthermore, the modern computer internet technology now offers the capability for sharing information on a scale that was not possible in the pre-computer age.

This heat monitoring system in space is analogous to the one that has been constructed to perform this function on Earth. One can expect that, in the future, other similar systems will be developed to conduct continuous, minute-by-minute empirical observations of temperatures on other planets and celestial bodies. This includes a sophisticated network of space-orbiting satellites and some surface-based sensors on some of the planets as well. Given such a wealth of data, and the increasing power of computer technologies, highly-sophisticated models of planetary weather now are being constructed. These are especially useful for making weather predictions not only on Earth but on Mars, as well, for example. However, throughout the rest of the solar system, all of the major planets are monitored by a less well-developed protocol, which still relies on orbiting spaceships and lander vehicles on the surface of the planets to acquire weather-related data.

Based on the continuing data that is being captured by all manner of telescopes and other sensors, we know that temperatures vary throughout the solar system, in much the same way that these phenomena vary throughout Earth. One reason for this variability of weather on the various planets and moons has to do with the variable strength of the magnetosphere on the planetary bodies. Generally, the larger and more energetic the magnetosphere, the more violent the interaction between the solar wind and the magnetosphere will be. The solar wind has a somewhat standard effect on unprotected bodies, including spacecraft, but when the solar wind interacts with planets that possess strong global magnetic fields, complex and dynamic climatic radiation belts can be generated. If we were to look at a temperature map of the solar system, we would find that there are variations based on "latitudinal" zonation too. Only in this case, the "latitudes" would represent the orbital zones of the planets and other bodies in the solar system. Therefore, the "latitudinal" variation can be seen as correlating to distance from the Sun, as does the mean surface temperature of the planets of the solar system.

———

The overall atmospheric phenomenon on Earth is called weather. The classic definition of the term, weather, refers to atmospheric conditions that are occurring at a given location and during relatively limited time periods. Within the planetary atmospheres in which they occur, these can be seen as discrete fields of solar winds, temperature, and atmospheric pressure. Each field occurs over a relatively short period of time and within a confined region. However, when they recur over longer periods of time, they can be placed into categories that relate to longer periods of time, such as seasons. On Earth, we are accustomed to thinking about weather fields in terms of diurnal periods or seasons. These are related to the various aspects of the Earth's surface with respect to the Sun, as the planet rotates on its axis and moves along it sidereal orbit around the Sun. So, it can be said that "weather" is closely related to time and space, and that a given weather field is relatively "short-lived."

There is a difference between planetary weather and the "interplanetary medium weather." The latter occurs throughout the heliosphere. The former, on the other hand, appears to be "local" and varies from planet to planet. More generally, interplanetary weather can be said to be a function of the solar wind, which is emitted from the surface of the Sun, and which is then normally propagated throughout the heliosphere with decreasing intensity and force, as the distance from the source increases. However, there are times when acute eruptions of energy occur, and these are referred to as solar storms which involve electromagnetic pulses. When the solar winds interact with each of the planets and other larger celestial bodies, a variety of local "weather" fields are produced. Thus, the actual weather conditions at any point in space and time are the result of the interaction that occurs between the solar winds and such planetary fields as magnetospheres and the atmospheres.

Space weather is also a term which is used by scientists to describe the weather-like phenomena that occur within the heliosphere as a whole. It is within this solar atmosphere that the solar winds flow throughout the interplanetary medium, while encountering celestial objects and specific fields in the heliosphere along the way. Occasionally, perturbations and anomalies that occur within the Sun cause disturbances in the normal flow of solar winds, which produce more intense solar storms, in much the same way that occurs within the atmosphere of the Earth. One class of these solar disturbances is called the solar flare. A solar flare refers to a sudden and intense brightening of a portion of the Sun's surface. Like many such intense, short-term events, it develops in just a few minutes, but it can last for several hours.

The material that is associated with these flares is an extremely high-energy mix of mainly protons and electrons. This plasma courses through the solar system at speeds of between 600 to 1,200 miles per second, which means it reaches the terrestrial planets within less than 30 hours. When the plasma meets the planetary magnetosphere, the planetary storms are produced. One of these is the magnetic storm, which is caused by solar flares and by "holes" on some areas of the corona. These holes were first discovered from imagery that was taken by the U.S. Skylab. They are significant because the holes enable particles to escape the corona with relative ease and at higher velocities. As a result, these particles "catch up" with the previously-emitted and slower particles from normal emissions. Then, when this agitated plasma reaches a planetary magnetosphere, magnetic storms are produced. Some of these are quite different than those on our planet, both in terms of intensity and duration. Consider the case of the Great Red Spot of Jupiter, which is really a storm which has been observed to be raging for almost 350 years, ever since it was first reported in 1665. This giant storm is awesome; the anti-cyclone high-pressure system rises to about 5 miles above the surrounding clouds. Its winds whip around a huge whirlpool at a speed of about 400 miles per hour; and almost three Earths could fit comfortably inside the maelstrom. Meanwhile, all of this space weather on Jupiter is being systematically observed and recorded by the Hubble Space Telescope.

These heightened and acutely intense bursts of solar-induced weather can have significant effects on military space operations, including the orbiting satellite systems. Of special concern are the fluxes gamma rays and X-rays. Based on experience that has been gained from space operations during the past fifty years, solar flares are considered to be potentially hazardous to both humans and their machines in space. More generally, the input of electromagnetic energy (e.g., X-rays, ultraviolet and infrared radiation, and radio emissions, as well as visible light) helps to create the "weather" regimes within which military space operations will occur. Although the actual incidence on any square foot of space is relatively tiny (about 1 part in 120 million) it nevertheless poses the same threat profile to military space operations as does the weather in the atmosphere of the Earth.

The overall variability of space weather also appears to be closely related to the sunspot cycle that is related to the rotation of the Sun. This a periodic phenomenon in which several important kinds of solar activity repeat themselves. The cycle lasts about 22 years, on average, and includes two eleven-year cycles of sunspots (regions of extremely strong magnetic fields). The sunspots display opposite magnetic polarities in alternate cycles, as well as pairs of peaks and declines during the same period of sunspots. These perturbations, which also include solar flares, pose a hazard to space operations. Some of the reported symptoms of these disorders have included the electrical charging of spacecraft, a drag on orbital variations of artificial satellite systems, and dangerous increases in ionizing effects on both spacecraft and astronauts. Other reported problems which have been associated with these solar weather events have included computer processing errors and the degradation of artificial satellite operations, both in terms of navigational and control processes.

One of the things that we have learned from of long-term observations by telescopic systems like the Hubble Space Station, which operates on an orbiting satellite outside the Earth's atmosphere, is the danger of spatial storms. These are related to the violent solar events that are characterized as flare-ups which can likened to violent wind events on Earth, such as hurricanes or tornadoes. However, in the case of the outer space, these kinds of events involve radiation storms, rather than wind flare-ups that cause such havoc to military aerospace operations within the terrestrial atmosphere. It should be noted here that the effects of radiation storms are not exclusive or unique to outer space. The only reason why they are not considered to be of immediate or acute concern to military aerospace operations on Earth is that our planet is generally protected from such phenomena by our geomagnetic field.

So, we see that solar flares, charged particles, cosmic rays, the Van Allen belts of radiation, as well as some other natural phenomena that occur in space, all contribute to what is called "space weather." This interaction varies from place to place in the solar system, primarily as the result of the variability of the relative distance from the origin of the solar winds, as well as variations in the structure and behavior of the atmosphere and the respective magnetic fields that might occur around a particular planet. Generally, it is this interplay of gases, plasmas, and fields that have developed in the interplanetary medium and around the planets that produces the "weather" in a particular place and time in the solar system. Just as we have experienced during military air operations within the Earth's atmosphere, there is great variation in space weather, both in time and in spatial terms. Also, the localized weather phenomena on Earth, such as tornadoes and storm cells have their analogs in space weather as well.

———

Meteor showers and micrometeorite debris occur throughout the interplanetary medium in the solar system; these can be seen as being analogous to the hail storms on Earth. Most meteor showers originate with the comets. Each time a comet orbits the Sun, it produces a great amount of small particles which will eventually be spread out along the entire orbit of the comet to form a meteor flow. If the Earth's orbit and that of the comet happen to intersect at some point, the Earth will pass through this flow for a few days at approximately the same time of the year, thus encountering a so-called meteor shower. Because the meteor shower particles travel in parallel paths, at the same velocity, they together appear to radiate from a single point in the sky.

It might be noted here that kinetic energy is what makes all moving objects in space a potential hazard for military space operations. We have seen a vivid illustration of such energy when it comes to a comet (such as the Shoemaker-Levy 9) striking a planet (Jupiter), but the same principle holds for any projectile striking an object that has mass, at a high rate of speed. Thus, we know that a small-caliber bullet with a high muzzle-velocity will cause more damage on impact than a slower-moving, larger caliber one. In the same way, it would appear that the smaller, but faster, meteor can cause equal damage over a more localized area as a slower-moving larger meteor. Meteoroids are part of the interplanetary environmental, and they sweep through Earth's orbital space at an average speed of 12 miles per second. The hazards that are posed by meteoroids is now well-understood and, therefore, has become a factor in the design of spacecraft and in mission planning. The effect of these meteoroids is especially acute within the increasingly crowded "air space" of the geospace domain. This is exemplified by the half-ton of meteoroid mass that occurs within 1,000 miles of the Earth's surface. The fact of the matter is that there is a lot of natural and man-made debris in the solar system, especially below 500 miles of altitude. Some of these meteorites are the result of the fallout from collisions of natural bodies within the solar system, which often result in clouds of meteorites that constantly rain down on the planets.

As humans continue to venture into space, the threat potential from meteorites will increase for spacecraft of all types, including orbiting satellites. There also will be a constant threat of impact from all the particles of rock of all sizes, space dust, and the debris from human activities in space. Indeed, some of the collisions will now involve the velocity of the spacecraft, as well as that of the meteor or other particles. The reality is that there is an ever-increasing assortment of debris and trash which is being left in orbit by the whole process of placing and maintaining of artificial satellites in orbit. Space debris (space junk) refers to the objects in orbit around the Earth that are the regolith of human space activities. These are the waste material of space operations which no longer serve any useful purpose. They can consist of everything from spent rocket stages and obsolete satellites to the minute detritus of erosion, and the fragments resulting from explosions and collisions in the orbital zone. At the same time, as the total mass of these "foreign objects" has grown, these objects often overlap the trajectories of newer objects; just like the automobiles in an oval race track. These "FOD" have long been the bane of runways on Earth, especially in the jet propulsion age, when jet intake cowls were prone to sucking up nuts and bolts that were left over from previous operations. As a result, some international diplomatic steps have been taken to ensure that the operators of the satellites will implement measures to bring obsolete satellites down into a safe recovery site on Earth. However, there are still many uncontrollable factors occurring which result in the cumulative build-up of debris in the artificial-satellite belt.

As we have seen, collisions of all kinds are a major source of space debris. Some involve only natural objects, while others are occurring between natural phenomena and artificial satellites or space vehicles. Some of these orbiting particulates are the detritus left in the wake of a satellite's lifetime; while others represent the debris from satellite explosions or impacts, orbiting fragments of rockets. But they all can cause damage to operating spacecraft. Consider that a particle as small as .1mm can cause surface erosion on the surface of an object, and cause even more serious damage to a spacecraft over time. And, it can cause life-threatening damage to a spacesuit. Even a slightly larger particle that is traveling at 6.2 miles per second has the kinetic energy of a cannon ball, and a 1cm particle would have the energy of a 50 pound safe thrown at the same speed (NASA Technical Memorandum 100 471). The hazard posed by these high-velocity particles can be partly mitigated by the use of so-called "meteor bumper" devices. Such technology is currently being used on such spacecraft as the International Space Station. However, this protective technology is still not able to protect certain vital components of orbiting spacecraft, such as solar panels and optical devices (that is, telescopes and star trackers). Therefore, it can be assumed that these components are still subject to ongoing wear and tear by contact with debris. The bottom line is that this class of foreign objects, as well as the natural debris that we call micrometeoroids, continues to be a major element in the military geography in space.

Explosions which are initially caused by kinetic energy or chemical reaction can be the catalyst for a succeeding chain of collisions and fragmentations. In such events there could result a cloud of shrapnel throughout a region of the artificial satellites and space vehicles in orbit. This phenomenon is known as the "Kessler Syndrome," which is named for the NASA scientist who first discovered the potential for this type of disaster in space. The primary collisions of the bodies in such a situation will produce many incidents of collateral damage to many satellites or space vehicles which are located within a distant proximity from the center of the kinetic-energy explosions. The upshot of such as scenario would likely be one in which a given Earth-orbit zone essentially becomes impassable and unfeasible for continued orbiting satellite operations. The orbiting satellite which is considered to be most at risk from such collision-events is the Envisat Satellite, an environmental monitoring system whose mass is approximately eight metric tons. It is now in an orbit whose mean altitude is 480 miles; which also happens to be a region where the amount of debris accumulation is the greatest, and still growing.

Natural meteoroids also have historically been a design consideration for spacecraft, so as to minimize the collision hazard risk. A meteor refers to a particle or small chunk of stony or metallic matter that travels in seemingly haphazard trajectories throughout the solar system. Some of these enter the atmospheres of the planets and vaporize. I see the threat that derives from meteor showers or individual meteors as being similar to that which results from anti-aircraft artillery projectiles (or "flak"). In one sense, the level of the threat (probability) is a function of the location of the spacecraft within the solar system. This probability can be determined mathematically through the use of models that have been developed for flak analysis within a given space event, for a particular space mission. Another is to constantly scan for patterns in terms of density within a volume of space. As part of the strategy of early detection, NASA scientists have determined that, given the nature of space missions today, the most effective way to deal with the potential risk of a spacecraft colliding with a meteor is through avoidance and evasion. In practice, this means that mission planning should probably deal with the risk of meteor showers in the same way that is done with respect to storms and turbulence within the atmosphere of the Earth. Should the trajectory of a particular meteorite in space pose a hazard to a mission, then detection and avoidance maneuvers is seen as best way to minimize the risk of a collision.

But, as is the case in atmospheric operations, there is always a search for ways to minimize the damage from collision events through the use of structural reinforcement or protective materials. One such countermeasure that has been developed is the so-called Whipple shield, which is named for its inventor, Fred Whipple. It is a type of hypervelocity impact shield that is used to protect manned and unmanned spacecraft from collisions with micrometeoroids and orbital debris whose velocities generally range between 2 and 10 miles per second. Unlike the monolithic shielding of early spacecraft, Whipple shields consist of a relatively thin outer bumper placed a certain distance off the wall of the spacecraft. This improves the shielding-to-mass ratio, which is critical for spacecraft components, but it also increases the thickness of the spacecraft bulkheads, which is not ideal for fitting spacecraft into launch vehicle payload space.

———

Another kind of hazard is posed by radiation storms in space; most of which are caused by eruptions on the surface of the Sun or disruptions among stars. These eruptions often produce mega-winds and violent sandstorms and static electricity charges, which affect entire planets within the solar system. The degree of protection from all this space weather that is provided by the Earth's magnetosphere and atmosphere will no longer be available in space. Therefore, operating outside this protective bubble in space is like going out into a severe thunderstorm without any protection from the wind and the lightning. Fortunately, as a result of the ongoing NASA missions in space, we now have a better understanding of the practical effects of the charged particles on astronauts in space. Their negative effects on man-made machine systems also are becoming clearer, and are being incorporated into the design of the hardware systems. The negative effect on communications by the radiation particles is another area in which considerable research and development has been done to counter these effects. Thus, for example, shielding and system redundancy strategies are being employed to minimize the effects of the interaction of the charged particles with the geomagnetic shield and the ionosphere, which can disrupt radio communications between space vehicles and ground stations.

A more acute hazard profile occurs in the case of solar flare eruptions. This sudden and intense brightening on a small part of the Sun's surface produces energy fields that consist of ultra-violet radiation and intense X-rays, as well as cosmic rays and less energetic particles. Since the particles travel at almost the speed of light, they reach the vicinity of the planets of the solar system some days later, the actual time depending on the planet's distance from the Sun. But the more constant solar winds also pose a threat to military space operations. This wind is free-flowing plasma, made up primarily of electrons and protons, and when the solar wind encounters a magnetic field, a shock wave develops. However, it appears that its effects on space operations or even the full scope of its nature is still not understood. With respect to a planetary magnetosphere, when the wind interacts with this shield, we know that it is diverted around the sphere, travels to the extent of the heliosphere, and eventually diffuses into galactic space. The solar wind is also thought to be responsible for deflecting both the tail of the Earth's magnetosphere and the tails of comets away from the Sun.

With respect to the Earth, one practical consequence of the occasional acute eruption of solar wind is that the Sun is episodically ejecting billions of hot hydrogen gas in our planet's direction. About three days following the eruption, this huge mass strikes the geomagnetic field of our planet. Within seconds following this collision of fields, immense electric currents surge and strike the electrical grids on the surface of the Earth; transformers are fried and relays melted. Such a degradation of a global power grid would then severely affect both civilian and military operations, in the atmosphere and throughout the geospace domain. The orbiting artificial satellites would also be subject to significant damage or destruction. Worse, if a satellite should be struck or even be diverted out of its planned orbit, it is possible that the satellite could crash onto the Earth's surface. At best, the satellite-based navigational and communication systems devices also would virtually be rendered useless. Computer systems, including the internet, no longer would be able to function normally, until human intervention could restore the various servers and other nodes of the hardware systems to their normal state.

A generator of "weather" within the magnetosphere that envelops some of the other planets of the solar system is the electromagnetic "storm." This storm is created by the interaction of a planetary magnetic field and the incoming solar winds. These might be likened to the collision of storm fronts on Earth, which result in a kind of maelstrom of wind, precipitation and atmospheric pressure. Indeed, based on a careful study of the NASA space operations over the last half-century, I have come to believe that the most significant "geographic feature" in space operations – after the Sun itself – might be the magnetosphere. This is the balloon-shaped body of charged particles that are produced by the interior core of some of the planets, and which are further shaped by the solar wind which is emitted by the Sun's surface. Initially, the magnetosphere deflects most of the radiation of the solar wind before it reaches an intermediate exo-space, such as an atmosphere which blankets some of the planets, most notably, the Earth. The deflected radiated particles then go around the target planet, forming the bulbous plasma around the planet, and finally recede so as to form a comet-like tail on the far side of the planet.

The reason why this geographic formation is so important in military space operations is that it is basically like a "cloud" of charged particles that can negatively affect both astronaut and spacecraft that is operating within it. Some efforts to gain a better understanding of the extent and nature of this threat to military operations is space have included Project Argus (1958). These were a series of tests to gain knowledge about the formation of radioactive belts and any possible military applications. The project involved a series of nuclear explosions of nuclear missiles over the South Atlantic Ocean by the U.S. Defense Nuclear Agency. These explosive events were monitored and measured by the Explorer 4 space mission to gather data about the radiation effects of the explosions from the perspective of space. The U.S. Atomic Energy Commission observed the tests gain better understanding of their physiology in space as compared to previous nuclear explosions on Earth. Satellites were also used to measure the electron density over time, and included the use of a magnetometer for measuring ambient radio noise.

In the case of the Sun's magnetosphere, it is defined by field lines that protrude from and loop back into the burning ball of gas. Sometimes these field lines snap and then join other nearby lines, thereby releasing energy that can then launch bursts of plasma known as solar flares. Large chunks of this plasma from the Sun's surface can then flow toward Earth and damage orbiting satellites, or even bump them off their orbital paths. Like the complex radiation formation that is known as the magnetosphere, the solar storm itself affects man and machine, particularly the electronic circuits that are the heart of so many military systems, on the surface of the Earth, in the transition region of "geospace," and in outer space as well. In January of 2012, the largest solar storm since 2005 occurred, and it generated some of the most spectacular northern lights in recent years. Equipped with space-based sensors that operate in the infrared phases of the electromagnetic spectrum, scientists were able to determine the source of that storm as being the Sun's magnetic field.

This flux of radiated energy and particles presents invisible field lines that protrude from and loop back into the super-heated ball of gas. Sometimes this field lines break and join with other nearby lines, thus releasing energy that can then launch bursts of plasma known as solar flares. Huge packets of plasma from the Sun's surface can zip toward the Earth over a period of about 20 days, and damage orbiting satellites or even divert them from their paths. These packets of plasma, known as coronal mass ejections, can also snap the Earth's magnetic field lines, thus causing charged particles to speed toward Earth's magnetic poles, ultimately producing the northern and southern lights. Even though the process of field lines breaking and merging with other lines (called magnetic reconnection) has such significant effects, a detailed picture of what precisely is going on has long eluded scientists. More recently, however, using high-speed cameras to look at jets of plasma in the lab, scientists have discovered a surprising phenomenon that provides clues to just how magnetic reconnection occurs.

With respect to the planets themselves, the gaseous exteriors of some planets also can pose a hazard to space operations to any spacecraft that attempts to transit them. The same holds true when or a probe descends slowly through them in order to study them or to attempt a landing on the hard surface of the planet. Consider the case of the layers of clouds and gases that cover the surface of the planet Venus. Within this most massive atmosphere of the four terrestrial planets, there is a mix of gases which includes more than 96% carbon dioxide and 3.5% molecular nitrogen. There are also trace amounts of other gases, including carbon monoxide, sulfur dioxide, water vapor, as well as argon and helium. As a result of several flyby and orbital missions, as well as a few attempts to inject a probe into the atmosphere of Venus, it has been determined (at this point in time) that this planet has the most hostile atmospheric regime for both humans and machines in the solar system. This aspect of planetary weather outside the Earth will likely pose significant challenges to any future "amphibious" military space operations in the solar system, wherein a "lander-rover" spacecraft is sent through the gaseous environments.

The potential effect of the atmospheric regime of Venus on military space operations also can be inferred by the experiences of NASA and other space agencies, as they have tried to explore and study the gaseous envelope around the planet. In 1962, NASA launched Mariner 2 in order to attempt to gather data about interplanetary space and on solar activity as it headed toward Venus. During this mission, the spacecraft took direct measurements and made several other observations during a 45-minute flyby. It was the first successful interplanetary spacecraft to continue past Venus and toward the Sun. From this mission, NASA has learned that Venus lacks a strong magnetic field, that its extreme heat is generated within its lower atmosphere or its surface, and that the radiation levels in the space between Earth and Venus are no more dangerous than the levels of radiation anywhere else outside the Earth's protective magnetosphere. Perhaps the most important lesson of the Mariner 2 mission was the conclusive evidence that interplanetary space is neither empty nor field-free. Rather, it is full of particles that have a wide range of energies (nasa.gov).

Jovian weather is predominantly global in scale, as opposed to the local weather that occurs on Earth. On Earth, huge spiral cloud systems often stretch over many degrees of latitude and are often associated with motion around high and low-pressure regions. They are much less zonally confined than the cloud systems on Jupiter and they move in latitude as well as longitude. Similarly, on Earth, local weather is often closely related to the local environment which, in turn, is determined by the variable nature of its land surface. Jupiter has no solid surface and, therefore, no significant topographic features. At the same time, the planet's global circulation is dominated by latitudinal currents. The upshot of all this is that there is great persistence in the currents and their associated cloud patterns on that planet. The Great Red Spot, as an instance, moves in longitude with respect to all of the planet's rotation systems, yet it does not move in latitude.

––––––––

As a geographer, I see the interplanetary medium as another example of a geographic region. Indeed, in some ways the IPM may be likened to the medium of liquid water that is found in the Earth's oceans. That is, it too is a region of thinly scattered matter and energy, which is interspersed with relatively massive celestial bodies that are composed of rock, dirt, ice, and water, and gas. Many of them are enveloped by a "thicker" gaseous medium which is held down by magnetic forces emanating from the cores of the bodies. In a sense, the bodies resemble the continents and islands on the surface of the Earth. As humans have ventured more and more into the domain of the IPM, it has become apparent that the interplanetary space is far from being empty. Indeed, it contains: electromagnetic radiation which also is known as the solar wind; cosmic rays; microscopic dust particles; and electric and magnetic fields. Beyond that, we are now discovering other entities there, such as "dark matter" and "dark energy."

Like most physiographic regions, the IPM also contains subregions. Some of these are called fields, which are comprised of charged particles in various states. These charged particles are packaged as photons and are propelled through the medium by electromagnetic interaction and the solar wind. As these charged particles move through the medium, they sometimes encounter planetary fields of charged particles which also are in flux. The interaction between these flows then creates bodies of charged turbulence (reminiscent of the turbulent waters in an estuary). Sometimes these flows are referred to as plasma, which is made up of electrons, protons and other ions that often combine with the solar wind. There are also flows of cosmic rays that originate outside the solar system. Some of these are the so-called galactic cosmic rays that are thought to be produced as a result of stellar processes, such as supernova explosions. Finally there are the so-called micrometeoroids which are often as small as a few hundred micrometers in size, but which nevertheless pose a significant collision hazard to spacecraft and their payloads because of the high speeds at which they travel. On impact, these tiny particles – whose speed has been measured in the tens of miles per second – can puncture a vital component of a spacecraft system, or create a cloud of transient ions that can short-circuit an electrical system. For this reason, protection against micrometeorite impacts has become a major element of space hardware design. For example, many spacecraft are being equipped with dust bumpers, or "Whipple Shields," to guard against damage to the superstructure from colliding debris. Spacesuits that are used in extravehicular activities in space also incorporate protective measures in their outer layers.

Space missions have been launched to study the particles of the IPM region. One such mission in 2004 flew past a comet and collected particles from its coma for study on Earth. Another space mission, launched by Japan's space agency in 2003, was able to capture small amounts of material comprised of fragments and dust from a near-Earth asteroid for laboratory analysis. Just as was the case within the Earth's atmosphere, the IPM contains both clouds of microscopic dust particles and fields of charged particles. The planets and some other objects in the solar system also can produce dust through such processes as crater impacts and volcanism. However, most particles are the result of surface erosion and collisions of asteroids, and from the comets which expel dust and gas when they are orbiting near the Sun. Another source of particulates is the solar wind, which contains a flux of particles, chiefly protons and electrons, together with nuclei of heavier elements in smaller numbers. This wind is sometimes described by scientists as being "collisionless" plasma. The solar wind also represents an extension of the Sun's magnetic field which is congealed into a highly-conducting fluid. This overall substance is then accelerated by the higher temperatures of the solar corona, or outer region of the Sun, sometimes achieving velocities great enough to allow particles and photons to escape from the Sun's gravitational field. The outflow of plasma also transports the magnetic fields of force that are present at the surface of the Sun, and then radiating them away from it. It is also responsible for deflecting the tails of the Earth's and other planetary magnetospheres and the tails of comets away from the Sun. Eventually, when the solar wind happens to encounter a planet's magnetic field, a local shock wave is created. The portion of the solar wind that does not interact with a planet continues to travel to a distance of approximately 20 AU; then it begins to cool and eventually diffuses into galactic space. The magnetic field lines that are carried outward from the Sun by the solar winds remain attached to the surface of the Sun, but, because of the rotation of the Sun, the lines are drawn into a spiral formation. Closely associated with the interplanetary magnetic field are electric forces that act to either attract or repel charged particles.

———

The emission of light and heat from the Sun also acts as the basic engine that drives what we call "weather" on the various planets of the Solar System. From its points of origin along the outer surface of the Sun, electromagnetic radiation (light, ultraviolet, and infrared) generally flows outward, without great variations in intensity, until it ultimately merges with the intra-stellar medium. Aside from its encounters with the celestial bodies and especially with the magnetic fields that envelop some of them, the main degradation in this flux is simply a function of distance from the Sun. A lesser amount of cosmic radiation from the stars also is combined with the solar radiation to create periodic acute fields. This results in a model of the solar system which consists of a uniform space, which is void of planets or other bodies. The total radiated energy from the Sun is propagated without variation, except for the dissipation which is a function of distance from the origin of the energy. If we add the planets to our model, then the total radiation can be said to be a "solar constant" of radiated energy from the Sun, per unit of time and area, on a theoretical surface that is perpendicular to the Sun's rays and at a planet's mean distance from the Sun. This value would be modified if we were to introduce atmospheric effects and seasonal changes in distance from the Sun. In fact, one definition of the solar constant is that it is the "climate" of the interplanetary medium, within which "weather" occurs to affect military space operations outside the atmospheres of the planets. Thus, it can be said that the geographic phenomena that we call "weather," on Earth, and which interacts with air operations within our atmosphere, also occurs throughout the rest of the solar system; however, in outer space the major elements are temperature and electromagnetic radiation.

Weather also occurs on other planets and bodies of the solar system. There are differences from Earths atmosphere, however; the main one being the fact that the weather on the other objects tends to be more extreme. One major reason for this is a function of the durations and intensities that occur in planetary rotation. Then there is the matter of angle of incidence (with respect to the ellipsis) of a planet's axis. This angle varies considerably among the planets, and Neptune, for instance, presents a longitudinal axis that almost lies flat on the ellipsis; on its side, if you will. The effects of gravitational influences, or tidal effects, from nearby planets and moons also can affect weather patterns. In addition to these external factors that influence weather on other planets, there are internal factors as well. These include such things as the local topography, emissions from internal heating core systems, and tectonic activities. The center of all these external and internal variables happens to occur in the atmosphere and/or the surface of the planets; this is what is referred to by the term, weather.

So, it has become apparent that a useful method for comparing weather elements, between Earth and the other planets is to compare the extremes that occur in the population. We traditionally have marveled about the extreme manifestations of weather that occur from time-to-time at various places on our planet. These include such phenomena as: the great concentrations of lightning, which light up the skies over the Patagonia region of South America; the extreme cold is that is experienced in places like Vostok, Antarctica, where temperatures have dropped to -129 degrees (F); the extremely high temperatures that are experienced in Death Valley National Park in California, which can reach temperatures as high as 136 degrees (F); the powerful tornadoes of notorious "Tornado Alley" of the southern plains of the United States, which extend from central Texas to Nebraska. The winds that are associated with these tornadoes have been measured at speeds of 300 miles per hour or more; and the hurricanes of the eastern seaboard of the U.S. and the Gulf of Mexico littorals which also generate winds of more than 100 miles per hour. But these extreme weather phenomena on Earth are dwarfed by those that occur on some of the other planets of the solar system.

So, weather also is a critical factor when designing a space mission, or conducting operations on a planet, moon, or asteroid. It is a major consideration in the selection of a landing site on these bodies that lie outside our planet. Indeed, the presence of dangerous dust devils was one of the most important factors for choosing a landing site near the north pole of the Red Planet for the NASA 2008 Phoenix Mars Lander. Lightning on Saturn poses a tremendous risk to the electronics of any future atmospheric probes on that planet. The Great Red Spot also presents its own complex of extreme weather which any lander or balloon would have to face. And then, there are the complex layers of thick clouds on Venus, which consist of a toxic mixture of sulfur dioxide and droplets of sulfuric acid. It is these clouds which block the view of the surface and thus make any attempts at landing on the surface almost like flying in the blind; they rain sulfuric acid, but the rain is evaporated by the heat and, therefore, it never reaches the surface; it simply rises again into the higher atmosphere and becomes part of the clouds again. Because of the obfuscating nature of the clouds at the visible-light spectrum, radar is necessary to conducting empirical observations of the surface.

The weather on Mars is basically driven by the Sun, in the same way as occurs on Earth. The Sun's heat warms the spinning, tilted planet in an uneven manner, and the heat is distributed from the relatively warm equator to the colder poles by a complex mix of atmospheric movements. Clouds of ice form high in the Martian air, driven by winds that can gust up to 50 mph on the surface, and more than 125 mph within the great dust storms that have been seen occurring on the planet. One of these, a global dust storm which enveloped Mars in 2001, as the southern hemisphere's spring was beginning, was captured by the Hubble Space Telescope. The storms covered the planet for nearly two months. Like storms on Earth, those on Mars usually occur at particular latitudes. Most are short-lived, even though they can cover vast areas of the planet. Earlier, in 1999, the Hubble Space Station captured images of a giant cyclone that is located near the planet's North Pole. It also occurred during the summer of the northern hemisphere, when the planet's weather is further agitated by increasing temperatures. These storm clouds contain mainly water ice crystals that have evaporated from the North Pole ice cap and then have refrozen high in the atmosphere. The Mars Reconnaissance Orbiter's camera has captured images of raging dust devils on the Red Planet. These dust phenomena can easily rise to altitudes of as much as 6 miles above the surface, and are driven by westerly winds. One such dust devil has been seen towering to 12 miles above the surface of Mars as it sweeps across the northern region of the Amazonis Plains. Generally, NASA mission reports indicate how the pervasive dust can cause problems to exposed hardware and space suits, but these storms actually have proven to be a positive factor in the operation of the rovers; that is, they have served to clear accumulated dust off the solar panels of the Spirit and Opportunity rovers.

Several manifestations of many interesting weather phenomena have been capture by imagery from spacecraft, during the last few decades. For example, a mosaic combining several images of the Great Red Spot of Jupiter has been produced from imagery that was captured by Voyager 1 spacecraft in 1979. Little is known, at this time, about the actual weather conditions on the so-called Jovian planets: Jupiter, Saturn, Uranus, and Neptune, mainly because there are no local "weather stations" that can record temperature, atmosphere, precipitation and winds at a given place on these planets. However, we can infer the broad outlines of those atmospheres, based on the data that we have garnered from telescopes and space probes. This gives us some notion of global weather conditions on those planets.

The atmospheric structures of the four Jovian planets are quite similar. The primary difference among them is that the atmospheres of the particular planets get cooler with increasing distance from the Sun. On the other hand, the structure of the atmosphere of each of these planets is similar in all four cases. For example, clouds of a particular composition always seem to occur at approximately the same temperature, but within them, each of these so-called Gas Giants presents distinct cloud layers. Generally, these layers are presented in accordance with the atmospheric levels at which various gases can condense into liquid droplets or solid flakes. Thus, on Jupiter and Saturn, the cloud layers include ammonia clouds or water clouds, depending on the atmospheric temperature. On the other hand, Uranus and Neptune cloud layers are mainly methane clouds. Methane can condense in the very cold upper troposphere of both Uranus and Neptune, but not in the warmer troposphere of Jupiter or Saturn. For its part, Neptune's atmosphere is banded, and includes an extremely high-pressure storm, known as the Great Dark Spot; it is similar to the Great Red Spot on Jupiter that is found in its southern hemisphere.

Some elements of weather phenomena are found on all four of the planets. But there also are many variations among the Jovarian planets. Jupiter's clouds, winds, and storms are the strongest and most active of the planets in this subregion of the solar system. Saturn has stripes of alternating color and wind direction, similar to those of Jupiter, but less vivid. Even though Saturn has an axis tilt angle similar to that of Earth, there are not the clearly defined seasons that might be expected. Some weather changes have been observed, but Saturn's internal heat keeps temperatures about the same the year-round and global. Although great bursts of lightning can occur on all the planets (except Mercury), on Saturn these electrical discharges are much more powerful than on any of the others.

Much of what we have learned about lightning phenomena in the solar system has come from NASA missions. As an example, the Cassini spacecraft is particularly well-equipped to detect and measure these occurrences because of its on-board Radio and Plasma Wave Science instrumentation. With it, Cassini is able to detect radio bursts which are emitted from lightning at great distances; as much as 55,000 miles above Earth. This is the instrument which also has measured the radio pulses of Saturn, as far out as distances of about 100,000 miles from the ringed planet. These have witnessed the so-called "Dragon Storm" which occurs in Saturn's "Storm Alley," wherein tremendously strong radio emissions occur. In one particular case, extraordinary strong radio emissions were detected in September of 2004. These are thought to have been the output from intense electrical discharges that were associated with these storms. When Voyager 2 flew past Venus in 1986, Uranus's northern hemisphere was facing almost directly toward the Sun, and its southern hemisphere was perpetually dark (the result of the fact that this planet is essentially tipped on its side), and also due to the planet's 84-year orbit around the Sun. Imagery has revealed the almost complete absence of clouds and no banded structures such as those that occur on the other Jovian planets.

With respect to military space operations, it is important to know about the general patterns of weather on the other planets of the solar system. However, the only "information," or intelligence that can prove useful in the conduct of military space operations is what geographers call "micro-weather," or local, real-time weather conditions. Thus, while data such as average planetary temperature are useful for general knowledge or for comparative analyses, it is the local package of temperature, humidity, pressure, and wind that is crucial for effective planning of military operations. As we saw earlier, it is the environment that is presented within every cell region, throughout the sortie that matters, at any given slice of time. Fortunately, NASA, the U.S. Air Force, and NOAA (U.S. National Oceanic and Atmospheric Agency), among other agencies, have been busy deploying a comprehensive weather monitoring system in the geospace domain in order to gather the necessary continuous and long-term data sets. These will enable military planners to better understand the weather processes and their manifestations within each planet. More specifically, the mission planning that is being carried out by NASA will be of great value in the mission planning for military space operations.

The Lithospheres of the Solar System

During a past life as a U.S. Air Force Medical Corpsman, I came to realize that there were various diseases that could be diagnosed by a physician, simply by examining signs and symptoms that were presented on the patient's skin, or epidermis. The various and sundry blemishes, boils, scars, and other "signs" which were evident, often were also manifestations of internal processes. Similarly, the apparent wounds, cuts, scratches, disfigurations, scars, and other external markings on the skin often told the patient's life story. This medical history included the developmental history and past diseases and traumas. In similar fashion, the planets and other bodies of the solar system tell their respective life story through the state of their outer skins, the lithospheres. Rifts, craters, volcanoes, and lava flows – to name but a few – that are presented by the planets and other celestial bodies in the solar system present a history that can be utilized by the diagnostician. The scientists who are trained to read these presentations are the geologists and the geographers. And once again, the knowledge, techniques, and models that have guided the study of the lithosphere of our own planet are now being applied to the other bodies of the solar system in order to better understand the history and present condition of their lithospheres.

Thus, what we know about the crust on Earth today has been accumulated from many different data which have been gathered during human history. The earliest exploration activities were probably done in search of water or minerals, and involved relatively shallow horizontal and vertical shafts. By the middle of the 19th century, humans were digging much deeper, both downward and horizontally, again in search of minerals and, now, petroleum and gas. The earliest methods for understanding what is occurring under the surface of our planet can be best characterized as being careful, empirical examination of the rocks and soils, but only as a side effect of mining and well-digging. The scientific observation and analysis of the part of the Earth that lies beneath the veneer also began in the 19th century, when the sciences of mineralogy, geodesy, and stratigraphy arose. These sciences still were mostly concerned with the minerals; that is, the inorganic elements or compounds that have fixed chemical composition. Other scientists, such as the geographer, geologist and the archeologist, as well as others, have been more likely to study the spatial aspects of the strata under the surface of the Earth, as well as the intrinsic attributes of the matter.

One of these scientists was Alfred L. Wegener, who was concerned with the crust of the Earth on a macro-scale. In 1910, he began to study the spatial movements of the continents over the past hundreds of millions of years. In effect, he was studying the historical geography of the continents that occur on the outer crust of the Earth. Because he lacked much of the modern technology for examination and analysis of the subject matter, the methodology that he used for developing his theory of plate tectonics was very much like that of other Earth scientists of ancient times. That is, he made multiple empirical observations, compiled a data base of his sample of observations, and then applied human brainpower of induction and deduction to arrive at a solution. In short, what he did is what all scientists do when presented by a set empirical data: he tried to make sense of it all. In Wegener's mind, he visualized a terrestrial crust that once had been whole, but which had subsequently been broken up into several "plates" over geologic time. From this he postulated that there is an internal global force which causes these plates to move about the surface of the planet. We now refer to this force as the convective movement of material from the core to the sub-dermal layer (the mantle). Partly as a result of this conveyor belt movement of molten material to the surface, "new" crust is being created, and "old" crust is being returned to the depths of the Earth. The upshot of this vertical convection is that the denser continents actually are "floating" on the more fluid mantle. Today, this process is known as "plate tectonics." There also are secondary tectonic forces, which include such events as earthquakes and volcanism, traumatic strikes by other bodies, the effects of tidal stretching and other deformation caused by adjacent bodies, and the persistent weathering caused by ice, water, winds, and temperature.

The initial reaction of his scientific efforts was a rejection of his theory by his colleagues, basically because the modern tools of scientific measurement were not available to him to buttress his claims. Ultimately, however, science and technology caught up with inferential logic in the later 1960s. Today, the theory of plate tectonics is generally accepted by the scientific community. The model itself is now recognized as a valuable unifying model which can explain all manner of geologic and geographic phenomena. Much of the detail work on the "tectonic plates" has been done in the latter half of the 20th century and into the present day, by scientists who have at their disposal the use of sophisticated imaging and analytical technology of the space age. Thus, for example, Earth-orbiting satellites, with their enhanced sensors, ground-penetrating devices are only now available to deal with the great amount of continuous streams of data that are being collected. These data have served to confirm much of the work of Professor Wegener.

What happened in the case of the development of the theory of plate tectonics was that science and technology finally advanced to the point where it could verify the original hypothesis. In the main, this is now possible because scientists today have at their disposal much more powerful technologies with which to analyze what the anatomy and physiology of not just the Earth, but the other planets and celestial bodies of our solar system as well. One of these is the seismograph, a device that can detect and measure any movement beneath a planet's or moon's surface. Like the electrocardiogram of the heart, this is a tool that can indirectly provide information of what is occurring in a planet's interior. More importantly, the seismograph provides a continuous record of the pulse of the Earth and other planets. Another example of modern technology that is available to the scientist today is the use of radar energy to peer into the depths of the surfaces, both on continents and ocean basins. This use of the so-called "sounding radar" enables the scientist investigate the makeup of sedimentary deposits that lie deep below the surface of the Earth. Beyond that, there now are computer technologies that make it possible to store vast amounts of data and imagery, to process it with spectrograph equipment, and to relay these over great distances, as needed. With the latter technology, scientists have been able to develop characteristic "signatures," which they can ascribe to certain types of rocks and soils, as well as their composition.

During the latter decades of the 20th century, the U.S. space agency, NASA, began to develop a wide range of tools and "suites" of instruments for use by space probes and surface rovers. Among other things, these were used to study the crust of the other bodies of the solar system, especially the Moon and the planet Mars. One of these is a specialized spectrometer, which measures the transmission of gamma rays from a radioactive source. Another is a magnetic device, which has been used by the rover, Opportunity, to draw samples from the dust on Mars, which have been determined to contain a magnetic component called titanomagnite. Also, a small amount of hematite has been found; an indication that there may have been liquid water in the early history of the planet. On the other hand, the Rock Abrasion Tool (RAT) instrument that is carried by the rover, has detected olivine, which is seen as a sign that a long arid period has occurred on the planet during its developmental history. Yet another suite of instruments that has been employed by the NASA rovers includes the Alpha Particle X-ray Spectrometer (APXS) and the Thermal Emission Spectrometer (nasa.gov).

Sometimes "spade work" which exposes what lies under the surface of the Earth is done by nature. It may occur in one cataclysmic event, such as when an asteroid or other body collides with our planet and leaves a crater. More often it is the result of millions of years of work that is performed by the dynamic systemic processes of a planet. One example of the latter is the work that a river system has done to expose strata that represents millions of years of geological formation within the Grand Canyon in the United States. Another would be the exposed strata along rift valleys and in mountain ranges. Water also has worked on certain types of rock, such as limestone, to form caves and sinkholes that also expose millions (if not billions) of years of planetary development. And, other times, the wind operates in conjunction with water to expose rock and soil strata that can be examined by scientists. All these forces enable the direct, close investigation of the crust of our planet and, now the other planets, moons, and asteroids in the solar system.

However, much of the work of the geologist and the geographer on Earth is still being done by humans in the field. So, in this sense, the technologies and devices that have been developing can be seen as artificial "enhancers" that extend the powers of the human body. However, this "hands on" approach is being found to be problematical on the other celestial bodies of the solar system. The essence of the matter is that atmospheres, and the surfaces of the other planets and celestial bodies of the solar system, can be said to be encapsulated with the term "alien." This is especially true with respect to such elements as oxygen, temperature, and pressure, and their effect on human life forms. On the other hand, human activity within hostile environments is really nothing new. Many of the environmental extremes that are found on Earth, such as the Antarctic or the Atacama Desert, or that which is found beneath the surface of the oceans, or within the domain of the volcanoes have been sites of considerable "high value" human activities (mainly resource extraction or military imperatives). The input factor that made these activities in these "alien" environments on our own planet possible has been technology (and money). Now, we see a replication of this technology-money model on the planet Mars, with the utilization of in situ geological work that is being done by the Curiosity rover.

In a sense, the Earth's crust is a baseline phenomenon that can be used to compare other planetary crust systems. Once the layers of the lithosphere had been sorted out into vertical layers, the Earth wound up with an overall relatively uniform crust; one that was enriched by lighter elements that floated on an upper mantle made of denser materials. Later, a more distinct horizontal differentiation also developed as continents began to form. That is why the continents comprise a wide range of rock types, including granitic igneous rocks, sedimentary rocks, and metamorphic rocks. The first continental rocks were the result of repeated melting, cooling, and remixing of oceanic crust, driven by volcanic activity above mantle convection cells, which were much more numerous and vigorous than those of today. Each cycle has left more of the heavier components in the upper mantle and concentrated more of the lighter components in the crust. The first iterations of continents grew as lighter fragments of crust continued to collide and fuse. The resultant thickening of the crust led to increased pressure on the bottom layers, causing melting at the base, and thus resulted in the formation of the foundational granitic igneous rocks. At the same time, weather accelerated the process of continental rock formation, and retained the most resistant components.

From our new vantage point in space, we have found that only the tallest of mountains on Earth can be clearly seen by the observer with the unaided eye from that perspective. Indeed, with the persistent cloud cover and other particulates that are present within the Earth's atmosphere, it is difficult to get a very clear view of the surface of the planet even with telescopic technology. It is only when the observer uses multi-spectral remote sensor systems, such as those that are sensitive to the radar and infrared portions of the electromagnetic spectrum, that truly effective observation and analysis is possible. At the same time, radar, infrared, or other similar remote sensing technologies are also valuable in appreciating the differences in relief that exist on the surface of the Earth. It is this sort of technology that makes possible long-term orbiting observation of the Earth's surface, which provides us with the ability to observe the diurnal, seasonal and yearly cycles of the other planets. Even then, a highly sophisticated analysis of the imagery by specialists, who can detect the nuances on the surface, is required to develop actionable intelligence about the object of study.

From this extraterrestrial point of observation, one can appreciate that the continents are not at all relatively flat "planes." Rather, there are considerable differences in the local vertical distances between adjacent parts of the continental surface. These include the features that rise to relatively greater heights than the surrounding terrain, which are described as "mountains" and their congregations known as "ranges." Also at this scale of observation, it becomes feasible to record the whole physical region, in one imaging run. Compare this with what the traditional geographer had to do, which essentially was solving a jigsaw puzzle, done by organizing the batches of spatial data as they became available. From such a mosaic, the maps and charts were created.

Also, from the perspective of a platform which is orbiting a planet, at a distance of about 200 miles, a first glance of the surface of the planet reminds the observer of the skin of a veteran warrior. One sees craters that have been gouged out of the planet's surface by repeated collisions with other bodies. The dry streams of lava indicate that volcanoes and other thermal processes have occurred there. The composite picture that emerges from all of the multi-spectral reconnaissance, and the geological and chemical analyses that have been performed, is that there has, indeed, been a war among the various bodies of the solar system. The net result of the external and internal forces that have occurred during these wars is the composite of the surface features that are seen today. Meanwhile, analyses of the samples of rocks and dirt – taken at various points along and within the surface crust – indicates that planets have gone through the many phases of development since their first coalescence from the original mass of gas. Therefore, it is apparent that the outer surface (as well as the subsurface) of the planets and other celestial bodies has been undergoing extensive and, sometimes, cataclysmic internal remodeling throughout the last four billion years.

————

An extra-planetary analysis of the mountains on the surface of the various bodies reveals some interesting patterns. One is the tendency of some mountains to occur in groups, and often in what look like chains of high protuberances relative to the surrounding landscape or seascape. Then there is the propensity for these mountains to occur along lines which correlate with the seams of the tectonic plates, which indicates the nature of their birth and developmental growth. Other mountains, often appear like lone sentinels over a surrounding plain, betray more localized mechanics, such as violent eruptions of material from the deeper cores of the body. At least on Earth, sometimes, these "volcanoes" emerge from the deep in the oceans to form an island, or a chain of islands. In any case, these eruptions are generated by the super-heated core of a planet. More generally, the volcanic features are a manifestation of the cyclic movement of material from the core to the surface, and then back down to the internal regions from whence it came. In the meantime, the tectonic plates are motivated by this internal energy to "drift" from place to place (in geological time) and, in the process, shoving and pushing huge areas of surface material that happened to be in the zones of subduction and collision. The upshot of all these tectonic processes has been the emergence of great "heaps" of folded and uplifted material which form of mountains.

At the same time, from the global scale that is provided by the perspective of a space vehicle that is repeatedly orbiting a planet along several different trajectories, a cluster of mountains can be appreciated as a whole and in context by the observer. For example, on Earth, one of these chains of nodes forms a "spinal backbone" along the western edges of the two Americas, generally in a north-south direction. These are collectively referred to as the Rockies in the North America, and the Andes in the southern landmass. A lesser (more mature or worn-down) chain of mountains run longitudinally, near the eastern edge of the North American continent. These are the Appalachians, a generally lower and less massive chain which forms an eastern buttress of this landmass. Actually, it appears to submerge under the Atlantic Ocean as it turns eastward on its extension to the northern edges of the British Isles. On the continent of Europe, there is a mountain chain whose name seems to symbolize the continent as a landform and as a cultural icon – the Alps. These mountains extend along a central east-west axis of Europe, from the Iberian Peninsula to the region of Vienna, Austria. The Himalayas are the next great mountain system; it traverses Asia and forms a divide between the Tibetan Plateau in the north and the plains of the Indian subcontinent to the south. The trajectories and relative locations of these mountain systems do not appear to be random; rather they seem to correlate rather nicely with the "seams" between the continental plates on the surface of the Earth.

As is the case with most of the surfaces of a planet, the mountain domain does not end abruptly as a rule. Rather, it blends into the domain of the plain via a series of smaller protuberances, such as foothills. These transition zones can occur on land or under the surface of adjacent waters. Examples of the latter are the mountains of Iceland, which seem to plunge directly into the fjords that point toward the North Atlantic Sea. Mostly, though, the transition to the lower flatlands at the feet of the mountains is gradual. Thus, in their description of some of the mountains, geographers have noted that very often they merge into plateaus and high plains regions. Sometimes these relatively flat, adjacent lands appear to have suffered some trauma, which has resulted in such features as rifts or volcanoes. The latter can be seen as manifestations or symptoms of the roiling that is occurring beneath the surface, which is fueled by the heat of the inner regions of the planet.

There are forensic clues that indicate that all mountains, no matter how tough their rock material, are destined to eventually be broken into pieces and ultimately, to particles as small as grains of sand. The geologist reads the regions of sand the way an archeologist uses the remains of a civilization to tell its historical story. The sands lie at the end of geological spectrum of size, with the mountains on the opposite the end. But, there is a connection: the creation of a grain of sand signifies the very end of a process which is called "weathering," in which mountains are gradually (over millions of years) deconstructed. No matter how hard and resistant a rock may seem, it is inevitably vulnerable to the insistent and persistent work of the forces of reduction. These include gravity, collision, wind, and water in all its forms, as well as the chemical reactions that are provoked into action by the water, plants, and even microbes and other animals.

Then, there is the action of changing temperature and pressure, which causes expansion and contraction, which also contributes to the breakdown of the mountains. According to the geologists, the particles of sand, and the formations which they create, are perhaps more interesting and more telling than the monumental boulder. The accumulated on Earth, for instance, contain oxygen, which is the most common element that is present in the crust of the planet. Rather than occurring as a gas, however, it is chemically bound up with other elements to form solid minerals, in much the same way that sodium and chloride combine to form salt. Thus, what happens, in effect, is that nature is doing much of the basic extraction work of the geologist who is chipping away at the strata of mountains to obtain samples for analysis. Incidentally, according to NASA literature, the sand may provide the most efficient of the samples of material that can be gathered from the surface of the other planets.

———

Another geographic phenomenon that can appreciated globally, and only from an extra-terrestrial vantage point, are the larger areas on the surface of a planet that show little or no evidence of rainfall or other precipitation. Following many thousands of years of empirical observation on Earth, we have developed models which explain their patterns of distribution, and we also have learned a lot about their anatomies and physiologies. We know, for instance, that there are some deserts which experience only a very few inches of rain per year, in a scattered and seemingly capricious manner. Then again, there are deserts whose temperatures rise well above 100 degrees (F), during the daylight hours, only to plummet to below freezing or even subzero temperatures overnight. Some deserts, are located relatively adjacent to mountains. This spatial happenstance means that those mountains tend to force the upper movement of moist winds, wring the moisture from them, and thus cast an arid "shadow" on their leeward sides. Such geographic configurations, as well as the relative distance of a land surface from a water source (e.g., oceans or large lakes), often produce deserts. Some of these deserts only receive water from thin bands of rivers whose original source of moisture occurs outside the desert itself. Dryness does not necessarily imply that a desert is always hot. Some deserts, like the Gobi desert of central Asia, experiences temporal periods of cold; others are always cold (Antarctica, for example). So, we can say that, generally, deserts can be both hot and cold (but always dry) places, which cover several large areas of the Earth's surface. And then there are those which are high, cold, and dry at various times of the day or of the year. We also should note that some have proportionately less sand than rocks, of various sizes. These rocky deserts bear a close resemblance to the terrain that is found on the Moon; that is to say, a regolith surface. The overall conglomerate includes rock fragments in a continuous distribution of particle sizes, including a fine fraction of dust-like objects. However, only on the planet Earth, as far as we know today, is there a biological component.

Thus, it can be seen that observation of the planet Earth from an orbit of 200 miles in altitude enables one to appreciate distinct patterns in the distribution of the deserts on a global scale. Beginning at the 20th latitude (N), there can be seen the largest desert in the northern hemisphere of the Earth: the Sahara Desert. There also are extensive deserts on the Arabian Peninsula, which also lie astride the same general latitudinal zone. Along this latitudinal zone there also are extensive deserts in Central Asia, Pakistan and India. Most of these have resulted from several factors, but mainly from the fact that high-pressure atmospheric systems over these surfaces are unable to carry much water, and because of their sheer distance of the lands from the sources of water. There is also an interesting pattern that is presented by deserts that lie along the western areas of several of the continents. These include: the deserts of the southwest part of the U.S. and northwest Mexico; and the great deserts of western and southern regions of Australia. The explanation for this spatial distribution pattern has to do with the colder ocean currents that flow along adjacent coasts. The Atacama Desert is especially noteworthy. It is a thin strip of land which is bordered on the west by the Pacific Ocean, and the Andes Mountains to the east. Due to the fact that it has not received any rain in living memory, it is considered to be a useful analog for the Moon and the planet Mars. In fact, NASA has used this particular desert to train potential astronauts, and to test the vehicles and other machines which will be used on the missions of exploration to many bodies in the solar system, including the Moon and some of the terrestrial planets.

———

In a larger context, one might view the collection of lithospheres of the planets and moons (as well as the asteroids and comets) as parts of an overall "uniform" region which is comprised of smaller, discrete subregions that have common properties. In short, what we are describing is a lithosphere of the solar system itself. So, in much the same way that the continents on Earth are parts that have broken off one proto-continent (Pangaea), and now form discrete parts of the overall lithosphere of our planet, the planets have derived from a common Solar Nebula. And, in the subsequent processes of constitution and clustering in the billions of years that followed, each planet has developed its own surface regime. A few of these bodies (most notably the so-called terrestrial planets) and some of the moons of the Jovian subsystem have been able to develop the classic planetary configuration which includes a series of concentric spheres that are emanating from a central core. Other bodies, however, have not been able to achieve this degree of development. Among the latter are non-planetary bodies seem to have clustered as a kind of field within the Asteroid Belt. These were large enough, however, to be able to remain within the inner portion of the solar region. Other non-planet aggregates remained so small that they finally sought gravitational refuge in the outer areas of the solar region. Some of these are today known as comets that populate the so-called Oort cloud; others grew large enough to reach near-planetary proportions and are known as dwarf planets (such as Pluto). And, like the continents, the planets and other bodies of the solar system have experienced the effects of tectonic reconstruction and weathering. They also have experienced varying numbers of impacts by external solar bodies and internal upheavals that have also created variations on their surfaces. As a result, the lithospheres, where they still exist, are variations about a common theme, geologically speaking.

According to current thinking, the planets originated from the solar nebula that emanated from the Big Bang (about 14 billion years ago) and coalesced around our star, the Sun, several billion years later. At this stage in the formation of what is now the solar system, the individual accreting objects began to differentiate in their growth and composition, depending on their distance from the proto-star and the kind of neighbors they had. Some of the planets and other bodies in the inner solar system have formed crusts, but there is variety in the form and composition of these planetary outer skins. Indeed, the geologists are now finding out that each planet in the inner solar system has a common basic geologic structure. As indicated earlier, the crust of each of the terrestrial planets, for instance, is the product of billions of years of both external and internal tectonic forces. The external factors include the varying solar radiation and cosmic radiation, the gravitational tidal forces produced by adjacent celestial bodies, collisions with other bodies, the nature of the secondary atmosphere (if any), and the presence and the strength of a magnetic field.

During the process of development of the solar system, strong winds associated with the proto-star – a huge ball of gas that had not yet become a full-fledged star – created a series of nuclear reactions. Eventually, these produced the gas and dust that propagated throughout the cosmos, and was seeded with the heavy elements, such as carbon, nitrogen, oxygen, and others. The upshot of this was the creation of vast inter-galactic region, whose limits were eventually reached only when gravity began to effectively counter the force of the explosive event. The point to be made here is that all of the bodies of the solar system have a common origin and similar developmental histories and, therefore they will present many similarities in their structure and composition. Of course, there are variances too, but these can be seen as being the result of varying developmental histories of the various bodies of the solar system. Thus, over an early period of billions of years, the initial ball of gas and dust began to cool down. Then began the process known as accretion; this is the term that describes the gradual adhesion of external parts or particles to a body. Eventually, this process of accretion led to a period of sorting out of elements according to atomic weight, by the gravitational force, from the outer regions down to the cores. The discipline that underlay this sorting process also assigned relative orbital locations to the growing bodies within the solar system: the planets, moons, asteroids, comets, and the planetesimals. The significance of this to our cosmographical analysis of the solar system is that we can begin with the assumption that all matter and energy in the solar system which we see today has a common ancestry and, therefore – that there is an identifiable logic to the structure and composition of all the planets and other bodies in our solar region.

Meanwhile, within this model of planetary development and continued reconstruction, we are finding that all of the major planets of the solar system have undergone more local tectonic processes that led to the present variety of planetary lithospheres. Another set of factors in the location and development of the bodies of the solar system are other longer-term ones. These are the revolutions around the Sun, and the rotations about the planetary axis. But these movements also are affected by the attributes and behaviors of other nearby planets, their respective moons, and of the transient asteroids and comets. More recently, we have seen continuing evidence of this planetary surface change as volcanoes continue to erupt on some of the bodies of the outer regions of the heliosphere. Also, comets such as the asteroid or comet that struck the Siberian region of Russia in 1908, and the Shoemaker-Levy 9 comet impact on Jupiter in 1994, show that there is a continuing series of assaults on the planets, which result in significant changes of their surfaces. Another example of these would be the meteor strike that is thought to have struck the Yucatan peninsula in southeast Mexico some 65 million years ago. Scientists believe that this huge (6 miles in diameter) meteor collided with Earth at a speed of perhaps 100,000 miles per hour. This event, it is believed, then ejected a huge cloud of dust and particulate matter that encircled the Earth and blocked out the rays of the Sun for a significant period of time, possibly years. This, in turn, stopped the process of photosynthesis, thus causing the extinction of herbivores, and then the carnivores. It is now thought this external event and its effects then interacted with natural processes like "continental drift" to create further changes to our natural home on Earth.

———

Today, geologists and other scientists are continuing to analyze the imagery and data that is received from the satellites, probes, landers, as well as their telescopes that can operate in several wavelength environments. From these studies, they are finding that the traditional concepts and methodologies of the science of geology work just fine on other planets and celestial bodies of solar system. So now, with the advanced observation and scientific work that has been occurring during the Space Age, there has been developed a growing catalog of geological and topographical features that occur throughout the planets of the solar system. From the perspective of the geographer, this catalog of phenomena that is related to the structure of the outer crust of the planets is interesting in terms of the descriptive geography of place. However, geographers also would want to focus more on the spatial distribution patterns that are perceived from the analysis of both imagery and quantitative data. Thus, in general, whenever a geographer is presented with a spatial pattern of any phenomena, whether natural or man-made, there is always the elemental question that comes to mind: why these patterns? Indeed, many of these kinds of questions are now being investigated by scientists on Earth who continue to receive imagery from the surfaces of the other planets and their satellites, especially with respect to Mars, as the orbiting spacecraft and rovers continue to do their work there.

———

Thus, as we examine the other bodies of the solar system we find that, for example, the geological history of Mercury evidently fits with the theory of planetary process of accretion and differentiation in the solar system. This planet, whose orbit is closest to the Sun, has had its original atmosphere worn away by the solar winds, whose anatomy practically consists of a dominant core that is covered by only a thin outer shell of surface, and only a tenuous atmosphere. The planet's rocky outer shell, including its surface crust and its underlying mantle, is only about 400 miles thick. Based on the data gathered by the Mariner 10 flybys and subsequent probes, we at least know that the surface on Mercury somewhat resembles that of our own Moon. Thus, we again see a dense distribution of larger craters, interspersed by small inter-crater plains. Such a terrain presents an eerie reverse-image of the familiar inter-montaine terrain of regions like the northern Andes of South America, with the basins taking the place of the mountains on Earth. And, like the mountains on Earth, some of the crater basins apparently present a significant light-shadow contrast with respect to sunlight. Some relatively smooth plains are also evident; these are thought to be the result of ancient volcanic activity, although some may be the due to cumulative deposition of material that has been ejected during impact cratering events. Another of the most notable features of Mercury's surface is the system of enormous escarpments. These can extend as much as 60 miles in length and a mile or two in height. Some cut through the rings of the craters and other features, so as to indicate that they have been formed by some kind of compression. It is thought that these scarps may have formed when Mercury's crust shrank and buckled as its interior cooled rather precipitously. Otherwise, it is apparent that some water ice has been delivered to Mercury over its geologic history, probably by a series of comet crashes. Also there is evidence of water that somehow has managed to collect in permanent shadow regions of the polar topographies. Such are the vagaries and possibilities of what might be found on Mercury and the other planets and bodies of the solar system. Finally, the planet shows evidence of having been struck repeatedly by planetesimals, over a period of billions of years. As a result, it is now about one-sixth its original mass and considerably smaller in diameter as well. For these reasons, Mercury is one of the densest major bodies in the solar system, second only to the Earth. In many ways similar to the Moon, Mercury's surface is heavily cratered, but it presents no evidence of plate tectonics.

The upshot of all this is that Mercury is today a relatively very small, but extremely dense planet. It is thought that, to have such high density, its core must be rich in iron. Its molten core occupies about 42% of its volume. (The Earth's core is 17% of the total volume). Nearly two-thirds of its mass is contained in its largely iron core, which extends from the planet's core to a radius of about 1,100 miles (or three-fourths of the way to its surface). Spectroscopic analyses of the surface of Mercury have shown evidence of traces of oxygen atoms, potassium, and sodium. It is thought that these elements are only transitory: accumulating during the night and then being dissipated by the brilliant morning sunlight. Again, these findings can only be classified as tentative, and will be the source of much more study by astro-scientists in the coming years.

———

Venus presents a planetary surface that reflects considerable volcanic activity. It includes abundant shields and composite volcanoes, like those that are found on Earth. However, compared to the Moon, Mars, and Mercury, there appear to be only a few small impact craters. One reason for this might be the presence of such a dense atmosphere that smaller meteors are burned up before they can strike the surface. While there are medium-sized craters on Venus, there are not as many as is found on the Moon or on Mercury. There also is little evidence of wind erosion of the surface, which might be due to the density of its cloud cover. On the other hand, the volcanic activity on Venus moves the crust vertically, rather than horizontally as on Earth. The lateral extent of all this vertical tectonic movement can be seen in the system of lava tubes that cover the surface. The longest of these has a length of 4200 miles, and has been named the Baltis Vallis. There also is evidence of some chemical and mechanical erosion. Among the most notable surface features are the "montes," which include the Freya, Akna, and Danu montes.

Rifts are amongst the most spectacular surface features on Venus. The best-developed rifts are found atop broad, raised areas; they radiate outward from their centers and they appear to be places where large areas of the lithosphere have been forced upward, thus splitting the surface to form great rift valleys. In many ways the rifts on Venus are similar to great rifts on other planets, such as the East African Rift on Earth or Valles Marineris on Mars. Volcanic eruptions also appear to have been associated with these rift features. More globally, Venus shows considerable tectonic activity. Thus, it has about 1600 large volcanoes and perhaps hundreds of thousands of smaller volcanic features. Actually, about 80% of the surface is covered with lava flows. The larger volcanoes are often shield volcanoes which are similar to some on Earth.

Regardless of how the physical geography and geology of planets and other bodies has developed, their description reads a lot like the description of geological phenomena on Earth, particularly in terms of the vocabulary that is being used. Thus, the nomenclature of the geographer and the geologist that has been developed on Earth is now being used to describe the surface and the crust of the terrestrial planets, such as Mercury, Venus, and Mars, and the Moon. Among these is the term "multi-ringed basin" which is the name that has been given to a type of geological feature that has been observed on various planets and their satellites in the solar system. This class of features presents a "bulls-eye" appearance and they can cover a wide area of thousands of miles on the surface. Also of interest to the astro-physicist and other scientists is the characteristic escarpment or cliff-like quality of the outer rings of the basins. These are believed to be the result of giant-impact events that have occurred throughout geologic time.

———

The Moon was the initial site for the in situ exploration of such planetary features. Much was learned about the nature of the craters on its surface and many new technologies for conducting such studies were developed. These lessons have now been incorporated into the next major exploratory effort on the surface of the planets outside the planet Earth. Consider the trio of land rovers that have worked on the surface of Mars in the early years of the 21st century. Another is the NASA-ESA Mars Sample Return Mission, which is scheduled to be carried out from 2018 through 2027. During this planned mission, another rover will be landed on Mars (January 2019) and will spend about two years collecting rocks. Using an integral rotary coring drill, approximately 40 cores, three inches in diameter, will be extracted and then transported to the original landing site (Popular Science, 2011) on that planet. Because Mars is only approximately 35 million miles from Earth, it transits at the margins of the Sun's habitable zone. This is significant because it this is a place where liquid water could exist; and that means life as we know it also might be found on the subsurface of Mars. Scientists already have found evidence of ice there from prior observations; and also have found evidence of seasonal water flows, as well as signs of ancient rainfall, lakes, and even oceans. All this has led many scientists to the conclusion that Mars may have been warmer at some time in its planetary history.

Mars apparently is in a state of a geological coma; still alive, but just barely. One manifestation of this state is the lack of horizontal moving of tectonic plates, which allows the topographic features to reach impressive extremes in the solar system. Thus, volcanoes on Mars can continue to erupt in place and reach sizes that are greater than on geologically-active planets. Another manifestation is the feature known as the Valles Marineris, which is a canyon on Mars that is deeper than the Grand Canyon on Earth and longer than the Mississippi River. The most notable topographical feature, however, is the Oympus Mons, a mountain which rises thousands of feet above the local relief; it is the highest mountain in the solar system. At an elevation of 17 miles above the average radius of the planet, it would tower over Mt. Everest on Earth.

One of the most significant surface features on Mars is its dust storms. They rage across the planet and obscure the entire surface. They are like those on Earth, but they are "more." That is, they are more global, more persistent, and more intense than any on our planet. The fact that the planet is arid to the extreme means that strikes on the surface by asteroids and meteoroids, as well as volcanic eruptions, will generate great volumes of dust which will accumulate and persist for millions of years. There are great swaths of bare Martian surface which often expose the underlying substrata that lie beneath them. This is not thought to be due to extremely violent local dust storms, called "dust devils," but rather it is the result of a long, cumulative process which has occurred over time. The more rare planet-wide dust storms, however, do exert enough force in the low-atmospheric pressure conditions to generate global dust clouds that can obscure the entire surface of the globe for months at a time. These can be observed to be as much as a mile-wide and rising up to about altitudes of 36,000 feet.

———

One of the major challenges in the development of military bases on Mars will be to counter the harmful effects of the dust environment. These include the sandblasting of buildings and equipment, similar to what has been encountered on Earth. But on Mars, the process will last longer and be powered by sustained 200 mile per hour winds. Perhaps the most significant effect of the dust storms on the Red Planet will be the great amount of static electricity that will be generated by the movement of the sand and dust there. The discharges of the static electricity are known to foul and damage all manner of filters, connective devices, and to disable any equipment that depends on electrical power. One possible countermeasure will be to go under the surface, as has been done on Earth to counter nuclear radiation. Fortunately, the geological stability and aridity that characterizes Mars presents excellent conditions for establishing underground settlements and military bases there. Thus, for example, there appear to be many caves and miles-long lava tubes which may offer protection from harsh surface conditions. Identifying actual caves with significant volume will be the highest priority targets for NASA (nasa.gov). One of the main reasons for this interest in the caves and other underground features has to do with the planned manned missions to Mars in the second half of the 21st century, which includes the establishment of space bases there.

Human operations on Mars also will require access to H2O molecules, which can then be used as sources of water, oxygen, and hydrogen fuel. If subterranean water deposits already exist, caves may provide the best access to these resources. Also, caves with a protective rock ceiling would provide protection against radiation and meteor storms. But most importantly, they would provide an environment where the wide range of diurnal and seasonal temperatures shifts could be more easily managed. Techniques for detecting and studying caves on Mars already have been developed on Earth. Most of these remote sensing and in situ techniques utilize thermal and visible imagery. Previously, these had been tested in places like the Atacama Desert in South America, and the results are now being used to examine cave-like features on Arsia Mons, Mars. As an outcome of these exploratory efforts on Mars, lava tube remnants, deep pit crater chains, and isolated deep pits called "anomalous pit craters" have been identified. Air temperatures in cave entrances also have been measured with thermal infrared imagery and have been found to be different from ambient temperatures. Future studies of the Martian caves will deploy temperature and barometric pressure data loggers at the cave sites in order to develop diurnal and seasonal variations in these metrics.

———

Another aspect of the planetary surfaces that likely will affect military operations is related to the volcanic eruptions and their secondary effects on an atmosphere and a surface. These can have lasting effects on military aerospace operations, mainly because of what they do the surrounding natural and cultural environs. These include the local atmospheric conditions, as well as those in areas that extend for many miles beyond the epicenter. An example of these phenomena occurred in 1991, when Mt. Pinatubo, which is located in the Philippine island of Luzon erupted. More to the point, with respect to military geography, these eruptions occurred in close proximity to Clark Air Force Base, a long-time center of military aerospace operations in the Pacific region. As it happened, the base was in the process of being turned over to the Philippine government for their domestic use. However, by the time the cubic miles of rock and ash from the volcanic eruption had been spewed onto the surrounding area, and the fluxes of pyroclastic materials had destroyed the land surface over an area that extended out to distances of over 30 miles, the sprawling air base had been rendered unusable by anyone. One can only imagine the impact that the eruptions and their corollaries would have had on an operating air base.

In the other regions of the solar system, most of the active volcanoes are found on the moons of the Jovian planets. Jupiter's moon, Io, is the most active body in the solar system because of tidal interaction with its parent planet. Volcanic activity also is occurring on several Jovian moons: Enceladus (Saturn); Triton (Neptune); and Europa (Jupiter). These have volcanoes that are currently erupting, or that have erupted sometime during human history. A great deal of what we know about these volcanic eruptions on the Jovian moons has been learned only recently, since the beginning of the Space Age. This is because, during the past half-century or so, telescopes have gotten much more powerful; some have been deployed in space, above the filter imposed by the Earth's atmosphere. Another source of new knowledge has come as the result of flybys and orbits in the Jovian subregion. As a result of these, volcanism on Enceladus is the best documented of the moons, having been extensively imaged by the Cassini spacecraft in 2005. Before that, Voyager 1 and 2 had been returning imagery of volcanism on the moons during the decade beginning in 1979; and the Galileo orbiter has been operating in the subregion since the mid-1990s.

A better understanding of how military space operations could be affected in the future by volcanic eruptions occurring on other bodies also can be gleaned from what has been learned so far by NASA space missions. The key objective of some of these missions is to improve the capabilities for determining the likelihood of earthquakes, volcanic eruptions and landslides occurring on celestial bodies, including Earth. From such studies, it is hoped that technologies will be developed to enable scientists to actually anticipate volcanic eruptions, instead of simply reacting to them. Indeed, there is a need for very fast reaction, within a few minutes, since it only takes five minutes for the first ash from an explosive eruption to reach cruising altitudes. Such procedures and technologies would be part of the overall desire to maintain situational awareness of all possible threat regimes.

―――――

The mission reports from the various NASA missions also have provided us with some understanding of the craters that occur throughout the solar system. These geographic features already have proven to be useful in the initial explorations of the planets. There are two broad categories of craters which can be found on the planets and other bodies. One is caused by internal volcanic processes which produce large bowl-shaped depressions. Typically, these are more than one kilometer in diameter and are rimmed by inward facing scarps, which can reach considerable relative heights. Another is the so-called impact crater, which results from impacts from asteroids, comets, or fragments of these on celestial bodies. Generally, impact craters are created when external objects, such as asteroids, comets, and meteors crash into the surface of a celestial body. When the solar system was young, geologically speaking, there were many more impacts which occurred during the period of accretion. And they continue today, although at a lesser rate. Ultimately, some of the planets have wound up with many impact craters, but relatively fewer volcanic events. On the other hand, other worlds have been saturated with many impact craters, but may have had little tectonic activity reshaping their surfaces.

Impact craters are created by the impact of celestial bodies, especially meteorites, asteroids, or comets. These craters usually are categorized in terms of size, progressing from small, simple depressions with raised rims (at least originally), to larger basins with central peaks and hollows. The most developed of these are the classic multi-ringed basins that have diameters of as much as 600 to 700 miles. For eroded or buried craters, the nominal diameter is only an estimate of the size of the original rim that existed prior to the changes caused by subsequent geological events. The rim of the crater itself is the part that extends above the height of the adjacent surface; it is usually presented as a circular or elliptical edge which rises to an upper lip. In those cases where there is no raised portion that is presented, the rim may simply refer to an inside edge of the curve where the flat surface merges with the curvature of the crater. Impact craters also may have a dome, which is a deformational feature. Usually it is a large elliptical structure that is formed by warping up of rock strata. The strata of the dome are pushed up towards the top of the dome, which is then eroded off, exposing the oldest rocks appear at the center of the dome. A more rare type of protrusion in the impact crater is the so-called shatter cone, which is known to form in the rock beneath meteorite craters or underground nuclear explosions. They are evidence of extreme shock and pressures occurring under the surface.

Crater sites can be found on Earth too, but they are more commonly seen on the other bodies of the solar system. The main reason for this may be that on Earth there has been undergoing continuous tectonic remodeling and weathering, which has obscured the original features of an impact crater. One of the most conspicuous impact crater sites on our planet is the Bushveld Igneous complex and the Witwatersrand Basin system which evidently were created during an early period of the Earth's development. It is thought that the mass and kinetics that were involved was of sufficient magnitude so as to produce subsequent regional volcanism. However, the impact crater which is located in the vicinity of the town of Vrederfort, in South Africa, for which it is named, is considered to be the largest example of this phenomenon on Earth. Other examples of this type of geological feature are found in the Black Hills of South Dakota, in the western United States.

Impact cratering also provides a window into a celestial body's history. The formation of multi-ring basins on the Moon in its early history, for example, can be just as instructive as the geological strata manifestations on our own planet. Evidence from several other bodies in the solar system also indicates that the effects of macro-scale impacts go beyond the immediate impact structure and serve to cause increased internal geologic activity over an extended period of time, and over a large spatial area. Thus, despite having occurred during the early period of the solar system's development, these impact events have continued to generate secondary and local geological effects on the inner planets and other bodies several billion years later. Furthermore, these powerful impacts still occur in the solar system, albeit less frequently, but with the same magnitude of effect on the bodies.

One example of this occurred in July, 1994, when the Comet Shoemaker-Levy 9 smashed into Jupiter's atmosphere and broke up into several fragments. The succeeding explosions were observed by Earth-based telescopes, and the orbiting Hubble Space Telescope. In addition, the spacecraft Galileo, which happened to be approaching Jupiter, was also there to observe that collision event. The fireballs that resulted from the explosion of energy could be seen to rise above the planet's clouds and large black smudges in the clouds. Spectroscopic analyses also revealed that the impacts produced concentrations of many chemicals that are known to exist on Jupiter, but to a lesser degree of concentration. These include water, hydrogen cyanide, and carbon dioxide (nasa.gov).

As an illustration of the importance of craters to human activities, one of the main objectives of the current NASA Lunar Reconnaissance Orbiter is to find safe landing sites on the Moon for future manned missions. Among the tasks that have been assigned to the LRO is: (1) to look for any natural resources that could be of use to military forces that might be based there in some hypothetical future; (2) to measure the temperatures at various points on the surface in order to determine the range of these that humans will have to deal with, over long periods of occupation and; (3) to make copious measurements of radiation levels at various points on the Moon in order to develop a three-dimensional model of the radiation environment that might interact with any military space operations. To this end, devices have been developed to study the lunar soil and regolith to determine the optimum conditions for the construction of a military space base there.

There also will be a search for any evidence of water on the Moon. This vital resource can also be exploited to provide oxygen for life support systems, as well as hydrogen for use as rocket fuel. Similarly, topographical studies will be done to determine the best site for a military space base on the Moon. This will involve the construction of more detailed three-dimensional maps of the lunar surface, as well as analytical profiles of the upper crust of there. On the other hand, the search for chemical and mineralogical "signatures" as indicators for the presence of microorganisms seems to be another guiding imperative in the development of sustainable human bases on the Moon.

Based on these studies of the Moon as a possible manned base in the future, as well as those that are being conducted with respect to Mars, it appears that the most promising environment for the establishment of a military space base on either of these bodies would be that which is presented within an impact crater. This is borne out by the decades of data that have been accumulated by all the landings on both the Moon and Mars. Thus, for example, the Gale Crater on Mars was selected as the landing site for the Curiosity rover, and the Endurance Crater was selected as the landing site for the Opportunity rover. The case for the utilization of craters in the solar system as micro-environments also is strengthening. Indeed, when I view the imagery and read the descriptions of the all these craters, I am reminded of the "geo-cultural oases" that have occurred on Earth, some of them since the earliest days of human history in arid environments that happen to be located near a river or other water feature. The craters on other celestial bodies are perhaps more susceptible to this kind of utilization than those on Earth, mainly because they have not been filled in by sediment or obliterated by a tectonic event as has been the case on our planet. And, of course, such craters are relatively flat and they are protected from the elements by those high ridges.

Imaging radar has charted the dark regions of the poles of the Moon and has found reflective indications of water ice in the permanently shadowed craters in its polar regions. Then there were the Lunar Crater Observation and Sensing (LCROSS) detections in 2008 and 2009 of water vapor and ice particles that were projected by the impact of the object on the Cabeus crater on the Moon. The Lunar Reconnaissance Orbiter has also relayed images of a potentially ice-rich crater on the north pole of the Moon. This permanently shadowed crater lies on the floor of a larger, more degraded crater. With no sunlight to warm the crater floor and walls, ice that is brought to the Moon by comets or asteroids could potentially collect here, at least on a seasonal basis. Finally, the NASA instrument aboard India's Chandrayaan-1 spacecraft (2009) has found evidence of water molecules on the surface of the Moon.

———

The Military Geography of the Lithosphere

The military utilization of Earth's lithosphere has been a familiar phenomenon on Earth for thousands of years. I was first exposed to the concept of subterranean warfare at a very young age, thanks to the circumstance of living just a few blocks the World War I Museum in Kansas City, Missouri. This large complex of buildings and open spaces was one of my favorite places to ride my bike and to explore in my early years. One of the exhibits which most intrigued me was the scale model of trenches that characterized the First World War along the Western Front. From the mockup of the systems of trenches in France, I could easily see why this terrain of mud and barbed wire would be a veritable killing zone for the machine guns on both sides. Later, in my studies of military science and tactics at De La Salle Military Academy in Kansas City, I could easily visualize what was being taught about the subsurface military complexes that were developed by both sides during World War I. Even then I understood how this "cultural geography" had made the frontal assaults by cavalry and infantry of the previous wars obsolete; especially with the advent of rapid-firing artillery and the iconic heavy machine gun. This was to be the state of affairs until the airplane and the newly-invented tank broke the stalemate towards the end of the war.

Later, during the Cold War, the United States also went into the Earth's crust to deploy one group of its ICBM (Inter-Continental Ballistic Missile), with its nuclear payload. Its purpose was to be able to withstand a nuclear first-strike by the Soviet Union, and thereafter be able to launch a retaliatory attack. This assured retaliation capability was used as a hedge against a potential first nuclear strike by the Soviet Union. In fact, this subsurface strategy was complemented by a similar subsurface strategy in the depth of the oceans, except that the submarine missile-bearing platform could also move about throughout the depths of the oceans. More concisely, the underground ICBM utilized the hardness of the subsurface to provided protection against a nuclear strike; while the submarine depended on stealth and mobility within the oceans to ensure its survivability. I learned these lessons of military geography first hand during the height of the Cold War, when I was posted as a young 2nd Lieutenant at Grand Forks, AFB in North Dakota. This was a classic example of a Strategic Aerospace Command air force base during the Cold War. It hosted a Wing of B-52s (long-range nuclear bombers), and a Wing of Minuteman II intercontinental ballistic missiles. Both had the capability to strike targets in the Soviet Union, even after a first-strike. The manned bombers relied on their ability to be airborne in response to early warnings of impending enemy nuclear strikes within a matter of minutes. For their part, the missiles were embedded in nuclear-proof cocoons (known as silos) within the crust of the Earth. The idea was that the ICBM silos would be able to withstand a nuclear explosion and be ready for a retaliatory strike on command. Indeed, the US, chose to develop this mixed-portfolio of warheads, which is referred to as the "Nuclear Triad" strategy as a means for enhancing the nuclear deterrent through mutually assured destruction. So, this triad of nuclear forces included the B-52s, the ICBMs in their hardened silos, and a fleet of nuclear-powered submarines. This can be seen as a possible prototype of a military space strategy in the future.

———

It can be argued that the Earth's crust is comprised, not just of rocks and dirt, but also the vegetation that covers the terrain, and the water systems on and within the surface. This total environmental system also is where humans do their "cultural" work, to satisfy a variety of their needs. Such a military "ecosystem" developed along the frontier between the Vietnams, and Laos and Cambodia during the Vietnam War. The name given to this region by the American forces was the Ho Chi Minh Trail. There the North Vietnamese Army utilized every aspect of the lithosphere to avoid detection and attack from the air. So, they utilized the triple-canopy rainforest to camouflage the movements of their trucks, tanks, and personnel through the system. They also took advantage of the triple-canopy vegetation and the subsurface to provide further protection against air attacks, as well. And, they also went underground to construct a series of military bases along the trail to support their combat operations in South Vietnam and in Cambodia. The essence of HCM system, however, was the sophisticated network of underground storage facilities for all manner of war materiel, as well as barracks, hospitals, and other such facilities.

So, it can be seen that, even when there is no significant cloud or other atmospheric obfuscation, nor vegetation that can be utilized to hide from air observation and attack, there is often an outer layer of rocks and soil that can be manipulated to provide protection from air attack. In modern times, the combat engineers who design and construct the underground military bases tend to utilize the natural terrain as the foundation for their cultural constructs. Examples of these subterranean bases include the complex and sophisticated underground redoubts that were constructed in the various Pacific islands, including Iwo Jima, by the Japanese military, during WWII. In that case, the underground defenses made use of the numerous caves that had been created naturally in the volcanic rock of the island. Sometimes these caves actually became long tunnels which connect with other tunnels and caves. This natural underground fortress was so extensive and hardened with the cooling of the original lava flows from the resident, extinct volcano, Mount Suribachi, that it took the two divisions of American Marines almost two weeks to finally conquer the island. The underground defenders held out all this time, with terrible cost of human life on both sides, even though the island had undergone weeks of preliminary bombardment by U.S. air and naval forces.

————

Mountain environments are a subset of the lithosphere, and they pose many special challenges for military aerospace operations. This lesson has been painfully learned by both the Soviet Union and the United States in Afghanistan. Put simply, the conduct of air attack operations in that mountainous environment is damn hard and dangerous. This is true even when the mountains are not extremely high; all it takes is a significant difference in local relief – what is called "rugged terrain." In the mountain environment, the visibility of the landscape changes drastically as the Sun makes it diurnal journey across the mountains and valleys. For this reason, shadows are another element in the military geography of mountainous terrain that can increase the difficulty factor for the military aviator. Even the relatively flat high-altitude plateaus with their narrow crevices can pose ever-changing visual "illusions" to the air reconnaissance or attack aircraft. The mountainous terrain also presents a kaleidoscope of steep slopes and other obstacles or conduits for military air operations. This makes it easier to concentrate air resources, but it also concentrates the lethal air defense weaponry. Another feature of the mountainous terrain which affects military air operations are the caves that are present in these environments. These too are obvious locations to be reconnoitered for the presence of targets that might be struck by attack aircraft – either manned aircraft or an unmanned air vehicle – but as we have seen, they are also ideal locations in which to deploy air defense weapons. The most prominent example of this has been the utilization of the RPG (Rocket-Propelled Grenade) by insurgents in Afghanistan against allied air forces.

———

The military dynamics of these topographical features on Earth are most familiar to us, but they likely also will prove to be significant on other bodies of the solar system. Thus, it can be said that valleys will continue to be a main venue for air attacks on the surface of the other bodies of the solar system as well. This is because valleys are where human populations are likely to congregate, and where their economic activities are likely to occur, regardless of the particular place in the solar system in which they are located. Valleys also are natural venues for "lines of communication" in terms of military aerospace operations, even on the other bodies of the solar region. At the same time, it could be said that the concept of the signature "pattern" in a valley is somewhat akin to a particular object's electromagnetic "signature," and therefore is useful in reconnaissance, targeting, and the laying down of ordnance. The above illustrates why it is crucial for military warfighters to become very familiar with the geography of the land surface of their particular area of operations. In every case, whether on Earth or out in the solar system, the universal problem set for military aviators is cast in two dimensions: how best to fly and to fight in a given geographic region.

With respect to the underground military installations which are ensconced within the crust of a planet, it can be said that they represent another dimension of the problem set; one which likely will be presented to military space operations as well. Indeed, this may become the most efficacious model for the establishment of military bases on places like the Moon or Mars. Many of the challenges of life-support for humans over extended periods of time in hostile alien environments can be seen to be the subject of much continuing study and experimentation in various locations within the crust of the Earth. There also is a considerable amount of work being done to develop the technology for dealing with these challenges in outer space. I would argue that the work that is being done by NASA to solve the problems of space exploration and long-term residence will provide many opportunities to adapt the technologies to the development of subsurface military operations on other places in the solar system.

The reason why such an adaptation is useful in the discussion of a solar system lithosphere is that all the bodies in the solar system have derived from a common ancestry – the solar nebula. Thus, through all the condensation, accretion, collision, gravitational takeovers, mergers, and other cosmic tectonic processes that has taken place, the Sun and the various planets have gone through several iterations of development and evolution during the past 3 to 4 billion years. Because of the commonality of cosmic genetics, all the celestial bodies in the solar system share many common traits and attributes. The iterations of cosmic conditions and the effects of "random selection" also have produced many differences and variations in the structure and physiology of the individual celestial bodies. And, like all siblings, some planets have grown up to be larger than others. Some are cold and dry; at least one is extraordinarily hot and dry. Still others have never evolved beyond the gaseous stage, but they have grown to a much larger size than the siblings who have developed into rocky planets. Another variation in their respective development has been in their central cores. Some planets and other celestial bodies have well-developed heat engines which effectively interact with the heat that is received from the Sun. Others either have no such internal heat source or it is not able to counterbalance the heat exchange. Those which have well-developed cores also seem to be the ones with the most effective cosmic and solar radiation "umbrellas" and atmospheric "blankets" so they can maintain more control over their planetary environments.

———

The Planetary Mantles

To understand what is occurring within our own planet, especially with respect to the relationship between its core and the surface, we have at our disposal a powerful model that has been constructed to explain the anatomy and physiology of our own planet. It essentially depicts a convection system whose components include the core, the mantle and the surface of a planet. It is thought that the system is set in motion when material in the mantle is heated by the core, and then transports that heat to the surface of the planet. However, the process by which the surface is heated by the interior regions is not a direct one. Rather, it involves the mechanics of convection and conduction. Thus, it is postulated that when the material at the inner portion of the mantle is heated by emissions of energy from the core, that material rises up toward the outer surface of the body. More often, the material that has risen toward the surface then cools and, therefore, descends toward the core. Occasionally, however, the heated material finds an opening in the outer surface, and the molten material bursts or oozes onto the landscape, where it is further cooled either by the atmosphere or by volatile materials, such as water, and is deposited as additional surface material.

This process is called convection, and it is the primary means of heat transfer in all types of media that are exposed to the heat source. It involves the movement of heat from its source to a location nearer the surface, where it cools. There are two main types of convection: forced and natural. Forced convection occurs when the medium involved in the transfer of the heat is moving of its own volition. In this case, the heat is disbursed by the movement of the air or fluid, but it is not actually causing the movement. Natural convection occurs when the medium that is transferring the heat is motivated by the heat itself. Both types of convection are thought to be happening within the mantle of the terrestrial planets, and some of the moons in the outer solar region. So, as the medium heats up, it expands. The resultant buoyancy causes the warmer fluid or gas to rise. These two types of convection often occur simultaneously, wherein the heat is causing the fluid to move throughout the medium. But the movement of all matter within this convection is also being caused by another force: mixed convection. Within the process of convection there also occur mechanisms that are known as conduction and radiation.

Conduction is one of the most common ways to transfer heat. It is done by transferring heat through matter from one atom to the next (this is readily appreciated when a poker is placed into a fire). More generally, convection refers to the movement of any molecules through any fluid state. It's the primary form of heat and mass transfer on a planet. Mass transfer refers to the movement of matter from one place to another, and resulting in a net change in the mass's location. Mass transfer through the mechanism of diffusion is a common phenomenon in planetary mechanics. It occurs because all molecules in liquid and gases are in continuous motion, and this causes concentrations of a substance to spread out until its mass is evenly dispersed. Other characteristics of matter, such as heat, pressure, and momentum also diffuse in this manner, often causing movement of the mass itself.

So, the heat that is generated by the immense pressures that are applied to the planetary cores also is generated, transferred, diffused and conducted outward by molecular transfer, through the mantle and onto the surface. If there is a functioning atmosphere, some of the heat is diffused into outer space. And, water also is released by the differentials in heat and some of the molecules that escape to the surface of the planet. On some planets, this water is being recycled from that which had earlier been deposited by asteroids or comets, and their meteorites, onto the surface of the planet. This often occurs when liquefied rock is extruded onto the surface by volcanism, cools and hardens, and releases its water vapor and other gases. Then too, igneous rocks and other material that is extruded, represents the recycled material that had once been deposited onto the surface by external asteroids, comets, and their meteorites. Thus, in one sense, a planetary mantle – in the same fashion as an atmosphere and the Earth's oceans – is a giant mixing system which also reminds me of an industrial processing plant or a chemical laboratory. So, just as occurs in these human processing facilities, a wide variety of separation processes occur within the planetary mantle; some occur as a result of mechanical means of transfer, while others are the result of chemical reactions as well. Distillation and crystallization are examples of the former, while the chemical processes used to create oil deposits are examples of the latter.

Interestingly, the similarities in chemical composition are greater than the differences in the structure of the mantles. However, there are variations in the chemical composition too. One reason is that trace elements will not be readily incorporated into the overall mix. And, they are more likely to concentrate in liquid material in the mantle and carried upward in solution, eventually being transported into the crust. As a result, the mantle is relatively depleted, and the crust is relatively enriched, in terms of minor and trace elements. On Earth and the other terrestrial planes, the mantle is important in determining the bulk composition of the planet. Part of it is due to the volcanic eruptions which have brought rock fragments to the surface and there is also the introduction and concentration of some elements. These so-called "hot spots" denote sites where plumes of hot mantle material are upwelling beneath the plates.

Although the mechanism of heat flow in the mantles is not completely understood, we are learning more from our continuing studies of the mantles in the solar system. As might be expected, we apply many of the lessons that have been learned on Earth to the other planets and moons. On the other hand, we now are utilizing the data that has been extracted from the other planets to further our overall understanding of this phenomenon on Earth too. Our knowledge of the mantle in the other bodies of the solar system is mainly inferential, but we have been able to develop working models for understanding the signs and symptoms that are presented to us. One such model seeks to explain the source of the heat of the interior cores. Thus, we believe that the Earth's interior heat has been left over from the process of original formation and subsequent accretions, as well as from the decay of radioactive elements in its interior. According to this theory, the pressure exerted by the upper layers then causes rock to partially liquefy at lower levels. This is the downward portion of the convection cycle. During the upward part of the cycle, this molten material then cools as it comes nearer to the surface of the planet, only to then plunge back down after it has cooled and condensed somewhat.

The Planetary Core Systems

What we know about the cores of the planets and some moons in the solar system also has been derived from of the empirical observation of signs and symptoms, and the application of rigorous inferential logic. Examples of such evidence are the volcanoes that are found on the surface of some of these planets. The hot gases, rocks, and molten lava that are found on the surface also indicate that there is some extreme source of heat within these planets. Another bit of evidence about the source of heat is that these volcanoes occur throughout the surface of a planet. In the case of Earth, volcanoes are even found on the floor of the oceans. A leap in the process of inferential logic then takes place when scientists begin to see certain spatial patterns of location and begin to correlate these patterns with what they have learned from other scientific studies – such as the theory of tectonic plates and drifting continents. So, now it can be further inferred that volcanoes are not phenomena that are produced from local subsurface processes, but rather a more global mechanism. After further empirical studies and analytical thinking, one probable explanation for the nature and distribution of the volcanoes might be that they are systematic surface manifestations of something that is occurring at the core of the planet.

The planetary cores function as a kind of central heating subsystem for these bodies. Generally, the essential components of any central heating system are: some type of furnace in which fuel may be burned to generate heat; a medium for transferring the heat to the space that is to be heated; and an emitting apparatus for releasing the heat, either by convection or radiation, or both. In the combustion of most ordinary fuels, carbon and hydrogen react with oxygen to produce heat, which is then transferred from the combustion chamber to a medium consisting of either air or water. Ambient air enters from below and in front of the radiator and, as it becomes heated, it rises vertically between the radiator sections and discharges into the environment, either in directed fashion or by general propagation. Convection differs from radiation in that the inlets and outlets are designed to properly direct a stream of warm air through a space, using the "chimney effect." In the particular case of the planets of the solar system, the core region can be thought of as the furnace which generates the heat.

Another visible phenomenon that provides the scientist with other clues as to the nature and function of the planetary core occurs miles above the surface of the planet. These are the magnetospheres, which are regions of charged particles which are constantly being re-energized by some mechanism, one that is hypothesized to originate in the interior of the planet. So, once again, based on what is known about electricity and magnetism already, it has been further hypothesized that the core is probably metallic and that it might be spinning in combination with a surrounding fluid medium to produce magnetic lines. Currently, scientists are developing models which attempt to connect what is being observed on the surface of the planet and the magnetosphere. One of these postulates that the core of a planet behaves like a dynamo and, therefore, generates electrical and magnetic phenomena that can be detected and measured at the poles of the sphere, and above the atmosphere. This is a very simple description of the dynamo theory but it serves to explain how we have arrived at what we know about the planetary cores and the magnetic fields.

So, we see that the planetary core also acts as a dynamo for the planet; that is to say, it is a dynamoelectric machine. As such, it also behaves like a generator in which mechanical energy is changed into electrical energy; mechanical energy being the sum of the kinetic energy, or energy in motion, and the potential energy that is stored in a system according to the positions of the parts. The mechanical energy of the Earth-Moon system is nearly constant as it is rhythmically interchanged between kinetic and potential forms. The energy which drives the dynamo effect within the core involves several different forms and processes. Chemical reactivity involves the loss of electrons in the outermost shells to form divalent positive ions, thereby exposing the next innermost shell with a stable configuration in each case of several electrons. Any number of electrons can be removed under conditions that can provide necessary energy, such as intense heat or interaction with powerful electric or magnetic fields.

Based on the above, it is postulated that, at the period of its initial formation, a planet's interior was generally evenly mixed in terms of chemical composition. As billions of years passed, however, the heavy metals sank down to the center, while the lighter elements rose to the top. Because this process is so slow, the planet might have solidified before this chemical fractionation could have fully developed. As a result, large and massive planets, such as Earth and Venus, remained molten long enough for iron and nickel to form within the core. On the other hand, smaller planets like Mars cooled much faster and therefore solidified before the heavier elements had a chance to sink to the core. Another postulated result of these differences in the rate of cooling was that certain elements, like iron, remained in abundance in the surface soil. Such an outcome is thought to be the reason for Mars's familiar red color, for example. The differential in the degree and rate of cooling also is thought to be responsible for relatively thin atmosphere that surrounds the large, rocky bodies of the inner region of the solar system; as contrasted with the Jovian planets which have a relative small, dense core that is surrounded by massive layers of clouds.

———

At the present time, we know more about our own planet's core for several reasons. For one, the theoretical models which had been developed prior to the study of the other planets of the solar system are now being vetted by the improved sensing and measuring technologies of the Space Age. These include more sophisticated seismic measuring systems that are tied to modern computer technologies. So, we now have a greater degree of confidence in the hypotheses regarding our own planet's core. Thus, we can state with greater certainty that the Earth has a solid inner core which has a radius of about 650 miles, occurring within a fluid outer core that is approximately 1,200 miles thick. Both regions of the Earth's core consist of mostly of iron and nickel, which are thought to have sunk to the center of the planet while it was still in a molten state.

At the same time, we also continue to refine our models of the structure and composition of the other planets' cores. Thus, we now have a more sophisticated and nuanced understanding of the formational experiences of the so-called gas giant planets. As an example, conventional planetary formulation theory of just decades ago held that Jupiter has a set of neat layers with a gassy envelope surrounding a rocky core which consists of heavier elements. However, later generations of evidence indicates that the inside of the "Gas Giants," such as Jupiter, are more like an undefined mixture of elements with ambiguous zonal boundaries. This newer model of a melting Jovian core buttresses similar mixing models for other gas giant planets in the solar system. This model explains how liquefied parts of a gas giant's core may have trouble reaching the outer envelope due to double diffusion convection, a process that is commonly found in the Earth's oceans. Thus, when salty water accumulates at the bottom of the ocean, its density keeps it from mixing thoroughly with the upper layers. In a similar fashion, the heavy elements in Jupiter's core may have trouble gaining enough energy to move upward and outward. Scientists don't know how much this hindrance will affect potential mixing inside Jupiter, and many questions remain to be answered about the melting process.

In many ways, the core system of Jupiter is the most interesting one in the solar system. Current models of the structure of Jupiter indicate that its core is solid and measures about 10 Earth masses. It is thought that it initially was formed as a result of the cumulative accretion of icy planetesimals. Because of its great mass, this core likely would have readily developed an atmosphere, as the planetesimals continued to release gases during the accretion process. Thereafter, as the mass of Jupiter's core continued to increase, it would have been capable of attracting gases from the surrounding solar nebula. The individual gaseous formations would have mixed and thus develop the present-day envelope that includes an unusually large amount of hydrogen-helium. Another upshot of the mixing process is the empirically-observed unusually high amount of the most abundant heavy elements, compared to other planetary gaseous envelopes in the solar system.

Scientists now also have evidence that Jupiter's core has been dissolving; this could provide insights into the structure of exoplanets outside our solar system. Based on the latest simulations, it appears that Jupiter's rocky core has been liquefying and mixing with the rest of the planet's interior. The results are also valuable in the quest to determine what is occurring inside the other giant planets. For example, scientists have hypothesized that similarly-sized gas giant-like exoplanets may well have internal structures that are similar to Jupiter. However, astronomers have found an exoplanet with approximately the same volume as Jupiter, but having about five times the mass (nasa.gov). Scientists still have to rely on analog inference and mathematical logic models of Jupiter's core to a significant extent, because the conditions on that planet are simply too extreme to fully apply the models that have been developed for Earth. (Phys.org/news/20-12-03).

The Hydrosphere of the Solar System

When the Space Age began in 1957, it was not known whether there was such thing as a "hydrosphere" in the solar system, outside our own planet. Since then, however, two major advances in the observatory technologies have made us more aware of the existence of such a Heliospheric water system that is operating throughout the solar system. One has been the development of evermore powerful telescopes and other sensors, some of which are deployed in space. The other are the spacecraft that are carrying "suites" of multi-spectral sensors, those that are capable of "seeing" in the other portions of the electromagnetic spectrum have journeyed even to the outer limits of the Kuiper Belt. The upshot of the matter is that we now have enough empirical data to deal with a true hydrosphere of the solar system. Like the hydrosphere on Earth, the water of the solar system occurs in all three states (gas, liquid, and solid) and varies greatly in terms of its spatial locations. As a geographer, I take special note of the spatial distribution of the "nodes" of water throughout the solar region.

One can reasonably assert that water is ubiquitous in the solar system; and it has been ever since the solar nebula began to cool and condense. Even today, theoretically speaking, planetary water can occur as solid, liquid, and gas form, at least in the form of water vapor molecules, everywhere throughout the heliosphere. Everywhere there is a basic truth that evaporation always transforms liquid water into water vapor, which is then able to move freely anywhere as a gas. Regardless of the particular state of water, however, it is always comprised of water molecules. These molecules, without an adequate planetary gravitational field, will move away from a planet and eventually be lost into space. However, even with the gravitational field, a molecule that is moving outwards in the upper atmosphere has little chance of colliding with other materials and would therefore be able to achieve the pertinent escape velocity to escape into space. Another factor that affects the likelihood and rate at which these molecules escape the gravity of a planet is atmospheric temperature in which they are located; that is, the average speed of gas vapor depends on its temperature.

Meanwhile, the distribution of water throughout the solar system, in all its forms, continues even to the present day. This is because, as a constant, the heat derived from the Sun and from other processes (such as nuclear decay and the pressures of accretion), continues to be an engine of change in the form that molecules take over time and across space. The total amount of water remains the same, even throughout the billions of years, within this closed system. However, the respective proportions of ice, liquid, and gas is constantly (in cosmological terms) changing. So, because of these dynamic processes, which have been ongoing for the last 4.5 billion years or so, the hydrological map of the solar system also has been changing. So, why has the Earth been able to hold on to its water allotments, while other rocky planets, such as Mars have not, at least to the same extent? Actually, at this point in geological time, each of the terrestrial planets is losing about a ton of atmosphere to space every hour, and some of this lost material is in the form of water. Also, how did the planets end up with vastly different quantities of water if they are all losing molecules to space at similar rates? The answer lies in the fact that today's rates of loss are not necessarily the same as they were in the geological past. The planetary magnetic fields could have made a difference, because in the past when the solar wind presumably would have been stronger than at the present time. In any case, water molecules are too heavy to just float out of the atmosphere, but hydrogen can. On the other side of the ledger, even if some small amounts of water somehow are lost, there should still be a net gain from the water that is constantly being inputted by icy comets and asteroids.

Again, why is Mars a dry, cold planet, while Earth retains a robust hydrological system? One reason for this difference is that the Earth's magnetic field helps to prevent the drying effects of the solar wind on the planet. Mars, however, lost its magnetic field when it was about 500 million years old, and it has dried out as a consequence of the constant effects of the solar winds. But even this explanation is still a matter for debate; whether a magnetic field is any kind of shield at all. The controversy derives from recent observations indicating that Mars and Venus both are losing oxygen ions from their atmospheres into space at about the same rate as Earth. More recent studies suggest that a massive impact could have shut down Mars's dynamo by warming the mantle layer, thereby disrupting the heat flow from the core to the mantle and shutting down convection. The fact that the crust of Mars's younger impact craters is not magnetized supports this notion. Another explanation is that an asteroid the size of Texas might have hit Mars about 4.5 billion years ago, thus producing the biggest impact in our solar system's history. Other scientists believe the Martian magnetic field might have been beaten into submission by repeated strikes from space (as many as 20,000). It is thought that no single impact could have short-circuited the dynamo that powered it magnetism. But a quick succession of 20 asteroid strikes could have done the job.

Another factor in planetary "water retention" mechanisms has to do with a planet's atmosphere. In the case of Mars, over the last four billion years the planet somehow lost most of its blanket of atmosphere, which is made of carbon dioxide. The carbon dioxide in Mars's atmosphere is basically a greenhouse gas, just as it is in Earth's atmosphere. A thick blanket of carbon dioxide would have provided the warmer temperatures and greater atmospheric pressure necessary to keep liquid water from freezing solid or boiling away. Perhaps an asteroid impact blew most of the atmosphere into space in one catastrophic event. Or maybe erosion caused by the solar wind might have slowly stripped the atmosphere away over a period of billions of years. The planet's surface also might have absorbed the carbon dioxide and locked it up in minerals such as carbonate. In the end, it is still not conclusively known where all the carbon dioxide went.

————

The water on Earth represents a classic regional system. On a global scale, it can be seen to be organized via several nodes of varying size. The largest nodes would be the oceans, followed by the seas and larger lakes. Like in all regional systems, the nodes are often interconnected by a network of rivers and straits. Some nodes and links of this water system lie under the surface of the Earth, while others move over the land. Another linkage is the convection system in the atmosphere that is powered by the Sun. The name that scientists have given to this region is the hydrological system; it serves as a transportation and processing operation that constantly causes the water to change states, including liquid, gaseous, and solid forms. So, now we will examine each of these global nodes and their linkages.

At this point, we should take note of the detailed understanding that we have of our oceans. This is the result of thousands of years of practical experience and careful scientific studies that have taken place during the course of human history. And yet, there is much more to learn. Indeed, some marine scientists continue to mirror the activities of the astronauts in space that have occurred during the past half-century. In any case, as far as is known today, the vast bodies of liquid water that occur on the surface of our planet are unique within our solar system. Approximately three-fourths of the surface of the planet Earth is covered by the largest bodies of water, the oceans. The volume of water that is contained in these oceans comprises about 97% of all the water that has been detected on Earth. For this reason alone, the cumulative and particular effect of the oceans on the land-based systems (both natural and cultural) is massive and complex.

As far as the geographic distribution of the oceans is concerned, there are four major oceans that are interspersed with large land masses. The largest and deepest of these is the Pacific Ocean; its surface area is about 64,000 square miles and its waters cover more than a third of the planet. In terms of depth, it averages 14,000 feet, while the greatest depth that has been measured is about 36,000 feet. It extends from the Antarctic region in the south to the Arctic region in the north, for a distance of about 9,600 miles. It is bounded to the west by the continents of Asia and Australia; to its east lie the Americas. The next largest ocean, the Atlantic, covers about one-fifth of the Earth's surface. It effectively separates the continents of Europe and Africa, and the Americas. One of its most interesting geographic features has been discovered by deep-sea probes: the zone of active creation of new surface along a seam which is located in roughly the middle of the ocean. The east-west distances of this body of water varies quite a bit, from its northern region, through its middle sector, and its adjacent land masses in its southern region. Then, there is another large expanse of water that covers most of the southern hemisphere, the Antarctic Ocean which surrounds the southern pole.

There is constant movement within the oceans, both vertically and laterally. The ocean currents make up a horizontal and vertical circulation system of ocean waters that is produced by the forces of gravity, wind friction, and water density variation, in different parts of the oceans. The direction and form of ocean currents is the result of a number of natural forces. These operate on the ocean waters in various ways, within the horizontal layers, and by exerting gradual pressures. Today, equatorial ocean currents are blocked by landmasses, and therefore the Circumpolar Current is the strongest current. Ocean circulation derives its energy, at its surface, from the circulation of the winds and heat circulation within it, which is driven by variations in the density of the waters that are the result of exchanges of moisture and heat in the atmosphere.

Along with the atmospheric currents, the marine currents and vertical cycling are the means by which heat is redistributed around the Earth. Most energy arriving from the Sun is absorbed as heat near the equator, and it is then redistributed to colder regions. Actually, about 40 percent of the heat that reaches the poles from the Equator comes via the ocean currents. As one can appreciate, this major pattern of circulation of heat in the oceans has a large influence on the Earth's climate. But such patterns can be said to be transient, in terms of geological time. Thus, as continents, oceans, and currents have shifted through periods of millions of years, major climate changes have occurred throughout the globe. On the other hand, the sea levels and relative extent of the seas has been affected by the cycles of warmer and colder periods on the dry surfaces of the Earth. As an example, there is evidence that the ocean froze to a depth of 6,500 ft in places during the series of so-called "snowball" events 750-580 million years ago, and possibly earlier, each event lasting for up to 10 million years.

Winds also affect every area of the oceans, but tend to be most noticeable, and have their main effects, on or near the coasts through their interaction with waves. Thus, ocean waves are another source of disturbances in the ocean that transmit energy from one place to another. The most familiar types of waves are generated by the winds on the ocean surface, but there are types of waves that are caused by other forces. These include the tsunamis, which are often caused by underwater earthquakes, and internal waves, which travel underwater between water masses. Ocean waves are mostly wind-generated and vary from tiny coastal ripples, to the regular, rolling swell of the open ocean, to monster breakers on world-famous surfing beaches. However, regardless of the causative process, all waves transmit energy and when the waves reach land, this energy maybe dissipated, either as a destructive or constructive force. Thus, on the one hand this energy causes erosion, but it also builds up features, such as beaches. Tides are another type of wave that is caused by the gravitational attraction of the Moon. As the additional load of water approaches a shore, the motion that is generated deep down begins to interact with the sea floor. Along with the waves, the tides involve widespread rises and falls in sea level, accompanied by horizontal flows of water. They occur all over the world's oceans but are more noticeable near coasts. The basic daily pattern of high and low tides is caused mainly by the Moon's influence on Earth. Variations in the range between high and low tides over a monthly cycle are caused by the combined influence of the Sun and Moon.

Another major variable associated with the ocean waters is density. Somewhat like the atmosphere, the oceans consist of distinct strata of water masses that increase in density from the surface downward. The density of any small portion of seawater depends primarily on its temperature and salinity. Any decrease in temperature or increase in salinity makes seawater denser, except when the temperature drops below 39 degrees (F). In the latter case, the seawater becomes a little less dense. It is also generally true that, in any part of the ocean, the density of the water increases with depth, because dense water always sinks if there is less dense water around it. The processes that change the density of seawater cause it to either rise or sink, and to thereby drive large-scale circulation in the oceans between the surface and deep water. Most important is water that is carried toward Antarctica and the Arctic Ocean fringes, which becomes denser as it cools and as a result of an increase in its salinity as a result of sea-ice formation. In these regions, large quantities of cold, dense, salty water continually form and sink toward the ocean floor.

Some secondary large bodies of water, referred to as seas, appear as secondary nodes that often appear as "peninsulas" which extend from the oceans. These can also be seen as marine bridges between that connect the landmasses, such as the Mediterranean Sea. This particular secondary body of water has linked the continents of Europe, Africa, and Asia for many centuries of human history. Another of these "constricted" large bodies of water seems to be one which has flooded a rift between Africa, to the west, and the so-called Middle-East region, to the east. A tertiary set of bodies of water are the lakes which are located within expansive continental land masses. Taken together, they are large enough to warrant consideration in measuring the total amount of water on Earth. Among the largest of these inland seas is Lake Baikal, which is located in the northeast region of the Asian landmass. This body of water contains about one-fifth of the total "fresh" (non-saline) water on the planet. Another of the large body of water, situated in the center of North America, is constituted by a complex of interior lakes, the so-called Great Lakes, which is situated in the center of North America. Other significant inland lakes are found along the Great Rift Valley in eastern Africa, and Lake Titicaca in South America. These may not be as large as the Great Lakes of North America, but they are important within the context of the cultural geography that surrounds them.

Then there are the rivers and streams that drain continental watersheds. These ecological systems also have served as a source of food for many great civilizations for as long as about 7,000 years. Just as important, these waterways have served as transportation and communication systems throughout most of human civilization. Some of the greatest of these, in terms of volume and length, as well as their cultural importance, include the Amazon of South America, the Congo and the Nile in Africa, the Danube in Europe, and the Mississippi of North America. More generally, these rivers represent the quintessential system in nature, which has proven to be useful as an analog and a metaphor for many cultural systems. Indeed, I have found the river system as a useful tool for understanding and describing all manner of natural systems in outer space. One example of this is the electrical and magnetic fields throughout the solar system.

But there are still other bodies of water on Earth, such as the subsurface water that presents a hydrological system within the crust of the planet. Based on the evidence provided by deep core-drilling, as well as readings from seismic explorations, there are indications of large amounts of water occurring beneath the surface of our planet. Some take the form of pools, while others form veritable rivers, and large quantities of water permeate significant areas of subsoil and rock, like a sponge. The most recent exploration and analysis of the underground hydrology of the Earth is only another step in a continuing historical effort to find water for human purposes. Thus, for instance, the art of developing water resources that lie beneath the surface probably began as early as 5,000 years ago in places like ancient China, Babylon, and Persia. At first, these were simple, shallow perforations into the Earth's land surface, probably adjacent to rivers or lakes. Later, however, as the tools developed and practical experience accumulated, deeper wells of as much as 1,600 feet in depth were dug. Construction of "qanats," which are slightly-sloping tunnels that are driven into water-containing hillsides, is thought to have originated in northwestern Persia (now Armenia) approximately 3,000 years ago. A similar form of water-extraction technology seems to have been utilized in China as long as 5,000 years ago.

So, in some places, where the porosity of the geological layer is propitious, there are underground pools of water which are confined by impermeable layers of rock. These are called "aquifers;" and water is drawn from them by pumping it out through a well, or from an infiltrating gallery of horizontal, perforated tubes. This type of natural underground reservoir is of particular interest to the military geographer for several reasons. Within the context of the Earth, these aquifers and the associated cultural systems are potential strategic targets for air attack. This is akin to the bombing of surface reservoirs and hydroelectric dams as fresh water may become an even more valuable resource than petroleum in the 21st century and beyond. Beyond that, it is becoming more apparent that significant "aquifers" exist on Mars; perhaps on the Moon. If this proves to be the case, there likely will develop the same competition for water resources on other planets in the future.

———

Based on extensive telescopic and space missions to the various parts of the solar region, it is now thought that water ice exists throughout the solar system, from Mercury to the farthest reaches of the Oort cloud and the heliopause. This form of water plays an important role on both Earth and Mars, where it is a dynamic part of their climate regimes. Also, on the Moon and on Mercury, water ice is thought to be trapped within permanently-shadowed craters, especially in their polar regions.

In the outer regions of the solar system, water ice is a major component of many of the moons of Jupiter, Saturn, Uranus, and Neptune, as well as that of Pluto, asteroids, and planetary ring particles. However, not all ice that exists in the solar system is made up of water. Other "ices" which are made up of other volatiles, such as methane, are known or thought to occur on some of the outer planets and on their moons. Some asteroids and comets also show evidence that they are made up almost entirely of ice; while others contain a mixture of ice and rock. Ice also occurs within polar caps, permafrost, or volcanic craters which are shaded from the rays of the Sun.

Some ice formations on these bodies are thought to have been present ever since the birth of the solar system; while others of more recent origins. Among the terrestrial planets, it is thought that Mars, which started out very much like our planet, may have supported large bodies of water billions of years ago, before it became too cold to support them any longer. Venus, too, might have had large bodies of surface water, but in its case, it would have become too hot to maintain them. The Earth's Moon, Mars, and Mercury have all exhibited evidence of ice on their respective surfaces, as well. Meanwhile, the comets and the asteroids are thought to continue to contribute to the total water inventory of the planets and moons of the solar system. These traveling bodies also produce the meteors and meteorites that also contribute to the water inventories. It is interesting to note that the composition of these cosmic fragments have been found to contain deuterium within them. In fact, meteoritic material has roughly the same proportion of deuterium as the Earth's oceans, and so the assumption has been that water probably came from these external sources.

According to other, more recent studies, it is the asteroids from the inner solar system that are the most likely source of a major portion of Earth's water. Other analyses of carbonaceous chondrite meteorites show that they are the key sources of other volatile elements, such as hydrogen and nitrogen. As reported in the Journal of Science (July 12, 2012), a team of scientists measured the abundance of different hydrogen, nitrogen and carbon isotopes in chondrite samples, and they determined that our planet probably accreted its water and other volatiles from a variety of chondrite parent asteroids. Other analyses of asteroids by space probes shows evidence of water embedded deep within their interiors, which is taken to mean that some of them formed with water as part of their total composition. One exception is the dwarf planet Ceres, which might contain a water-ammonia ocean beneath its icy exterior.

Within the domain of the Kuiper Belt, there are comets that still retain much of the original water ice that they acquired during the early formation of the solar system. Until now, all of the comets that have been studied have been Oort-Cloud objects. These are thought to have been formed early in the solar system's history, in the region of the Uranus and Neptune. Eventually, however, they were shunted to a more distant zone of the solar system by the vagaries of gravity, and they settled into their current configuration, although they have continued to bump into each other. The reservoir of Earth ocean-like material in the Oort-Cloud region is much larger than was thought earlier and it encompasses cometary material which we had not previously been recognized.

New information indicates that primitive asteroids and comets are really siblings. Therefore, it is likely that both have delivered water to the objects within the inner regions of the solar region. However, it is not known exactly from which region of the heliosphere, and by which dynamic mechanism, the delivery of water has taken place. These questions are the subject of continuing scientific studies, especially with respect to the comets. Thus, while these objects are well-known for their present high ice content, there are some recent studies of comet dust that show that they also contained liquid water at some point during the distant past. Also, through the use of a kind of molecular fingerprint technique which measures the proportion of deuterium (a rare form of hydrogen), it has been found that the Comet Hartley 2 contains more water any of the other comets that have been observed. This strengthens the theory that suggests that much of the Earth's water might have come from the impacts of comets on our planet. Other Kuiper Belt objects, such as Pluto, are also thought to have oceans hidden beneath their surface, and are the subject of continuing studies.

Evidence of water ice also has been detected within niches on polar caps and craters on several other planets and some moons. These niches seem to be permanent or seasonal shadowed areas which remain cold enough to sustain the ice. The seasonal ice caps take place due to the varied solar energy absorption, as a body rotates on its axis, and as it revolves around the Sun. Additionally, in geologic time scales, the ice caps may grow or shrink due to global climate variation. Ice caps are generally in the range of subzero temperatures on the other bodies of the heliosphere. The polar ice caps on the other planets and moons consist primarily of water ice, but there also a few bodies whose ice cap is made of other volatiles, such as carbon dioxide, methane and nitrogen. As an example, frozen carbon dioxide makes up a small permanent portion of the South Polar ice cap on Mars. And, in both hemispheres of the Red Planet, there is a seasonal carbon dioxide frost deposit in the winter, which sublimes during the spring. Fluctuations in Mar's orbit also are causing the southern residual ice cap to undergo sublimation inter-annually.

Other celestial bodies that are attracting a great deal of attention from scientists these days include the moons of the outer planets. It is thought that these are the most likely places where oceans of water (or other volatiles) might be found. At the present time, the existence of most of these oceans remains largely hypothetical. However, the one that is most likely to exist is the one on Jupiter's moon, Europa. Imagery that has been returned from the Galileo space mission shows indications that Jupiter's largest satellite has a surface covered by a layer of ice, and that it may have a very active ocean beneath the surface. They also provided some indirect evidence of the presence of liquid water under the surface of Triton, the largest of Neptune's moons. There, the surface appears to have an outer crust made of frozen nitrogen; it overlays an icy mantle, which is thought to cover a core of rock and metal. The moons of Saturn and Uranus also appear to contain ice, but their greater density indicates that they less in their interiors than Triton.

To the extent that they exist, these lunar oceans likely are maintained by various internal heating processes, particularly radioactive decay within their cores. The one wild card is Saturn's moon Enceladus. It has been observed ejecting water out of geysers and cryovolcanoes, but it is unclear in what precise form the water occurs. There is some evidence that cryovolcanoes are ejecting liquid water, but it is possible that only vapor and ice are making their way to the surface. In any case, the presence of these formations is a strong indication that there is liquid water and that it is likely nearer to the surface, perhaps as close as 6 miles down.

The Hydrological Cycle

One model that has been developed to explain the global movement of water on Earth is the so-called hydrological cycle. It attempts to explain the mechanism by which water is distributed throughout the planet, in its three states. What makes this model so useful for studying the hydrology – not only of Earth – but the other bodies of the solar region, is that it has been validated by innumerable empirical observations. These have been done at virtually all places of our planet, from all spatial perspectives, and over wide spans of time. In fact, it can be said that throughout human history, there are many examples of the practical testing of hypotheses that can be derived from this model. As an example, the human food production systems that have been developed in various places throughout history, in ancient times and into the modern era, have sustained the hypothesis that there are good water supplies under the surface of the Earth.

Also, it has long been understood that the ocean and other large bodies of water are the main source of the rain on which their agricultural systems depend. The mechanics of the hydrological system, both locally and globally, can now be empirically observed, especially since the advent of the geosynchronous-orbiting artificial satellite. We now can even measure the precise inputs of solar energy that powers this water system. By the same token, countless school science projects continue to validate the hydrological cycle. That is, that sunlight is converted to heat, which then causes the vaporize water to form clouds, and that these clouds then cool and consequently deposit the condensed water onto the surface of the Earth. Then the difference in surface temperature and, therefore, in atmospheric pressure, cause the water-laden winds to eventually cool as they are forced upward by mountains, and ultimately deposit rain over the lands of the region. The essential hydrologic cycle is replicated in various ways and over many regions of the Earth.

One element of the global water cycle is manifested by the ocean winds, which today are monitored by an instrument called a scatterometer that is onboard a NASA satellite. It is from such orbiting satellite observations that we now can appreciate the fluxes of air that form a web over the entire globe and within the subsets of it. These flows of provoke the movement of move masses of air and water over the surface of the globe. Within these macro horizontal movements of water within large bodies of water, there are smaller cyclic currents that resemble eddies in a river. Meanwhile, there are the vertical cycles of water that are occurring within the oceans, in which layers of colder water descend and warmer waters rise. It is the interplay between these macro and micro movements of air and water over the surface of the planet that determines the spatial distribution of water that is deposited on the surface. If this global surface were a plane, the geography of water on a planet would be much different than it actually is. In fact, the surface is not a plane; rather it is a rough surface which includes many peaks and lower plains and valleys. As a result, the deposition of water on the surface is more complicated.

Equally complicated is the continuous circulation and churning of seawater, both across the surface and within the depths of large bodies of water. The engine for all this movement in bodies of surface water is solar radiation. (Less well understood, is the part that is played by geothermal eruptions on the floor of the oceans). The various surface currents that are generated by this dynamic movement have profound effects on climate in many parts of the world. Two of the more important of these are El Nino and La Nina. These are responsible for periodic climatic disturbances, and they help generate the extreme weather phenomena which are known as hurricanes and typhoons, as well as the periods of drought.

Ultimately, the global patterns of air movement over the oceans result from variable solar heating of the atmosphere, and the Earth's rotation. However, this macro pattern of winds is locally modified by linked areas of low and high pressure (cyclones and anticyclones), which continually move over the surfaces of the oceans. Near the coasts, additional onshore and offshore breezes are common. These are caused by relative differences in the capacity of sea and land to absorb heat. It is also solar heating that causes the air in the Earth's atmosphere to cycle vertically in various places on the globe, most notably in sets of giant loops called atmospheric cells. The cells cause air to move in a north-south direction over the surface of the planet. However, this pattern is altered somewhat by the Coriolis Effect, which is a consequence of the Earth's spinning on its axis. Because the Earth turns continuously underneath the airflow as it travels, the air appears to be deflected from its straight north-south course. This effect is an apparent one; it isn't a true force, and no actual force is exerted on the wind. In the northern hemisphere, the Coriolis Effect causes all air movements to be deflected to the right of their initial direction. In the southern hemisphere, they veer to the left. The winds that are produced by differences in atmospheric pressures, and are modified by the Coriolis Effect, are called the prevailing winds.

Then there are the local Hadley cells which are produced by warm air that rises near the equator, cools in the upper atmosphere, and descends to the surface around the subtropical latitudes, at about 30 degrees of latitude. In the tropics and subtropics, the air movements toward the equator in Hadley cells, are deflected to the west. These are known as the trade winds: they comprise the northeasterly trades in the Northern Hemisphere, and the southeasterly trades in the Southern Hemisphere. At higher latitudes, the surface winds in Ferrel cells deflect to the east, producing the westerlies; while in the Southern Hemisphere, these winds blow from west to east without meeting land. Those around latitudes of 40 degrees south are known as the Roaring Forties. In the polar regions of the Earth, the winds deflect to the west as they move away from the poles, and these are known as polar northeasterlies and southeasterlies. Year-round, the winds over most oceans are trades or westerlies; an exception to this occurs in the northern Indian Ocean which has a monsoon climate. There, a seasonal switch in wind direction occurs as a result of the relative heating of the ocean and the landmass.

Another local system of winds, called onshore and offshore breezes, is generated near coasts, especially in sunny climes. Onshore breezes develop during the day when the land is relatively hotter than the adjacent waters. Thus, as the land warms up, it heats the air above it, causing the air to rise; cooler air then blows in from the sea to take its place. In the evening, the reverse process occurs. That is, the land quickly cools down, but the sea remains warm and continues to heat the air above it, producing an offshore breeze. By the same token, in any area of ocean where air sinks (often at subtropical latitudes), a zone of high atmospheric pressure, or anticyclone, develops. Conversely, where warm air rises, areas of low pressure called cyclones or depressions occur. These often develop near the equator and subpolar latitudes as well. Both cyclones and anticyclones create linked, circulating wind patterns, which continually move about. In the Northern Hemisphere, there is a clockwise movement of air around an anticyclone, and a counterclockwise motion around a cyclone. This pattern is reversed in the Southern Hemisphere. These local pressure systems can also affect the general pattern of prevailing winds. Thus, cyclones move more swiftly over the ocean, usually producing rapid changes in wind strength and direction.

———

The hydrological cycle model also can be useful for analyzing water systems throughout various phases of the history of the solar system. Thus, the original hydrological cycle of the solar region can be seen to have originated billions of years ago in the nebula cloud that was emitted by the Big Bang event, when the proto-matter was provided with the initial "dose" of water. Thereafter, the water was redistributed over a period of billions of years, as the matter accreted and was reconfigured by the dynamics that created the solar system geography that we see today. Meanwhile, it is the Sun that has provided the energy that drives the overall proto-hydrological cycle. Since the development of this original water cycle of the solar system, there has continued to operate another "internal" cycle. It is manifested by the continuing additional "inputs" of water to the planets by the asteroids and comets. Then, these inputs are processed by the various bodies. Thus, the water within the planets is transported and driven by the energy of the Sun and the internal regions of the planets themselves. The output from this internal hydrological cycle is still not completely understood, but apparently water in both liquid and ice evaporates everywhere throughout the solar system. So, it would appear that the mechanics of the hydrological cycle throughout the solar system operates essentially in the same way as the Earth's hydrological cycle. Ultimately, however, the planetary cycles are both variable and complex. The reasons for this can be expressed in one word: "location, location, location." Thus, the Earth is located far enough from the Sun so that its rays do not burn off the moisture, as is the case with Mercury. On the other hand, it is not so far away from the gravitational influence of the Sun so that a giant planet like Jupiter can complicate its existence and its temperature variations. Again, it is not so far away from the Sun that it remains an icy planet like Neptune.

———

The hydrological cycle model also can be applied to an analysis of the present movements of water throughout the heliosphere. Once again, the initial distribution of water was the result of the dispersal mechanism which occurred during the original "precipitation" event which followed the early condensation of the proto nebula cloud. In any case, it seems that all the planets received a given ration of water at the outset. Then, over the next billions of years of formation of the solar system, the individual planets have continued to receive additional allotments of water from the asteroids and comets that happened to collide with them. Scientists have paid particular attention to the number of such hits on the planets Earth and Mars, which are thought to have many similarities. This would have constituted the early "historical geography" of the hydrosphere of the solar system.

Today, there are indications of the presence of water in many places throughout the solar system, based on data that has been returned from space missions and telescopic observations. As an example, the Mariner 10 mission employed spectral measurements to determine whether Mercury has condensed water, most likely ice, within the permanently shadowed craters near the poles. On Venus, the interaction of that planet with the solar wind results in a gradual, but continuous loss of hydrogen and oxygen to space from the planet's upper ionosphere. This process has resulted in an equivalent, gradual loss of water from Venus. Over the course of that planet's history, the cumulative amount of water that is dissipated could have been equal to a few percent of the Earth's oceans. On Mars, the water vapor in the atmosphere appears to be in contact with a large subsurface reservoir in the soil. Subsurface layers of ice are believed to occur at higher latitudes. The very low subsurface temperatures there probably prevent the ice from subliming into the atmosphere. Farther out, within the Jovian subregion, there is evidence of an ocean of warm water beneath the crust of several of the moons. One difference between the hydrologic cycle on Earth and that of the whole solar system, however, is that of magnitude. That is to say, the movement of water in the overall solar region must be measured in terms of much greater lengths and over periods of time that are measured in terms of billions, or at least millions of years. The main reason for this is the size of the total region, which is measured in Astronomical Units (about 92 million miles each), in a heliosphere whose boundaries are approximately 100 to 150 AU. With this in mind, we have learned a lot about the movement of water throughout the solar system in the past half-century of space exploration.

In analyzing the hydrological cycle of the solar system, we can appreciate the existence of a "water budget" there as well. Thus, the planets and other bodies continue to "sublime" water back into the interplanetary medium (which is analogous to the Earth's atmosphere). This would seem to support the hypothesis that the ancient water that is stored in the depths of a rocky planet is sometimes cycled back to the surface via the volcanic eruptions and lava extrusions that occur through the fissures. It also supports the idea that the Martian interior originated from the same planetary building blocks as those that formed Earth. In this scenario, chondritic meteorites that contained small granules of minerals are a main source of water for terrestrial planets. This is further supported by scientific reports in 2012; these indicated that rocks on Mars contain volatiles, such as water, which were not recycled back into the planet's interior. Indeed, the primordial water the researchers found in the meteorites from Mars did not show any imprints of the recycled water from the planet's atmosphere. However, one such meteorite was relatively rich in elements such as hydrogen, while the other one was not. All of these findings have served to provide forensic evidence of the commonality among the planets, which supports the idea of a system-wide hydrosphere within the heliosphere. There are many other areas in which the Earth's hydrological system, while it appears to be unique in some ways, is nevertheless useful as a template for the analysis of other such systems on other bodies of the solar system. Most likely, we will probably find that certain subsystems and elements of our own hydrological system will prove to be useful as analogs for understanding the role of other water systems, whether they be in gas, liquid, or solid form. Water, in all its states, is also one of the most important elements in the solar system, and it is also essential to the vitality of all life forms on our planet.

The Military Geography of Water

With respect to military space operations, water in all its forms presents many opportunities and challenges. As an example, I witnessed the effect that water has in low-temperature regimes during my tour of duty at Grand Forks AFB, North Dakota, at the height of the Cold War. In a place where winter temperatures often dipped below zero (F), the base's complement of B-52 strategic bombers had to be ready for takeoff at all times. So, external heaters and chemicals were used to keep the ice from forming on the wings of the aircraft. This is but one example of how water, in all its forms, can affect the operation of military aerospace systems. Even the electronic systems of aircraft on the tarmac or the space rocket on the launch pad can be degraded by water.

Platforms that operate at the higher altitudes of the Earth's atmosphere, or in space itself, can avoid the deleterious effects of water during most of their sortie cycle, but those that must return to the surface, via the atmosphere, are still subject to the negative effects of moisture, in one form or another. On the other hand, most military aircraft generally operate within the sector of the troposphere, where "weather" occurs. This is the particular workshop of the atmospheric military aviator. In this sector of the atmosphere, water can interact with temperature to manifest itself in all three of its states. Thus, if air temperatures fall to a certain temperature in a given region of air, water can appear as hail, sleet, or snow – all of which have their particular impact on military aerospace operations.

Above the freezing point, there is opportunity for water to appear in the states of gaseous or liquid, and negatively affect military aerospace operations. In its various manifestations of water create many types of problems. In terms of air navigation, water can affect visibility, although this problem largely has been dealt with by radar and satellites. However, in the target area, water can still cause problems in the detection or the striking of a target. Thus, water can present itself in the form of clouds, which can be thought of as the "terrain" of the atmosphere, and like mountain ranges, can produce dangerous shears and other sudden changes in the air that are generally referred to as turbulence. Such turbulence can unsettle an aircraft platform as it engages in strikes against other air targets, or ground targets. The clouds also can provide "cover" for aircraft which are being stalked by other aircraft in an air battle. By the same token, they can serve to obfuscate the enemy targets on the ground, as well as the enemy targets in the air. Thus, clouds can be seen as a double-edged sword in many forms of aerial combat. The bottom line is that clouds can be seen as a source of instability to military aviators, for good or ill.

Water also can have negative effects on the conduct of electronic warfare. Thus, any of the sensors and weapon systems that rely on electromagnetic radiation will face the problems of attenuation, diffusion, and degradation of their signal. As a result, the "signal" to the weapon guidance and control subsystem will be degraded or completely blocked. Or the sensors may not be able to detect the target in the first place. And, even if the target detection and the weapon guidance subsystems are effective, the fusing subsystem may not operate as expected because of the degradation of the mechanism that relies on atmospheric conditions to detonate. Even the onboard avionics of a platform can be negatively affected by excessive humidity over the course of a sortie.

A VIEW OF FUTURE MILITARY SPACE OPERATIONS

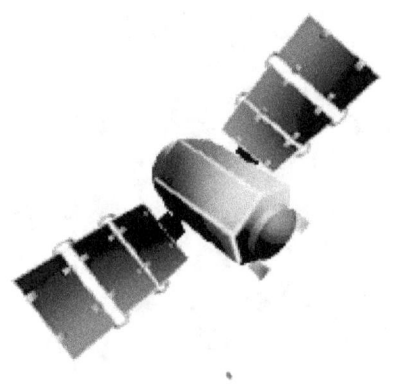

FROM PLOWSHARES TO WEAPONS

It is axiomatic that United States military operations, whether on Earth or in space, are based on national doctrine and the derived strategic objectives – as laid down by civilian national authority. However, we are at a stage of technological development which makes such a statement of doctrine and strategic objectives no longer the absolute domain of nation-states. So, now it is possible that even so-called "terrorist" groups will develop such statements of doctrine to guide their future military space operations. In any case, if we refer to the doctrinal and strategic statements of the U.S. Air Force and the Navy, we can discern the framework for the conduct of U.S. military operations in space, in the 21st century. It also might be inferred that the basic model for the command structure of military operations in space will very likely be similar to those on Earth; only the venue will have changed.

On the other hand, the "national command structure" that is utilized by the nation-states on Earth may evolve into an alliance command structure, much like the one that has developed within the North Atlantic Treaty Organization (NATO). The main imperative for such a geo-political command structure in the conduct of military operations in space is the fact that the solar system is so vast. This alone implies that no one country will have the resources to replicate the nation-state model in military space operations. One possibility is that the model for the militarization of space may even be similar to what has evolved in Antarctica, in which a more cooperative strategy is in place; one which is designed to ensure the security of all stakeholders. Already, we are seeing a trend toward multi-national cooperation in space exploration and scientific studies. The International Space Station and the Cassini-Huygens missions are two of the most obvious examples of this phenomenon. In any case, since the end of the Cold War era, the United States has begun to develop a command structure for conducting military operations in space. Thus, the U.S. Air Force Space Command has been organized; and it has been assigned responsibilities with respect to the areas of the solar system that lie outside the Earth's atmosphere. In the following listing of the stated objectives of the U.S. Air Force Space Command one can infer that there many different kinds of space sorties that will occur in pursuit of them. Thus, according to military planning documents which have been published on the internet by the U.S. Joint Chiefs of Staff, the outlines of the nature of military space operations can be inferred.

Like any traditional military operations plan, these documents begin by defining the mission objectives and describing the systems that would carry out these objectives. Proposed strategic objectives would likely relate to: (1) space-force enhancement; (2) development of space support systems and infrastructure; (3) control of space capabilities; and (4) development of space force application capabilities. These represent a statement of intent that should be familiar to anyone who has had experience with U.S. Air Force operational planning on Earth. Therefore, it is reasonable, I believe, to assume that the achievement of many of these objectives, especially the imperative to achieve and maintain space domain superiority, will require many sets of ongoing sorties that will parallel the air operations designed to achieve of air superiority on Earth. Indeed, as I analyze the stated strategic objectives, I am struck by how similar they are to the mission statement of the U.S. Air Force Tactical Aerospace Command that was defined during the Vietnam War. These include the imperative for attaining air superiority to control the battlefield, the development of aerospace combat support systems, and the development of a force structure that is capable of applying the required force against enemy targets. These should be familiar to anyone who has worked in air intelligence or operations in a combat area of operations.

But, if a united command structure, like NATO, were to be developed in space, then a multi-national military space force might be expected to emerge. One of the most interesting things about the NATO is the way in which it has been able to evolve in response to changing geo-political and military conditions. It has changed from its original configuration as a unified military force, whose main mission was to be a counterforce to the strategic threat from Soviet Union and its Warsaw Pact allies during the Cold War era. However, since the end of the Cold War in 1991, the NATO system has been transformed in several ways, usually in response to changing geopolitical realities. One of these has been the emergence of the European Union, which now presents many of the elements of a traditional nation-state. For one thing, this emerging political union (which now includes many of the former Warsaw Pact countries) has established a legislature. And, the outlines of a European judiciary system also are being sketched. However, the most relevant development, as far as military space operations are concerned, has been the configuration of NATO into a multi-national military force, which also includes the United States and Canada.

So, we can see that a NATO is now embracing a new cooperative and global military force structure to deal with the realities of globalization and "democratization" of technology that has been developing on Earth in recent years. (Consider that the development of an effective long-range rocket is becoming easier and cheaper to accomplish than ever before). Also, it is interesting to observe how this multi-national military force has been developing a relationship with the United Nations since the end of the Cold War. A recent example of such a unified NATO military force structure, operating under a United Nations mandate, was seen in the military operations in Serbia during the 1990s. Such an organizational structure also was utilized during the Gulf Wars, and in Libya. Should this model of unified action be followed in space, we might begin to see the emergence of theaters (or military space domains) as an organizing mechanism for assigning areas of operational responsibility to such a military space force. It is also probable that these military space domains will be ordered spatially in the form of concentric orbits around the Earth. If so, I would suggest here that the military space orbits of the future might resemble the operational routes that are taken by naval forces on Earth. This is because I see outer space as being analogous to the oceans on Earth, and the celestial objects as being like the continents and islands that seem to exist among the oceans. Thus, the trajectories that are planned and executed for the completion of NASA missions in space may bear many of the characteristics of the courses and the sea routes of the naval forces on Earth.

I think that maybe the best expression of overall military strategy in space can be derived from the writings of Alfred T. Mahan, a highly influential U.S. naval historian. Writing about naval operations, he stressed the interdependence of the military and commercial control of the seas and asserted that control of seaborne commerce can determine the outcome of wars. He also stressed the imperative of sea power in national historical supremacy. This military doctrine and the national imperative, he argued, depended on continuous improvements in military technologies. Another American naval thinker, Rear Admiral Bradley A. Fiske, in his work entitled <u>The Navy as a Fighting</u>

Machine (1916), noted that the Confederate ironclad, the Virginia, was able to handily defeat several wooden Union ships; only to be fought to a standstill a day later by the Union ironclad ship, the Monitor, which mounted only two guns on a turret. The difference, he argued, was the more advanced technologies and tactics that were employed by the Monitor, including the turret-mounted naval guns, which enabled it to fire continuously without respect to the orientation of the vessel. Another contribution to military science by these naval thinkers was the concept of the "constants." Among these are the realities that: (1) unlike ground combat, the principal aim at sea is to put the fighting machine, rather than the individual human, out of action; (2) modern machines are sensitive to damage; and (3) it is a longstanding constant that naval battles, once joined, are fast-moving and decisive. Therefore, I see future space warfare as being most analogous to naval warfare on Earth. That is: the processes of space warfare are essentially the same as in naval warfare. In both cases, the efficacious delivery of firepower to the precise target point is the most important objective. To reach this critical point in a battle, however, a lot of preparatory events must occur, including, what in naval parlance is called, "scouting." This involves such matters as the gathering of information (intelligence) by reconnaissance, surveillance, cryptanalysis, and other means. The information that is gathered must then be collated and otherwise analyzed for utilization by the tactical commander. The more strategic objective in space warfare, however, is to maintain a highly-responsive command and control presence in the space domain.

This, in turn, would require the continuing assimilation of all relevant intelligence data for processing. By "processing," I am referring to a particular phase within the input-process-output cycle, which includes: (1) the filtering and validation of the raw data for presentation to the "programs" that make up the actual production of actionable information; (2) the processing function involves many different algorithms, Boolean decision logic, and other manipulations of the validated data (these can be in the form of digital or qualitative measurement tests); and (3) the presentation of usable intelligence in a form (verbal presentation, imagery, or as electrical charges that cause actions by automated subsystems) that enables the user to take a desired action. All of these processes and events are part of what is called command and control of military forces. This is the locus of the decision-making which determines what actions are called for, and directs forces to act accordingly. At this point, it should be noted that the processes of space warfare are increasingly being done by computers of one sort or another. This trend can be seen today in the move toward remotely-piloted vehicles and virtually autonomous platforms. For this reason, one of the most critical elements in space warfare will be what is called telemetry, which is the common method for communications between spacecraft and ground stations, usually via radio signals.

U.S Naval Doctrine, as expressed in a document posted on the U.S. Naval War College website (usnwc.edu), provides a formal structure to this perceived analogy between naval and space warfare. "Engaging in and winning battles in the maritime domain and preventing conflict through presence offshore" is a statement of purpose that could easily be applied to the inter-planetary domain in the solar system. In the same vein, the stated imperative of maintaining the "...continued prosperity" of the U.S. and its national partners is "tied to the maritime domain..." sounds very much like a reference to a future inter-planetary domain in our solar region. The assertion that "our freedom of the seas is secured by naval forces" is another tenet of U.S. naval doctrine and it, of course, could also apply to the necessity of maintaining freedom of action by space forces throughout the inter-planetary medium of the solar system. Finally, just as the littoral regions of the oceans are relevant to this freedom of action on Earth, the surface of any planet can be seen as being a littoral to outer space. Indeed, the surface littoral zone is where virtually all launch facilities and detection and tracking stations will likely continue to be located for the foreseeable future. This means that ground military forces also will continue to play an important role in the military space environment, at least during the next few centuries. As such, they will be an echo of the U.S. Air Force Security forces that provided asset protection for the SAC facilities throughout the world during the Cold War era. Like the U.S. naval forces on the Earth's hydrosphere, the space forces will need to possess the staying power to protect and sustain operations, as long as is necessary to project and sustain operations in the new domain.

Although the naval paradigm is especially relevant to military operations in space, U.S. Air Force doctrine documents (af.mil) also make the point that the essence of military operations – whether in the Earth's atmosphere, inter-planetary space, or in cyberspace – is not just about particular platforms or weapon systems. Rather, it is about effects and desired outcomes. Doctrine is the blueprint for warfighting in any milieu. It focuses on the desired outcome from a particular action, and it is based on the idea that a desired outcome can be produced by many different configurations of force structures, and platforms and weapon systems. Also, it is stressed that warfighting is not tied to certain physics. That is, differences between operations in the Earth's atmosphere, space, and cyberspace are simply differences in military environments that require exploitation of different sets of physical laws. Therefore, to achieve a common purpose, air, space and cyberspace capabilities must be applied in an integrated fashion. Also, it is of paramount importance in military operations, anywhere in the solar system, to properly use the given medium in order obtain the optimum warfighting effect.

Therefore, it is always the case that U.S. military forces, in any medium or environment, are formed and deployed with an eye to furthering certain national objectives. The U.S. Navy of the 21st century defines certain sets of objectives to guide the use of naval power on the oceans and seas of the Earth. These objectives are ultimately defined in terms of stated policy which guides the action of military commanders. According to the Naval War College documents, these currently include the dicta to: (1) limit regional conflicts with forward-deployed decisive naval power; (2) deter major-power war; (3) win the nation's wars; (4) foster and sustain cooperative relationships with international powers; and (5) prevent or contain local disruptions before they impact the global system. These policies could also be applicable to the U.S. Air Force and the U.S. Army in the 21st century. Consequently, strategic policies are seen as being national strategic policies that are formulated by the National Authority (U.S. Congress and Commander-in-Chief) and converted into strategic objectives; and then into operational and tactical orders that deal with unfolding global events.

————

A naval force only has control of the sea when it is so strong that its rivals are unable to attack directly. Such dominance implies either supremacy or at least significant supremacy of the oceans and seas of the planet. In this sense, it is the equivalent of air supremacy in the atmosphere. By extension, in the case of its analog – the interplanetary medium – it would be like having supremacy or dominance throughout the space that lies within the boundary of the heliosphere. With the command of space, a nation-state or an alliance of nation-states could ensure that its own military spacecraft (and merchant spacecraft) could move around the solar system at will, while adversaries would be forced either to "stay in port" on Earth or some other celestial body – in much the same way that the German navy was bottled up in its North Sea bases during much of World War I. Another advantage of having space supremacy would be more freedom to conduct "amphibious" operations against hypothetical enemy bases on celestial bodies in the solar system.

At the present time (in the early decades of the 21st century), the United States enjoys paramount control of interplanetary space, mainly because of its sustained investment in exploration and technological development that began in the middle of the 20th century. But it is not total dominance by any means. Indeed, it continues to exist by virtue of the U.S. investment in existing military weapon systems that are relevant to the geospace domain – which includes the artificial satellite belt and the surface "littoral" of the domain on Earth. Indeed, it is along this littoral that many nation-states (including not only the U.S., but also Russia, China, India, France, and even North Korea) are developing the ballistic missile capability to either damage or destroy the artificial satellites that orbit the Earth. Even if a nation-state or a coalition of states choose not to directly challenge the supremacy, or at least superiority, of the United States in space, many states and "non-governmental" agencies (sometimes referred to as terrorist organizations) have the money and other assets to someday launch a missile with a nuclear warhead into the orbit of the artificial satellites.

————

Military forces in space also might be deployed as "expeditionary forces," which would presumably be custom-designed for a particular mission. The idea would be to develop the precise military weapon systems and support systems to achieve a given objective, in response to a particular event or set of events. However, because of the vastness of space even within the solar system, it likely will be necessary to develop several regional military space commands (domains) as a more workable framework for the application of military force to achieve a desired outcome. The military space domain that is most active today is the so-called "Geospace Domain." It is bounded by the surface of the Earth at one end, and extends outward toward the outer limits of the artificial satellite belt – anywhere between 200 miles and 22,000 miles above the surface of our planet. There we find a growing number of space stations, orbiting telescopes, and other spacecraft that are operating within this domain, and can be seen as part of the "cultural geography" of this region, or spatial system. Among the natural nodes of this spatial system are included the Earth and its Moon, the planet Mars, the Main Asteroid Belt, and several transient asteroids, comets, and meteoroids. As with all functional geographic regions, several other nodes of significance are gradually developing, and they are being interconnected by communication and transportation linkages that are being established. It also might be said that the various nodes and linkages are developing functional specialties to further the efficacy of the system. An example of the nodal specialization would be the Earth's Moon which, it appears, is being groomed as a possible staging site for conducting space operations within the Jovian domain.

As to the cultural geography of this region, many nation-states and non-governmental organizations already have deployed a substantial network of artificial satellites in low-orbit about the Earth. Some of these platforms already are providing communication, navigational and meteorological support to the Earth-based missile defense system and to other military forces that are deployed along the "lower littoral" of the Geospace Domain. These are not logistical or supply satellites in the strict sense of the term, but they are "life-lines" on which military space domain control operations would depend in future space conflicts. We also are witnessing the beginnings of an effort by governmental and private organizations to exploit the natural resources that occur in the solar system for commercial and strategic reasons. This phenomenon can be seen as a motivating factor in the development of space dominance in this region. And, these trends can be seen as precursors to some kind of military presence in space. In any case, the following is an outline of the kinds of military strategic and tactical activities that might be expected even during the early decades of the 21st century.

At the present time (at the turn of the 21st century), the greatest threat to the orbiting artificial satellites comes from the long-range ballistic missiles on the surface of the planet that are being deployed in increasing numbers by the various nation-states and, potentially, by non-governmental organizations. In any case, the next generation of military-purpose satellites likely will consist of smaller, but more efficacious satellites that operate in highly specialized orbits. This will effectively serve to increase our asset presence in geospace, and therefore make it more problematical to defend against attack. As with any asset base that must defended, the options for doing so will depend on the resources available and the nature of the threat profile. In the case of the artificial satellites, the practical problem of defending them reminds me of the historical problem of defending the B-17 bombers as they attempted to strike targets in Europe during World War II. In the same way, one option for defending the artificial satellites would be to arm them so they can defend themselves with onboard weapons. There are many problems with this approach, including the danger of causing "friendly fire" or collateral damage to our own assets. Also, current international treaty agreements proscribe the deployment of certain kinds of weapons onboard the satellites. A second option, which probably will be found to more effective, is to develop a "combat air patrol" or escort platform system to provide security for the satellites. This platform would operate in the same fashion as the P-51 fighter of World War II, or the F-4 Phantom in Laos, during the Vietnam War. If history is any guide, there will almost certainly be other weapons and tactics that will be developed to solve this basic problem of defending assets in space.

To counter the potential growth of the surface-based missile threat on orbiting assets, more sensitive and precise anti-missile systems are being developed. These will be able to detect dimmer and shorter-duration heat events within the near-space region in the future. The newer satellite-defense systems also will be able to provide more efficacious notice of missile launches and more precise impact-point predictions. One of the elements of these anti-missile systems will likely resemble the current GPS constellation (now comprising 31 orbiting satellites), which provides time, location, and velocity data. And, they will be designed to be jam-resistant, redundant, and hardened against nuclear radiation.

One metric that can be applied to the nodes of the solar system is their distance from the Sun. Actually, there are no fixed distances between the nodes of this functional region, rather there are the changing distances that occur along the orbits around the central node. For instance, within an elliptical orbit, the point at which a planet is closest to the Sun is called its perihelion, which implies that distance involved will always be dynamic. Although the mean of these metrics is often cited in tables of planetary data, for military space operations such a value would not be useful "information." Instead, it would seem that the most useful measure of distance from the Sun would be the specific distance of a spacecraft at a given point along a particular orbit, and at a specific point in time. In the same manner, the particular distance at a given point in time relative to one's vantage point would represent true data that could be inputted into a navigational algorithm to arrive at a target rendezvous point. By the same token, the mean gravity of a planet would be of less value to the navigator or mission engineer than the escape velocity value for a particular body. Escape velocity is that which is sufficient for a body (or a molecule) to escape from the gravitational center of the body of origin without undergoing any further acceleration. At the surface of the Earth, for example, escape velocity is about 6.96 miles per second (disregarding the atmospheric resistance that is involved).

Table 3

The Parameters of the Space Domain

Relevant Natural Factors That Affect Launches throughout the Solar System

	Mean Distance-Sun	**Mean Gravity**	**Escape Velocity**
	AU	**cm/squared**	**km/s**
Mercury	0.4	370	4.3
Venus	0.72	860	10.4
Earth	**1.0**	**980**	**11.2**
Mars	1.5	372	5.022
Jupiter	5.2	2,312	59.5
Saturn	9.5	896	36.0
Uranus	19.2	869	21.3
Neptune	30.1	1,115	23.5

One light-year is equal to approximately 63 AU (1 AU equals 93,000,000 miles). If 1 AU equals 1 inch, then 1 light-year equals 1 mile. Given the propulsion technology that is available today, a rocket travels at a speed of about 17,000 miles per hour. Therefore, given that the distance to Mars is 139,500,000 miles, a rocket launching from Earth can reach Mars in about 82 hours (or 3.41 Earth days). A one-way trip to Jupiter would take approximately 18 days. Ultimately, to reach Neptune, the farthest planet from the Sun, would require a one-way journey of about a half a year.

———

Another aspect of the escape velocity that is relevant to military space operations is the particular escape velocity of the individual planets, moons, and asteroids which might be used as launch sites. In planning for the various space missions, NASA engineers have calculated the exact value of the escape velocity based on two factors: the mass of the body that hosts the launch station and the distance from the center of the body to the space vehicle that is being launched. The following table, which is derived from nasa.gov data, lists the various surface escape velocities for some of the planets and bodies of the solar system. This simple listing of relative escape velocities in Table 3 indicates that it is more efficient (in terms of the quantity of energy that is necessary to lift a given weight load) to locate a launch site on an asteroid, as opposed to a launch site on the Earth, Mars, or even the Moon.

One model which attempts to explain, scientifically, why this is so, is the so-called "rubber sheet" model of gravity. The essence of this model is that gravity well, or the gravitational field, is like a rubber sheet that is spread out throughout space, and which surrounds every celestial body. A "gravity well" is a conceptual entity that occurs in a gravitational field. The idea is that the more massive the body, the deeper and more extensive will be the gravity well that is associated with it. Thus, the Sun has an extensive and deep gravity-well. On the other hand, asteroids and small moons have much more shallow wells. Anything on the surface of a body is considered to be at the bottom of its gravity well, and the energy required to escape from this well is much greater for the more massive bodies.

———

The Military Space Domain

Potentially, the largest, all encompassing, military space domain would be the solar system itself (within the context of the Milky Way galaxy). In practice, its theoretical outer boundary is defined by the potential for projection of military power at any point in the future, given the state of the military technologies of the day. This capability can be inferred from the space operations that are being conducted today by such space agencies as the U.S. NASA and the European Space Agency (ESA), among others. I am referring specifically to the robotic space probes that have gone beyond Pluto and onto the zone of the heliopause itself. The central node of this military space domain would not be the Sun, but rather the central operating node, such as the Earth, given the most likely circumstances. The secondary nodes in this system would be the particular military bases and resource nodes, such as certain asteroids or the Moon. Perhaps another military space domain would be centered on Mars and its secondary nodes might be bodies in the Main Asteroid Belt. By the same organizational logic, one of the moons of Jupiter might be established as the central node in another military space domain in that subregion of the heliosphere. Most likely, the "natural" nodes of these spatial systems would be joined by hundreds of artificial bodies that have been constructed by humans and sent into space.

The Geospace Domain is the only active military area of operations in space at this time. Its configuration reminds me of the region of the Roman Empire that developed along the north and south littorals of the Mediterranean Sea. This seems to be a useful paradigm for analyzing the region of the artificial satellites. Its location in space, between the rocky inner planets and the outer gas giant planets, can be compared to the medium of water known as the Mediterranean Sea – which lies between the northern littoral of Europe and Asia Minor and the southern littoral of North Africa and the Middle East. By the same token, the artificial satellites can be likened to a fleet of naval vessels that is deployed along various orbits within an interplanetary medium. Continuing with the analogy, we can see that, like the individual naval vessel, each satellite performs a particular function, or set of functions, in order to promote the objectives of the whole "constellation" of satellites. In the same way that a fleet in the Mediterranean Sea might have been oriented toward both littorals of the sea, the constellation of artificial satellites is designed to focus attention on both the Earth and outer space.

In this geospace military space domain there currently are being conducted operations, of one sort or another, not only by the United States, but also China, Russia, the European Union; and more are expected to enter in the coming decades. Thus, even though military activities within this zone are proscribed by international convention, the Chinese already have pushed the issue by actually firing a missile to destroy one of their obsolete satellites. In a sense, the geopolitical situation with respect to the zone of orbiting satellites appears to present many of the same characteristics of the "mutually-assured destruction" condition that stayed the hand of the adversaries during the Cold War. In any case, by the turn of the 21st century, several thousand artificial satellites have been placed into orbit around the Earth, mostly at a distance of approximately 200 to 500 miles above the surface of the planet. Most of these have been launched by governmental entities, but a few have been sponsored by private sector organizations. The missions that have been assigned to these space robots can be sorted into several categories, to include: communications, navigation, remote sensing, and science satellites. Within these major categories there are those that serve a military purpose. These are the satellites that are variously referred to as "dedicated systems," whose missions are invariably classified. However, they can categorized be as performing the following general functions: intelligence-surveillance-reconnaissance (ISR), communication and navigation, as well as airborne reconnaissance, targeting, and ordnance guidance and control.

A review of the literature of the military space community also points to the zone of orbiting satellites as the most likely theater for military operations in space at the present time. As of 2012, the number of satellites which are orbiting the Earth was on the order of thousands. A review of the various summary descriptions of the orbiting satellites which are under the operational control of the U.S. Air Force Space Command reveals that they are performing many traditional combat support functions from their locations in space. It should be noted that some of these are support services that are still being provided to ground forces in places like Afghanistan. A vital area of combat support which is being provided by orbiting satellites is that of communications, wherein the satellite serves as a relay station. Meanwhile, the Global Positioning Satellite system provides navigational, timing, and velocity data to all forces. Another area of support is the provision of meteorological information to all military units throughout the world. Early warning of possible ballistic missile attack on US forces or on the homeland is also provided by an orbiting satellite system. And, other satellite systems provide a more strategic detection and warning function; these operate in the infrared portion of the EM spectrum to conduct surveillance of space, that is: to maintain total "situational awareness" in space.

So, it would appear that a de facto military situation has developed in which these satellites will have to be "defended" against potential threats from either nation-states or non-governmental organizations. I would argue that this situation has gone beyond the realm of speculation. It now can be characterized as being a matter for serious contingency planning by some major powers, including the United States, Russia, and China, at the present time. One illustration of this fact is that today (in 2013), there exists a formal, fully-functioning U.S. Air Force Space Command whose mission, broadly speaking, is to defend United States interests in space. In many ways, this reminds me of the mission statement of the U.S. Navy and the navies of the other major powers during the late 19th and early 20th century, with respect to international waters. These naval forces were deployed with the mission of showing the flag and to apply military force when and where necessary – to ensure the secure flow of natural resources to the home nation-state, including petroleum, copper, and bauxite from sites in Africa, Asia, and the Americas. Agricultural commodities such as bananas, coffee, sugar, and rubber also were deemed as resources whose secure flow to the major powers required security by naval forces. More recently, the most critical natural resource happens to be petroleum.

It appears that the same dynamic is beginning to unfold in space. In the Air Force Magazine, July, 2011, there is an article titled "Five Roads to Space Dominance," authored by Robert S. Dudney, he outlines the mission of the modern U.S. Air Force Space Command (and it appears to be very similar to that of the naval fleets of the 19th century). The article describes the current solar system military geography as a "vast, but also crowded and dangerous place." It goes on to say that the geospace domain is quickly being populated by a growing number of satellites that are being deployed by nation-states and private entities. This region of space was initially dominated by two major actors: the United States and the U.S.S.R. (now Russia). But now, at the turn of the 21st century, about a quarter of the nation-states of the Earth and several private entities are operating some kind of spacecraft there. At the same time, even peripheral nation-states, such as Iran and North Korea have the capability to jam satellite links.

———

It is likely that there will be two major military space domains by the end of the 21st century. The first is the geospace military space domain has been developing since the middle of the 20th century. Its mission is primarily focused on the belt of artificial satellites that have been placed into orbit since 1957. However, it also includes nodes that are located on the surface of the Earth, and within the atmosphere. The most important of these is the complex of long-range ballistic missiles that pose the greatest threat to the orbiting satellites. However, there also are the launch sites, such as the one that NASA operates at Houston, Texas, that constitute the military map of this domain. Also, one must consider the anti-satellite missile systems are now carried by airborne platforms within the stratosphere.

The other military space domain can be expected to develop sometime during the 21ˢᵗ century. Its main mission, probably, will be to provide security to resource extraction and processing sites on other planets and asteroids that will be located in a zone between Mars and the Main Asteroid Belt. Perhaps there also will be a network of artificial space bases that would be situated on an orbit that would allow them to operate as intermediary links between the Mars Military Space Domain and the Geospace Military Space Domain. What makes such a hypothetic even more feasible is that these are regions of the solar system that already are accessible to humans today, given our current state of technological know-how. And, consider that the major and minor nodes of such a military space domain network already are being connected by identifiable linkages – that is to say, the electromagnetic signals and the plotted trajectories of past, current, and planned missions into space.

Also, a "mental map" of these regions as potential areas of useful natural resources in space is even now being developed by humans. Such a map would include many of the industrial and energy minerals and elements that are the focus of intense competition by the nation-states of on Earth today. At some point in this current century, there likely will be some economic development of these space-based resources; probably by the private corporate sectors of the nation-states. The hypothesis that is being proposed here is that is that the "flag will follow the corporate logo" in space – as has been the case during the past 300 years or so on Earth. In this book, I argue that the pattern is likely to repeat itself with respect to the rest of the solar system; that is a "New World" for future human colonization, if you will. The following is an inquiry into the military system that will be developed in the solar system to provide security for the cultural systems that will be established by humans there.

On a more detailed level, it can be seen that the first generation of nodes of these military space domains are already being developed. Consider the space launch facilities and the mission control centers, as well as the tracking and communication relay stations have been constructed to launch, track, and communicate with the growing number of space-based stations in the inner solar system. At this point, it should be realized that the military and the NASA aspects of the system already are coinciding in many ways. Thus, NASA has incorporated many of the U.S. Department of Defense assets and concepts in the early attempts to build the agency's infrastructure. Thus, in the early years of the NASA space program, the entire rocket program was transferred from the U.S. Army to the new space agency. Also, many military test pilots were reassigned to NASA to become the first astronauts. One could even say that many of the defense contractors of the day were developed as space contractors by the NASA system as well. On the other hand, many weapon systems that have been developed by the military services during this period of time have borrowed many technologies and engineering experiences from NASA.

The U.S. Air Force, for its part, has also been developing a somewhat parallel space system concurrently with the NASA system. After having formed the U.S. Air Force Space Command, the military has developed an extensive ground-based and satellite-based infrastructure with which to conduct the full range of military space operations, which currently dove-tail with existing military aerospace operations on Earth. The Global Positioning System orbiting satellites provide launch and ranging support to U.S. Air Force rockets, such as the Delta IV and the Atlas V, which are used to launch dedicated military payloads into orbit. Once in orbit, the USAF satellites and other space vehicles are provided ground-based mission-control support by one of two tracking and control facilities, one in Florida and the other in California. Another USAF system includes two control nodes and nine worldwide remote tracking stations which ensure responsive and effective satellite support to warfighting forces on a global basis.

On another dimension, there are plans to deploy thousands of kinetic-energy satellite interceptor platforms (weighing about 100 pounds each) which would be placed into low Earth-orbit, along with their associated ground-based tracking systems, and their laser and charged-particle beam weapon systems. These systems have been steadily improving in terms of their efficacy and reliability. It may be that a future generation of mini-satellites will perform space-based interceptor missions against potential threats to the system posed by enemy spacecraft, or even surface-based missiles. Another system might be designed to intercept incoming warheads in space. As a main component the overall U.S. national missile defense strategy, it would be aimed against ballistic missiles, including ICBMs. Generally, such space-based systems are referred to as Exoatmosphere Kill Vehicles (EKV), whether launched from the ground or a ship at sea, or even a high-flying atmospheric platform.

To manage all these assets, the U.S. Air Force has created an overall major command organization whose function is to integrate the various ground-based and space-based nodes into an overall system. In essence, the U.S. Air Force has created a parallel to the NASA model for space operations, except that it is dedicated to national defense purposes. The USAF model includes nodes on the surface of the Earth whose mission is to provide space launch capabilities, and then to provide the control and communications to manage the space vehicles and other elements, once they are operating in space. Specifically, these include such elements as the Space and Missile Systems Command (Los Angeles AFB, CA) and the GPS Systems Directorate. These are responsible for management of the space superiority systems, which involves the management of warfighters, expendable launch vehicle rockets, spacecraft launches, and ballistic missile and aeronautical testing. All of these activities serve to highlight the emerging military imperative with respect to the Geospace Domain.

The above provides an outline of the structure of an actual military space domain. An analysis of how it is structured geometrically shows that the military geography of space may be best studied in terms of orbits and trajectories. Similarly, the military aerospace regions can be described as being a web of sorties that connect nodes. Some of the orbits and trajectories can be said to be the result of "natural" forces, while others have been "constructed" by human intervention. However, all of these are subject to the influences of natural forces. With the eventual exploitation of mineral and other natural resources on asteroids, the Moon or the planet Mars, there will be sorties that are influenced by the laws of economics. This economic activity will create its own "geometry," which will be manifested by continuing transits between Earth and the Moon, and between the Earth and Mars.

The model that I would use for the hypothetical military space domain that would be responsible for maintaining the security of an economic system would be the naval military protection of trade routes during the 19th and 20th century. Consider the coaling stations that were used by the British navy in its operations to provide security for the ocean sea lanes of the British Empire. It so happened that iron ore, which was needed for the Industrial Revolution in Britain, happened to be located in external locations. These included the Labrador deposits of Canada, the deposits at Serra dos Carajas in Brazil, the Transvaal Basin deposits of South Africa, and the Hamersley Basin of Australia. However, the iron and steel works of England were mainly located in the United Kingdom. The only way to connect these economic nodes was through the use of a merchant marine. This produced a situation where a military naval fleet was necessary to secure the iron-ore resource production nodes as well as the sea lanes that were used by the merchant marine to deliver the raw material to Great Britain. The fuel for the vessels was provided by so-called "coaling stations" which were strategically located at various points along the shipping lanes, including Barim, an island in the strait of Mandeb off the coast of Yemen, and another in the south of Chile. Such an economic geographic system appears to be forming in the 21st century as well. So, it seems somewhat likely that today, in the 21st century, we will see embryonic efforts to establish a similar economic system that will be based on resource extraction on Mars, the Moon, and some of the asteroids and comets. Also, consider that the next industrial revolution may be based on different primary resources than the first one; silica rather than iron, for example. In any case, regardless of the nature of the input resource, a transportation system will have to be developed to tie together the nodes of this future economic system.

Command and Control

It is undeniable that reliable and continuous communications between all the nodes of a space system is vital to its effective functioning. NASA has developed a Telemetry and Control System to carry out this important function. Here again we see an example of a functional region. In this case, the telemetry "parameter" (information field) is the medium of exchange. This electronic packet describes the status, configuration, and health of the spacecraft and its payload, for example. These packets of electronic information typically are downlinked to the Command and Data Acquisition (CDA) Station, which in turn relays to the Satellite Operations Control Center (SOCC). On the other hand, this space communications and control system interfaces with the NOAA CDA Station and with the NASA Deep Space Network (DSN), which is available as a backup. The ground interfaces during orbit are with DSN, Air Force Indian Ocean, and NASA Wallops CDA Stations. The telemetry data can be transmitted in either analog or digital form, but each analog signal is converted into a digital signal. As can be seen, these space communications and control systems can easily become extremely complex and, therefore vulnerable to both natural and human electromagnetic attack.

One major systemic problem that is emerging is that the space communications bandwidths are being overloaded in a variety of ways. First, the data acquisition subsystems are continuing to grow in terms of the quantity of data that is being inputted per cycle. Simply put, the input devices are "byteing" more than they can chew; that is to say, the quantity of raw data that is being inputted at the beginning of each computer cycle is too much for the processor subsystems to handle during one cycle. In traditional computer systems, this can be overcome by just dumping the raw data into temporary storage locations or buffers, which is acceptable if the data that is being acquired is not time-sensitive. Another solution is to increase the size and/or speed the main processor, or to deploy an array of smaller processors in a form of an array. However, if the system must operate in virtual real-time fashion, then the raw data that is acquired has to be processed serially, in a short amount time. An example of the latter would be an early-warning and reaction system that must be able to detect, warn, track, and react within seconds or even microseconds.

It is feared by many experts that this looming problem of space communication systems overload is going to reach crisis proportions as the rate data-gathering increases. As is the case with the internet and other communications systems, the problem is essentially one of inadequate bandwidth. This is partly because transmitting over more frequencies allows a signal to carry more information. But, it is also due to the reality that the radio waves that pierce the atmosphere are not robust enough to carry the growing gigabits per second loads. One option for solving this problem is to filter the data onboard the spacecraft, then send home only the most "interesting" data. But, this only works if there are never any surprises that require "managerial overrides." Another solution that has been suggested is the use of laser telemetry, which would increase the speed of delivery of data batches, and thereby enable satellites and space probes to minimize an overload of the system by switching from radio to laser communication. Then, spacecraft could beam their data down to a network of dedicated light telescopes. However, a very wide-spread global network of such telescopes would be needed in case clouds should obscure some of the telescopes.

Nevertheless, it is thought that laser communications might be a practical solution to the overload problem, and may prove to be the dominant technology in the 21st century. Many private firms and the military-industrial complexes in several nation-states are already working on long-distance laser communications. However, to implement such a next-generation communications system, an infrastructure first would have to be constructed. Thus, for example, a network of high-powered lasers and specialized ground stations would be needed; then the channels for transmitting the information would have to be allocated, and relay stations for shuttling data between data-gathering probes and satellites, and Earth would be required. All of which means that there is a need to increase and enhance the infrastructure for a new laser communications network. At the same time, there is still much that can be done to upgrade the existing radio communications systems.

Still another approach to the problem of data or information overload in computers has been to employ "micro-miniaturization" technology. Essentially, this has involved the minimization of the path along which electricity courses through each component of a computer. The icon of this revolution has been the micro-chip, an extremely small, thin wafer of silicon material. The circuit, which used to involve literally as much as miles of wiring, now has been shortened to nanometers through the implantation of the tiny circuit within the wafer. While this technology can be applied to the miniature circuits of the processors within a spacecraft, it is not feasible when it comes to inter-planetary communications in which the size of the communications field may be measured in Astronomical Units or even light-years. That is why so much research is being focused on ways to increase the speed of the signal itself, through the vast distances of space. Finally, some scientists envision a time in the more distant future when the communication signal can be pushed to the limits of the speed of light, through a combination of "repeater" stations; a kind of "gravity assist" by objects in space, and ionic propulsion enhancements.

––––––––

The Sortie IPO Loop

All types of communications are intimately related to some kind of control and coordination, particularly with respect to some set of expected outcomes. One of the most effective visual portrayals of this process is the input-process-output loop (IPO), which usually is presented in either two or three-dimensions. However, within the basic "outer" loop, there are other "inner-loops" which make up the total sortie IPO. As an illustration of these processes, consider as an example the following mission narrative that documented a portion of the Apollo 13, as presented by NASA (nasa.gov). Thus: "...at 46:40 [into the trajectory], the crew routinely switched on the fans in the oxygen tanks briefly. A few seconds later the quantity indicator for tank number two went off the high end of the scale, where it stayed... a master alarm had indicated low pressure in the hydrogen tank."

This is an example of an IPO cycle that might be occurring during a space mission. In this situation, a switch is turned on by a human, thus sending an electrical signal to the actuator mechanism that controls a specific function. So, in this scenario, the receiving sensor mechanism detects a level of electricity that is beyond its assigned (expected) limits, and therefore, sends back an electrical signal that something is wrong with the programmed loop. Later, an investigation board reviews the entire history of the oxygen tank number two. This includes every step, from fabrication to launch, which has been recorded in the detailed documentation that follows every piece of equipment, as it proceeds along the series of IPO loops, from its time of manufacture to the launch pad. In one particular event, the conclusion of the investigators was that there "had not been a random malfunction..." Rather, there had been a "somewhat deficient and unforgiving design [of the sensor-actuator-output system.]" This is just one instance of the myriad of IPOs that occur throughout the launch-cruise-payload delivery sortie in space operations.

One of the main subsets of the main loop occurs when a mission spacecraft inserts a lander, rover, or other spacecraft into the atmosphere and/or onto the surface of the target celestial body. Consider, for example, the Curiosity rover which recently has been deployed onto the surface of Mars. Here, another set of IPO loops occur within a subset of a NASA space mission. These involve the Alpha Particle X-Ray Spectrometer (APXS) and the Chemistry Camera (ChemCan) sensor systems with which the Curiosity rover has been equipped. In one particular type of loop, the robotic arm of the rover will fire a laser to vaporized and "scoop up" materials that have been extracted from a planet's rock, dust, and other particles (otherwise known as regolith or "soil"). In its gaseous state, this sample will be analyzed by an onboard spectrograph, which will then produce details about the microstructure of the rocks and other material, and the chemical properties and characteristics of the minerals that occur in the regolith. Here again, there is an IPO loop occurring within the overall mission loop. In this case, a scooper that is attached to a remote arm of the planetary rover deposits a sample of the planetary surface for input into onboard processors. The semi-processed data is then transmitted to Earth-based stations.

Back on Earth, another IPO loop occurs when a Rover Environmental Monitoring Station (REMS) acquires the data from Mars, in order to generate a daily report of atmospheric weather conditions on Mars. The data that enter into this particular IPO loop include atmospheric pressure, humidity, ultraviolet radiation (from the Sun), wind speed, wind direction, ground temperature, and air temperature. Also, the monitoring of changes in UV radiation will provide information that is necessary to assess the habitability of the near surface environment on Mars. Throughout this loop, the signals will have to be relayed via a satellite, thus extending the delay time for the data to actually reach the Earth; considerably more than the nominal 15 minute delay if the signal were to be transmitted directly between Mars and Earth.

———

Also known as feedback control systems, IPO loops include the following nodes: (1) input; (2) processing; and (3) output of actuating and control signal. Like all geographic regions, the IPO loops can be seen to occur in a spatial context. That is, they also have an internal spatial structure or configuration, which consists of nodes and linkages. The nodes of this loop system are of three main types: sensors, processors, and actuators. It is the sensor that acquires and measures the desired data from the surrounding environment. This raw data is passed on to one or more processing units, which initially "validate" the inputted data by sending it through a series of filters. The valid data is then processed further by a unit that manipulates it into a form of "information" to be outputted to the "user." The "user," may be a human or a controller-actuator mechanism; and the "information" is then passed to a "user" actuator (human or electromechanical) in order to make the work happen; that is to say, to achieve the desired primary objective of the system. The loop that occurs within a system will usually be very complex in military space operations.

It is generally the case that most of the data that is acquired for scientific purposes is gathered by sensors of one kind or another. Some of these are captured in situ, while others are acquired by remote sensors. Some of these sensors are located on the surface of the Earth, and they include a variety of powerful telescopes that can operate in many wavelength regions of the electromagnetic spectrum. Other sensors are deployed in space, some as far away as the outer limits of the solar system. The space-based sensors can be placed onboard many types of spacecraft (or platforms) including, orbiting satellites and planetary orbiting spacecraft, space probes, planetary landers and rovers, as well as space stations and space-based observatories. There are even sensors that have been embedded into the surface of the Moon and several of the planets. In any case, the sensor can be said to be the most crucial element in any IPO loop. If the sensor is not able to detect and collect the desired raw data in the first place, the processing and output functions will effectively be put out of commission. This sort of input failure actually occurred during the Pathfinder mission to Mars, when the scooper attachment to the remote arm did not filter out the regolith as was required for scientific experimentation.

Sensors also differ in terms of the particular property of the environment that is to be detected and measured. There are sensors that are designed to work in various phases of the electro-magnetic spectrum, and there are those that operate with acoustic; thermal; mechanical; and chemical sensations. Sometimes these various kinds of sensors are integrated to form a "suite" of sensors that combine to form "sensing systems," which are more efficacious in combination. Also, the integration of a combination of chemical sensors with a custom actuator eliminates the need for frequent calibration. And, solid-state transducers are getting smaller too. On the other hand, almost any modern system is equipped with a combination of sensors and actuators to both observe the environment, and to influence this environment as well.

———

An example of such an integrated sensor system can be seen on many modern spacecraft. These typically are equipped with temperature, pressure, speed, chemical sensors, and various types of controllers and actuators. The most important actuator types in a spacecraft include linear and rotary electromechanical converters, displays, and electrical converters. As an example of output of from such sophisticated sensor systems, in a military context, might be the refined "signals" that would be transmitted to space interceptor platforms. This would include such information as the location, rate of speed, and distance to a target. This algorithm might also be accompanied by even more acute input-processing-output sequences between the ordnance delivery subsystems and the enemy airborne targets. However, even the most elegant software cannot overcome the mundane problem of spurious signals in the loop.

Consider the case of the Apollo 11 mission. The mission summary report that has been produced by NASA includes the following cases of problems where a faulty sensor may have been a contributing factor: (1) spurious indications of leaky valves in the launch rocket; (2) lost communications linkage between the spacecraft and mission control on Earth ("communication was in and out for long periods of time"); (3) Alarms which began to go off repeatedly, even though there was no actual problem, "forcing a human override of the process, all because a tiny onboard computer was overloaded with data"; and (4) the switch for a circuit breaker, which was essential for starting the ascent engine (from the Moon) snapped off because of a spurious electrical overload, forcing the astronaut to "reset the breaker with a ball-point pen" (nasa.gov). All of these events involved anomalies in the nested loops that occurred in each of the components of the subsystems. "Spurious indications," for example, usually mean that the sensor in question read the data incorrectly in the first place – maybe due to faulty calibration or flaws in the original manufacturing design. By the same token, problems with a communications linkage, either in the upload or download phase, might involve external factors (such as a spike in cosmic radiation) or it could reflect problems with the transmitter or receiver subsystem at either end of the linkage. The same could be said for any of the myriad of IPO loops that are occurring during a space mission.

———

Input devices that are used in space missions can include the: keyboard; pointing device; joystick; digital scanner; image scanner and, of course, the ubiquitous sensor itself. What they have in common is that they are introducing "raw data," which then is processed and thereby converted to useful "information." Physicists say that information is energy that can be used to do "work." In terms of military space operations, information needed to make actionable decisions. Information may be passed along to a human or mechanical actuator, to produce a desired outcome. These "desired" outcomes can be caused through ad hoc action (or preprogrammed algorithm) by man or computer. Whether the decisions are made by a human or a computer processor, signals then are passed to the subsystem (human or machine) to make the desired "work" happen. This input-processing-output cycle is reiterated throughout the systems involved, until a final outcome is produced.

In his book: <u>Imaging Saturn: The Voyager Flights to Saturn,</u> Henry S.F. Cooper, Jr., describes the arrays of sensor devices which monitored the operational "environment" during those twin missions. These included: (1) magnetometers which studied planetary and interplanetary magnetic fields; (2) cosmic-ray detectors; (3) plasma detectors for studying the solar wind; (4) and an instrument for studying low-energy charged particles. Other instruments which were used for analyzing the geographic environment included: an infrared interferometer-spectrometer, radiometer, photo-polarimeter, and a radio dish. All of these sensors have been designed to capture environmental properties which occur at certain electromagnetic wavelengths, as the Voyager spacecraft moves through the regions of space, and in the environs of the Jovian planets and moons.

––––––––

What should appear to be obvious from this discussion of IPOs and closed loops is the growing intricacy of space operations, in general, and military space operations in particular. As we review the detailed and complex nature of the myriad of spacecraft's primary and secondary systems, as well as the redundant subsystems, we can appreciate the common saying that: "if anything or something can go wrong… it probably will." These probabilities for malfunction are inherent in virtually every aspect of the space sortie. Taking space-launch operations as an instance, consider that launch vehicles include many sensors and control mechanisms that are associated with guidance and control. In all these circumstances, it seems that the level of complexity in military space systems will only continue to grow as we begin to rely more on automated and robotic solutions to the problems of military operations in space.

Another area of study for scientists and engineers is found in the quest to reduce the size of the system components. Thus, in recent years, nanotechnology has been studied as a way to make spacecraft systems smaller and lighter. While the main motivator in this effort is the desire to drive down the cost of launching rockets and their payloads into space, there is also the objective of being able to embed more sensors into the spacecraft systems. Indeed, it appears that the "nano-sensor" has the potential to enable the construction of not only smaller, lighter platforms, but also to pack more sensors of various kinds throughout these systems. It is also hoped that additional sensors will enhance the internal and external situational awareness of the total operating environment. One application of such a nano-sensor technology would lead to the installation of smaller, more capable environmental monitors and smoke detectors in future spacecraft. As an example, NASA's nano ChemSensor Unit was embedded into the Earth-orbiting MidSTAR-1 satellite (2007) to enhance awareness of the chemical environment (nasa.gov). On longer manned missions in space, arrays of custom-calibrated nano-sensors may be put to use to monitor harmful chemical contaminants, which tend to build up gradually in the crew's air supply. It is hoped that nano-sensors would be able to detect minute amounts of these toxins and alert the crew of a potential problem. Similarly, to combat the stress of prolonged microgravity, nano-sensors could be built into a special gravity suit to protect astronauts in such situations.

———

The NASA Flight Control Room and the Space Shuttle Flight Control Room operations are examples of close-loop processes that occur within the overall loop in space operations. In virtually every case, these systems are defined by a central node: the "user." This term refers to ultimate consumer of the loops output. Within the IPO loop, the information itself can be converted to digital form and outputted onto a display screen or console; or it can be passed onto another IPO system as input for further refinement before being outputted to the user. In the case of NASA space operations, each console in the Flight Control Room can be seen as both a user and transmitter of information for a given space mission. It is also the case that the Space Shuttle Flight Control Room and the International Space Station Flight Control Room are basically identical in their equipment. Therefore, most data related to flight control of the International Space Station or the Space Shuttle can be viewed in either room. This is an example of utilizing the technique of redundancy to ensure that communication loops remain active. In each of the rooms there are large display screens at the front of the room, and several remote television cameras mounted in the room to provide live broadcast of activities. The television monitor in the main room shows pictures, or "raws," as they are arriving from the spacecraft. At the same time, the astronauts or the robotic pilots of the ISS and the Space Shuttle also are users and transmitters of data and information within the loop. Sometimes the output from a loop becomes input into another loop, where the IPO process is repeated. The "recipient" of this first output can be a human decision-maker, or an electro-mechanical actuator.

So, it is clear that each step toward greater autonomy in space systems begs several basic questions. One has to do with the ratio between human control and automation that will optimize the probability of achieving a desired outcome. Another issue that arises in determining the optimum ratio of automation is where to place the various decision-making nodes within the loop system. Thus, for example, if a spacecraft is fully automated, the question that arises is: should the decision-making locus be on the spacecraft itself, or should it be in the flight-control center on Earth? Aside from the question of how much autonomy of decision-making should be relinquished to autonomous control systems, there is the matter of the levels of priority in the various actions that are done onboard a spacecraft. That is, should an autonomous spacecraft "do science" and only "call in" if it encounters some issue that lies outside its preprogrammed algorithms? Any attempt to construct truly automated onboard systems will need to address such issues early in their overall software design. The ground system automation, if any, will need a reliable and accurate flow of information as it interfaces and interacts with the onboard systems. And, it must be equipped to modify the programs instructions; that is, to exercise "managerial override" when necessary. Of course, in the end, the decision of where to locate the "controller" function will have to consider the matter of the relative efficacy of the telemetry that is available.

The actuator device in a subsystem is the component that causes "work" to be done; that is, it makes something happen within the overall system. An example of this is the relay, which is an electro-mechanical device (such as a solenoid), that provides remote or automatic control of the current. When an electric current is passed through the device, it becomes magnetized. In its magnetized state, it is used to draw a core or plunger into the solenoid and the motion of the plunger is then used to actuate switches, relays, and other components of a system. As can be readily appreciated, the efficacy of the work that is caused by an actuator is basically a function of the quality of the information that it receives from the sensor system. For one thing, an actuator system typically is expecting a very particular signal, or range of signals, from the sensor. If the metric that is being used is temperature, for example, the actual temperature must be as "expected" by the actuator (according to the computer program instructions that it has received). One obvious conclusion that one can draw from the precision demands and complexity of the loop system is that there will be many opportunities for dealing with unexpected variables. This means that either even more automation will have to be added to the system, or redundancies will have to be added, or that human intervention (override) will be needed.

This leads us to the matter of how and where in the process the actual "programming" is to be done. These lines of code are the language that present the total environment to the computer processor, and provide instructions for conducting tasks during a mission. Both of these must be constructed for a particular mission, and typically will require the writing of a vast number of lines of programming code. As an example, consider the programming that was done by teams of experts that worked on the Voyager missions (as described by Henry S.F. Cooper, Jr. in the book, Imaging Saturn). The imaging team alone included "astronomers, geologists, meteorologists, and physicists." In general, during a typical NASA space mission there will be hundreds of humans writing thousands of lines of code, telling the various systems of the mission what to do, when to do it, and how to do it, all in exquisite detail. Therefore, as I have discovered in my computer-programming career, the computer has to be told everything, as though it were a five-year old child. So it appears that, for now, humans will do most of the actual "new" writing of the computer instructions in military space systems. However, as time passes, it is likely that there will be more "canned" program code that will be available off the shelf. Such "legacy coding" has been growing in volume as the number of space missions has accumulated over the past half-century. The upshot is that, future computer programming for space missions will increasingly be done by other computers.

Overall, some very sophisticated methods have been developed for managing the complexity of information that is passed on to the entities which are responsible for making high-stakes, time-critical decisions. After defining the set of expected values of information, the actual values are compared against it, and the appropriate decisions can be made. Such measures also can be used to enhance the efficiency of the computer displays used for monitoring complex systems. Imaging itself, is a process whereby impulses from television cameras aboard a spacecraft are encoded by a computer into a series of bytes and transmitted back to the user, either on Earth, or at some other location within the operational region. Upon receipt by the user, the impulses can be decoded by another computer, and reconstituted into pictures for display. Once again, we can see an IPO loop in action.

During the years of the Cold War (between 1947 and 1991) the scale of the military aerospace region (MAR) became truly global in scale. It also was a dynamic region wherein operations occurred unceasingly – day-after-day and year-after-year – until the Cold War was declared to have ended. Within this area of operations, the U.S. Air Force Strategic Aerospace Command, the North American Air Defense Command and other elements of what came to be called the "Nuclear Triad" carried out perpetual operations during that period of history. The "Nuclear Triad" rubric referred to the U.S. force strategic bombers, silo-based ICBMs, and submarine-launched missiles that were poised to launch a retaliatory nuclear strike in the event of an enemy first strike nuclear attack. Another essential subsystem in the overall nuclear deterrence strategy was the so-called "DEW" (Distant Early Warning) system of radars, command and control centers. These were designed to detect and identify any strike threat. This information was then relayed to the North American Aerospace Defense (NORAD) system. Within seconds, the necessary information was passed on to SAC, and the ICBM-bearing submarines. This can be seen to be a strategic IPO loop which operates on a global scale. Another loop that included the NORAD was the one in which threat information was passed to aerospace interceptors, which would be directed to defend against strikes against Canada and the United States.

The shortening of the IPO loop also has occurred within the computer. This was manifested in the time-distance shortening of the so-called "computer cycle." In one sense, this involved the micro-miniaturization of the distances which a particular programmed instruction had to travel in order to be processed by the computer. In great part, this shortening of a circuit in computer processing was due to the revolution in computer engineering in which "wiring" was being replaced by micro-chips. These silicon wafers, on which a micro-circuit was imprinted, were very small; therefore, the number of instructions that could be processed within a given time-period increased. So, the computer has become more "powerful." One manifestation of this micro-miniaturization of the cycle in military applications has been the creation of smaller and lighter hardware systems in every phase of military operations; both on Earth and in space.

––––––––

Another manifestation of this "tighter" IPO cycle has been seen in the area of aerospace reconnaissance. When World War I began, the cycle would begin with the photographs that were taken on board an aircraft. These were visible-light cameras which managed to capture a brief expose' of enemy order of battle. Later, the aerospace reconnaissance was widened to include industrial sites, lines-of-communication, and other manifestations of their ability to wage war. However, when aircraft came to be used to attack platforms against ground targets, the most crucial focus of air reconnaissance was on the air defense guns that were deployed nearby. (Later, these included anti-aircraft artillery and surface-to-air missile sites and, most particularly, the radar-sites that controlled their fire). By the time I went through photo-intelligence training in the early 1970s, the technology of aerial reconnaissance was pretty much where it had been during World War II. During that period of military aerospace history, air reconnaissance still depended primarily on the visible-light photography that was produced by onboard cameras. The most significant advance in the nascent science of aerial photography, at that time, was in the technical development of the camera systems themselves. Another was the development of more skilled photo-reconnaissance specialists, which brought the process of target detection to a more precise level. The results of this photo-analysis were then hand-carried to an "intelligence/operations shack" where the photography was processed and analyzed by intelligence and operations personnel. At that point in time, the lag time between the taking of the photographs of possible targets, processing by photo-intelligence analysts, and the development of targets, was still a matter of hours, if not days.

This paradigm continued into the early stages of the air war in Vietnam. There, as the science and technology of electronic data transfer matured, the first electromagnetic IPO system was developed and implemented. It would be in the form of the so-called Igloo White program, which was described earlier in this book. Thus, the process of capturing the images of enemy target systems through remote sensing continued to become more sophisticated; the camera systems and other remote sensing systems produced ever higher-definition images for analysis by the imagery intelligence (IA) function. At the same time the air reconnaissance platforms were able to comprehend greater swaths of territory, and cover greater distances than ever before. The ultimate platforms became the reconnaissance U-2 platforms and the satellites that were launched into orbit, some 200 to 500 miles above the surface of the Earth.

The upshot of these developments in military technologies was that, toward the end of the 20th century, the IPO loop entered into real-time territory (at least on Earth). That is, the time between the input of data into a system by sensors, the processing time, and the output of information to users now was approaching the speed of light. This would be the realm of electronic sensing elements, controllers and actuating devices. It also would be the era of satellites, the space shuttle, space probes, and any of the multitudes of processes to which power is applied. The output of this space-age IPO loop would become the input data for all manner of devices, such as motors, valves, solenoids, screws, pulley systems, chain drives, and other manner of mechanical and electrical components.

———

Along with the reconnaissance function, the IPO loop also began to shorten with respect to airborne weapon systems and the delivery of ordnance. One practical expression of smaller/lighter weapon systems and weapons control systems that I witnessed in Laos were the airborne sensors and actuators that controlled weapon systems such as missiles and cannons. These became increasingly compact and efficient, due to the greater incorporation of computer-controlled loop within components of the systems. Consider the case of the AC-130 Specter gunship, which proved to be extremely effective as a ground-support system during the Vietnam War. By the time I left Laos in 1972, the AC-130 gunship was sporting Gatling machine guns, 20mm, 30mm, 40mm cannon, and even 105mm howitzer artillery with which to rain down firepower on enemy ground troops and tanks. More significantly, these airborne weapons were being controlled by sensor-controller-actuator systems which utilized not only the radar portion of the EM spectrum, but the visible-light and infrared portions as well. All of these advanced weapon systems were made possible, in part, by the "tightening" of the input-processing-output cycle of the computer subsystems, as well as the sensor-actuator devices.

Another manifestation of this "tightening" phenomenon can be seen in the shortening of the IPO loop with respect to air ordnance delivery technology during the past 60 years. An example of this was the development of increasingly "smart" fusing subsystems of airborne ordnance. Thus, during the course of the air war over Laos, I remember that the "frag list" (scheduled air sorties) included information, not only of the particular platform (i.e., the aircraft) and the ordnance each carried, but also the type of fusing that each type of ordnance carried. One class of "smart" fuses was designed to could detect and follow the laser path down to the target. This pathway would be "painted" by a laser emitter that would be carried onboard another "pathfinder" platform, or by the attack platform itself. These, of course, were the precursors to the integral laser systems that are now embedded on munitions – such as the Hellfire missiles that are carried and launched by today's remote-controlled Predator drones.

Interestingly, during the course of the Vietnam War, the air defense system problem was turned on its head. In this case, the U.S. Air Force and the U.S. Navy were the attackers. So, in the interdiction campaign over the Ho Chi Minh Trail, an American program called Igloo White was utilizing seeded acoustic sensors to acquire data regarding the movement of the enemy forces along the surface of the trail. This raw data was then uplinked to airborne computers carried by EC-121 aircraft, which also carried communications equipment for down-linking semi-processed data to ground-based computers at a base in Thailand for final processing. There, at Nakon Phanon airbase, the "ground "truth" data was transformed into usable air intelligence that could be passed on to the attack aircraft. Usually these would be A-10 truck-killers and AC-130 gunships that were designed to wreak havoc along the NVA logistical line of communications. Of course, the "prey" in this deadly game would attempt to do everything possible to evade or nullify the radar energy that was being emitted by the predator. This was simply another expression of the EM warfare battle scenario of measure-countermeasure loop that operated within the larger loop.

It was during the Vietnam War that virtually all the parts of the electromagnetic spectrum came to be widely utilized on a large scale to solve problems of target identification and weapons guidance. Indeed, during my tour in Laos (1971-1972) as a member of the Airborne Command and Control Center, I observed the several uses of various portions of the EM spectrum, including the infrared phase in our MAR. There also was a more sophisticated utilization of the visible light phase. In one instance, the A-26 bombers which worked the Ho Chi Minh Trail utilized electro-optical technologies to enhance the available visible light at night. Infrared energy was also used to conduct the interdiction operations at all hours of the day, because the "heat" that is radiated by all objects occurs regardless of the daylight conditions. Also, infrared radiation can be quantified in terms of a "heat profile" or "signature" for various types of objects. Another use of the electromagnetic spectrum involved "excited" and "shaped" light molecules, known as LASER (now called directed energy). This technology was used to guide ordnance against lucrative targets in North Vietnam. Infrared energy also was used in many situations during the Vietnam War to either enhance detection of the target and/or to provide guidance to ordnance.

The same can be said for the specialized military aerospace operations that were given the name, "Wild Weasel" in Laos, during the Vietnam War. This was a technique for detecting and attacking air defenses through the use of the radar portion of the spectrum. That is to say, specialized hunter aircraft, such as the F-105 Wild Weasel platforms would utilize radar to detect any tell-tale emissions from fire control radar on the ground. Once locked on by the F-105's specially calibrated sensors, the F-105 could launch a "smart" air-to-ground missile that would simply follow the electromagnetic path back to the emitting fire-control radar and, of course, the air defense missiles or artillery which it controlled. Another tactic was to utilize the F-105 platform as an electromagnetic "pathfinder" who would relay the relevant data about the offending radar to another warfighter platform. This was yet again another example of how onboard computers were used to shorten the IPO loop in that situation.

———

Maintaining the Security of the Space Lanes

The foundations of a system of space lanes in the interplanetary medium have been under construction (at least in de facto fashion) since the launching of Sputnik I into orbit around the Earth by the former Soviet Union in 1957. And, the process has continued during the placing of other satellites into orbit by the United States and other space entities during the following half-century. In addition to the orbits, there have been many trajectories that have been "blazed" by the various space missions of NASA and other space agencies during the past 60 years. Like the sea lanes (or routes) on Earth, these space lanes can be seen as being linked to nodal points, or bases. At the present time most of these bases are distributed along the surface of the Earth. However, given the trends that are occurring with respect to space exploration, we can reasonably expect that there will soon be other bases that will be located on other planets, moons, asteroids, and even artificial satellites and orbiting space bases.

It is likely that traffic intensity within these space lanes will increase in the future. One reason is the a major trend in space operations today; that is, the nascent movement toward "privatization" of certain aspects of overall space exploration enterprise. At least one private company has successfully launched a shuttle-type mission which has resulted in a successful rendezvous with the International Space Station. It involved the use of a launch rocket and the spacecraft that was designed and constructed by a private corporation. One of the ramifications of this success has been a more insistent call by private corporations to assume more of the responsibility for the research and development of space systems in the future. Along with this development, there seems to be more immediate movement in the area of the establishment of mining and other resource extraction infrastructure. All of this can be viewed as very concrete step in the development of such primary industries within the geospace economic region. Of course, with the creation of these nodes, there will be linkages that will tie them into the Earth's economic system. Further, we can expect that other economic networks will be developed to integrate these primary resource systems into the global economy on Earth. This will have to be done through the development of a network of trading routes, or space lanes in geospace.

Space lanes – the trajectories and orbits that are being developed and understood as being "optimum" for traversing the interplanetary medium of the solar system – also are functioning as conduits for the movement of humans and materiel. So, if history is any guide, it would seem that these space lanes also will be utilized in a hypothetical military area of operations within the solar system. Here again, much of the heavy lifting has been done by NASA and other governmental agencies in the initial development of the present space lanes. This can be seen in the NASA space shuttle operations, wherein spacecraft deliver supplies, equipment, and structural parts to the

International Space Station, shuttle replacement crews, and ferry astronauts who perform repairs and modifications to the cultural nodule.

If we use the development of the ocean trade routes since the 15th century on Earth as a model for the geospace region, we can expect that the recurring trajectories and orbits of the exploratory missions will become the established "shipping lanes" in space. This expectation is derived from the same rationale that produced the selection of the earliest ocean shipping routes. Thus, certain pathways across the oceans came to be considered as "optimum," as a result of the cumulative experiences of thousands of marine navigators. In the era of the sailing ships, these established ocean routes or lanes were mostly subject to the discipline of the winds, the currents, and the well-known regions of lethal storms and doldrums. Later, with the advent of the steamship, the discipline of the natural environment was somewhat ameliorated through the use of technology and greater scientific understanding of the natural forces that operated in each marine environment. Subsequently, as the new routes became better known and rationalized, the cultural factors of ocean travel became more paramount in the selection of established ocean routes. Basically, these cultural factors can be categorized as: political, military and economic. Ultimately, the relative function of each of these factors came to be codified in maritime law and convention. In any circumstance, there has always been an imperative to use military power on the ocean to enforce the freedom of movement on the high seas, in accordance with the maritime system that has been established.

———

Travel along, and between, shipping lanes today is made possible by navigation systems like the Global Positioning System, which consists of a constellation of orbiting satellites. These serve as a navigation aid for both civilian and military "users," both inside the atmosphere and in space. However, the most-used method for solving the problems of navigation and positioning in the solar system today is the inertial guidance system. Such an inherently integral system will be absolutely vital in any combat operations in space, where the threat of enemy attack can come from any point along any 360 degree front within the air space. This reflects the fact that an attack can come from three dimensions. But in space combat, there is also the fourth-dimension of time that becomes part of the space defense problem. Given this degree of spatial uncertainty, it may be that the best strategy for defending the orbiting satellites is to employ the technique of "barrage anti-aircraft" tactics that were developed during World War II. More recently, this kind of barrage fire has been referred to as "swarm" interceptor tactics against an enemy threat to the orbiting satellites. Such swarm tactics in space will quite likely involve the utilization of "nano-interceptors," which would behave like bees defending their hive.

The most immediate threat to today's GPS constellation of orbiting satellites (as well as other communication and surveillance networks) comes from cyberspace attacks against the computers and electrical components that are vital to the orbiting satellites. Another eminent threat comes from the ICBM-type missiles that are already deployed on the surface of our planet, and which have actually been used to destroy satellites. The main problem of providing security for these assets essentially involves the development of "air defenses" in space, as well as on the surface of the Earth. The fact of the matter is that the threat profile also may come from attacks on the surface components of communications/logistics networks by so-called "rogue" nation-states or terrorist organizations that have the sufficient level of resources to strike them by some "improvised explosive device" (IED) method. It also is clear that relatively powerless nation-states, such as North Korea and Iran, possess the capability to launch rockets with nuclear payloads up to the altitude of the artificial satellites. (It is estimated that even a small kilo-ton nuclear warhead could devastate an entire section of the orbiting objects).

Therefore, it can be said that the main source of the threat to the artificial satellites at the present time still is the surface-based, long-range, ballistic missile. In this regard, it should be noted that China already has fired a ground-based missile to strike one of its dead satellites (in 2007). This has been one of the more immediate events which have spurred both the U.S. and Russia to develop prudent counter-measures to answer the potential threat that is posed by China with respect to their own space-situation. Another kind of threat is that of direct attacks on the launch bases and the flight control sites that provide command and control over the rockets and the satellites in orbit. Thus, we can expect that military surface operations will always include a robust security force that can quickly respond to any threat or actual attack on these ground assets.

While the placing of certain kinds of weapons in space is currently prohibited by international accords, many civilian and military "support" satellites have been put into orbit around the Earth to perform such functions as reconnaissance, which can be considered to be an integral part of military aerospace operations. The current treaty essentially prohibits the placement of weapon systems in space, and prohibits any attacks on orbiting satellites and other space hardware from weaponry on the surface of the Earth. However, it does allow the gathering of data that could be useful for military aerospace operations within the Earth's atmosphere and, hypothetically, in space itself. It also says nothing about the conduct of basic scientific research in space, even though it might produce outcomes that would prove useful to military research and development, and even actual military applications. Thus, for example, the rarified space environment has proved useful in the development of microchips that can be used to produce all sorts of modern military aerospace avionics, weapon systems, and other such systems.

———

The development of weapons and tactics that might be used within the geospace transition zone began, for all practical purposes, with the Inter-Continental Ballistic Missile (ICBM) systems during the Cold War. These weapon systems were designed and developed with the realization that the Earth is a sphere, and that its geometric properties could be capitalized upon to shorten the distance to nuclear strategic targets by ICBMs (e.g., between Grand Forks, North Dakota and Moscow). This could be done by placing the intercontinental ballistic missile in a very low orbital path just above the edge of the atmosphere. However, in terms of warhead type, the most likely weapon to be used against missile or satellite attacks on our assets would be one that causes an explosion through the kinetic energy of the impacting projectile. This type of weapon apparently is deemed to be a better choice than those which are based on either: nuclear, electromagnetic, or even conventional explosive energies. This is because of the risk of collateral damage from a weapon system, like that which resulted from the anti-satellite missile which was used by China to destroy one of its own satellites in 2007. Evidently, that incident has left a lot of space debris which could cause problems for subsequent space operations in that region.

So, the main response to the prohibition against the arming of orbiting satellites has been to develop surface-based anti-satellite weapons, which are designed to degrade or destroy an orbiting satellite. Currently, only the United States, Russia, and the People's Republic of China are known to have operational anti-satellite weapon systems. Thus, on September 13, 1985, the U.S. destroyed an obsolete American satellite, using an ASM-135 ASAT missile. Then, on January 11, 2007, China destroyed an obsolete orbiting weather satellite. A year later, the U.S. again destroyed one of its own malfunctioning spy satellites with a RIM-161 Standard Missile 3 anti-satellite missile. Russia, too, has revived its own anti-satellite program which, in addition to missiles, also includes developmental work on the MiG-31 Foxhound as a high-altitude launch platform for an anti-satellite system. Aside from the three major players, India has announced its intention to developing lasers and exo-atmospheric "kill vehicles" that could be combined to produce an anti-satellite weapon system; it is planned for implementation by 2014.

One anti-satellite tactical approach, which is a much more risky for all concerned, involves the use of high-altitude nuclear explosion (HANE), which historically has taken the form of nuclear explosions that took place at altitudes above 18 miles, but were still within the Earth's atmosphere. These have been nuclear weapons tests, which have been used to determine the effects of the blast and radiation in the exoatmospheric environment. These occurred in the early sixties and involved the detonation of 3.8 megaton warheads, the highest of which occurred at 335 miles above the surface of the Earth. So far, the only nation-states to actually detonate nuclear weapons in outer space are China, the United States and the Soviet Union. However, this approach has proven to be unacceptable, because of the very large radius that is associated with nuclear events. Indeed, it reportedly has been found to be nearly impossible to prevent indiscriminate damage to other satellites, including one's own satellites. Another problem area is the emission and propagation of gamma rays which penetrate the environment and collide with space molecules, depositing energy to produce huge quantities of positive ions and electrons that damage electronics with a force greater than that of the solar storms.

At the present time, the response to these threats to the orbiting satellites is being confined to passive measures and deterrence through implied retaliation. These include such techniques as developing redundancies in the most vital systems, the application of shielding materials that will provide a minimum level of protection against hostile electromagnetic radiations, and other methods that are familiar to the traditional electronic warfare on Earth. More active, but still defensive, techniques also are being developed at the start of the 21st century. Basically, these are borrowing from the "smart" electronic countermeasures that are being developed for atmospheric aerospace operations. These basically are intended to "reconfigure" the received electromagnetic signal and return it as a spurious reading to the adversary. This also involves the use of nanotechnology to embed arrays of tiny sensors in the skin of the satellite to perform these countermeasures dynamically, in real-time fashion. However, one can predict with a significant level of certitude that, by the middle of the 21st century, military satellites will have the capability for performing active self-defense countermeasures. The implied deterrence countermeasures might include the use of laser or other directed energy systems to strike the attacker. These weapon systems could be supported by advanced pattern recognition and high-powered electronic propulsion systems to be able to respond quickly to the threat area. At the same time, satellites also would be equipped with sophisticated counter-measure systems, such as chemical or electromagnetic "aerosols" and decoys to jam the adversary's offensive systems.

Still other self-defense tactics against attacks on satellite systems involve the building of layered redundancy within satellite systems, such as multiple processors or power systems. This strategy may also include the development of mechanisms for quick-reaction repair and restoration of the sensor platforms and networks to normal operations. This is a well-known protocol, which has been developed to combat sudden surges in a system, due to such natural calamities as lightning or sudden electromagnetic pulses of cosmic radiation. With respect to satellite-based computer, every data processing manager has learned from hard experience the necessity for developing efficacious "backup" systems to safeguard not only databases, but the software that runs the systems, as well.

Perhaps the main underlying and indispensable, aspect of the satellite defense equation is the need for situational awareness. Consider that, as of 2012, a new satellite defense system has been in place to defend against any threats to our Earth-orbiting satellites. This new system has been given the name of "Self-Awareness Space Situational Awareness System" (SASSA) by the United States. Its existence was made known in a very public announcement, which was probably intended to serve notice on other space powers (e.g., China) of our intention to defend these satellites, by any means necessary. More to the point is the advertisement of our possession of the military capabilities to do so. SASSA seems to be using the model which was developed to deter and, if necessary, respond quickly and decisively to any actual attack on the assets which are being defended. This was the strategy which was developed by the U.S. to deal with any Soviet first nuclear strike attempt. The model includes an early detection and warning system that is designed to alert friendly forces to mount an immediate and powerful counter-attack in order to, first, destroy the incoming aircraft or missiles, and then to mete out destruction of the enemy's systems. The latter represents the steel-fist of deterrence which makes a first-strike against the system suicide operation.

––––––

All of this derives from one of the outcomes of the space missions; that is, the discovery of significant sources of minerals in the inner solar system. From this, it can be expected that there will be a future imperative to maintain the security of these assets throughout the interplanetary medium, and at extra-terrestrial resource extraction sites. Mars, for example, is thought to be the site of significant, and accessible, mineral resources (discovery.com). The best places on Mars for exploiting valuable ores seem to be the volcanoes, especially their lava flows. Also, the impact craters are also thought to be places where ore extraction operations could be even more feasible because of their location relative to known landing sites on Mars. In a Sky & Telescope (August 2012) article, the subject of mining for precious metals on asteroids also is discussed. It appears that a commercial venture has already been formed, with the intention of developing plans to extract minerals, as well as water for space travelers, from near-Earth asteroids. Such an idea is not new; but until recently it has only been an interesting speculative matter.

Although it still remains an unstated consequence of such mining operations in space, the experience on Earth indicates that, eventually, military space operations may be necessary to provide security for such resource-extraction operations. Indeed, in his book The Economic Consequences of Peace (1920), John M. Keynes foretold that the next world war that would be instigated by the nations who suffered most from the results of the peace accords that ended World War I. In Europe, he said, it would be Germany who would attempt to regain access to coal and iron resources which it had lost as a result of the Treaty of Versailles (1919). During this same period of "peace," the Japanese Empire also found itself shut out from access to virtually all the energy and mineral resources that it required for its industrial growth. The upshot, of course, was that both Germany and Japan essentially went to war to gain access to the energy and mineral resources that had been denied to them by the existing world order (as they saw it). For the Japanese, this included the resources that were controlled by the British, Dutch, and the French in Asia. Meanwhile, the Germans focused their eyes on the coal and iron resources which had been taken from their erstwhile territory, as well as energy resources in Romania, among others. The point is that historically nation-states have found it necessary, and have formulated rationales, for going to war with other nations over the control of natural resources that are deemed to be vital to the interest of the coveting party.

Now, in the 21st century, we begin to see the outlines of the geography of natural resources – mineral and chemical – in the extraterrestrial solar system. It almost certainly will be that these will be the object of future attempts to exploit them economically by various governmental and non-governmental organizations of Earth. At that point, one can only say that if a global system of exploitation and allocation of these resources is constructed, it may not be necessary to replay the experience of the industrial revolution, with respect to the competition for control of these natural resources in the 19th and 20th centuries. However, even if there is a true consensus by the nation-states of the Earth on a peaceful model for developing these other planetary resources for the common good of mankind, there is still the danger that "rogue nations" or "privateers" may take advantage of the ease of entry in terms of acquiring the resources and weapons to cause havoc within the accepted transportation lanes.

As indicated earlier, "military space operations" effectively began with the race between the United States and the Soviet Union to place an artificial satellite in orbit around the Earth in the 1950s. Such tacit "military operations" in space continued throughout the latter stages of the Cold War and in the following decades, officially under the aegis of "civilian" space agencies such as NASA. With this in mind, I see the continuing space explorations, as well as all the scientific and engineering activities that are being conducted by NASA and the other national space agencies as, in effect, contributing to the building of "military capabilities." This situation reminds me of the "clandestine" preparatory activities of the major powers (including Germany, Japan, Great Britain, and the United States) during the years prior to the commencement of actual military operations in 1938 when Germany invaded Poland, thus starting World War II.

In any case, there likely will be a need for developing quick-reaction capabilities for responding to any of the threats in the space domains. Even when military space operations occur within the interplanetary medium or within the domain of planets and other bodies, there still will be a need for the development and deployment of the needed forces to secure assets on the surface of the Earth, or some other planetary surface. This will probably take the form of special operations, similar to those that have been carried out by the U.S. during the turn of the 21st century. Typically, these have been employed against a non-governmental entity, such as the insurgents in Afghanistan and Iraq, or against an organized "terrorist" entity that possessed the resources to mount audacious "flash" attacks on a global scale. Some experts have argued that the most effective response to this kind of a threat is not large-scale military operations against the supposed enemy base. Rather, it has been the evolution of "special operations," such as those launched the Delta Force, the U.S. Navy Seals, and other special operations forces that have been most effective. This argument has been buttressed by the success of "sniper" attacks by unmanned attack platforms like the remotely piloted Predator platform. The upshot of the matter has been the realization that small attack forces that are disciplined and powerfully equipped are best countered by small counter-attack forces that are more disciplined and powerfully equipped with advance technology. It is "nano-warfare," if you will.

Meanwhile, it can be seen that the lines of communications that are continuously developing in the solar system. At the present time the visible manifestation of this are the orbits along which a variety of artificial satellites are operating. Like all lines of communication it must be defended against any attack on the platforms that travel along these space lanes. The main threat to these platforms comes from (1) the anti-satellite missiles that are deployed on the surface of the Earth and (2) the threats from the computer cyber attack. To defend against these threats, there have been developing several countermeasures. One of these focuses on the Earth-based long-range missiles and the support bases. The other is based on a more passive defense strategy. It involves the shielding of computers that control the movement and operation of the orbiting artificial satellites from cyber attack from other computers. In general, all these actions can be seen as part of a greater security paradigm, which is the maintenance of the security of space assets.

———

Asset Security in Depth

The above scenario reminds me of conversations that I had with my friend, Colonel Karl J. Woelz, in 1972, when we both were just returning from a tour of service in the Vietnam War. Like me, Karl had begun his career in the U.S. Air Force as an enlisted airman, and we were both later commissioned and assigned to our initial duties. Karl was a security police officer, and I was as an administrative type, at Grand Forks AFB, North Dakota, in the early 1960s. One of the things that struck me as most relevant to the subject of military space warfare was the conversation we had about security for military assets. Karl was one of an emerging group of young officers in the field of force security who recognized the need for fundamental changes in the concept of protection of assets. This was a consequence of the Vietnam War that was ongoing amidst the Cold War. Indeed, by the time then 1st Lt. Woelz had been assigned to Nakon Phanom AB in Thailand in 1968, during the height of the Vietnam War, the traditional role of the U.S. Air Force Security Police already had evolved well beyond their main Cold War role as protectors of the nuclear B-52 bombers and the ICBM missile sites of the North American continent. They had now become a proactive security "warfighters" who aggressively pressed the security of the envelopes around the military bases throughout Southeast Asia, in a variety of situations. The latter became of vital importance during the period of the Vietnam War, when U.S. air bases were scattered throughout South Vietnam and Thailand, conspicuously in the midst of enemy guerrilla forces and agents.

Back then, in 1967, I saw Karl as a no-nonsense, perfectionist; one who could quickly size up all the parameters of a problem and then develop the optimum solution to the problem. He also was a natural leader who knew how to communicate his ideas and to motivate his airmen to carry out the mission. Thus, he excelled in his duties to provide security to the B-52s and Minuteman II ICBMs which were stationed at Grand Forks AFB during the Cold War years. Even in the midst of the Strategic Aerospace Command doctrine of "point defense" of assets (individual B-52s and Minuteman II ICBMs), he was one of a cadre of junior officers who worked in military security that foresaw that the true role of the "Air Police" (a descendent of the Military Police of the pre-1947 "brown shoe days" of the U.S. Army Air Corps) would be transformed in view of the changing nature of warfare during the Vietnam era. That is, in addition to the point security of specific U.S. Air Force assets, the "air cops" would evolve into a special operations force that would aggressively defend the entire hinterland of a military base in a hostile environment.

So now, in the war in Afghanistan, there has been a continual adaptation of the security forces to their new special-forces type mission. Brigadier General Robert Holmes, the theater director of security forces and force protection has characterized these changes as a "refocus" on how people train and fight (usmilitary.about.com). His statement reflects the reality that we are no longer engaged in the Cold War. Rather, we are now involved in a global war on terrorism, where the "front-line" occurs wherever the U.S. has a base or other installation asset. Therefore, the asset security forces have to alter their mentality and practices for the reality of today. Essentially, security forces will have to continue to focus on preparing for their warfighting mission at "forward locations" throughout in every region of the Earth.

A case study of this new asset-security mission can be seen at the Bagram Air Base in Afghanistan. There, U.S. Air Force security airmen are routinely going "beyond the wire" to conduct proactive security operations throughout a larger envelope around the base. They can be seen carrying out patrols aboard a Humvee, and manning an M-240 machine gun that is mounted on the roof of the vehicle. Like their Army counterparts, they travel along roads that are often seeded with IEDs (Improvised Explosive Device), and where the risk of attacks by guerrillas with RPGs and AK-47s is high. Indeed, this is one of the most dangerous areas of the airfield that must be aggressively patrolled by the Air Force warfighters. They patrol not only the road, but also the villages that lie alongside them. In fact, the typical two-man patrol's primary mission is to ensure that the base and Air Force assets remain safe. Beyond that, these airmen are providing a secure environment for not only Air Force, but also Army, Navy, and Marine fixed-wing and rotary wing aircraft at Bagram AB. Even in the case of an actual attack by enemy forces on the base, the Security Airmen are trained to operate in a similar fashion as the Green Berets in South Vietnam (Air Force Magazine).

NASA: A MODEL FOR MILITARY SPACE OPERATIONS

So, we might say that all of the various space missions of exploration and scientific study that have been carried out by NASA over the last half-century also have, in effect, served as research and design for the military space "sorties" of the future. These operations began in 1959 when the Mercury Program was initiated, and it was followed by the Gemini Program in 1963. Since those early days of NASA space operations, more than 200 such missions having been completed. Perhaps the best-known project has been the Apollo Program, whose ultimate objective was to place a man on the Moon, but which was accomplished only through a stepwise series of sorties that prepared the way. Even before the lunar project began in 1961, the U.S. Air Force also had conducted many precursor sorties of its own. These were intended to explore the unknown of manned operations at the very limits of the stratosphere. This was the era of the test pilot and the experimental high-altitude balloon sorties which, of course were designed to build and enhance military capabilities for the future.

When NASA assumed most of the responsibility for such research and testing in space, during the early 1960s, their sorties were designed to develop knowledge regarding the new environments, and to develop human and other systems for conducting operations in space. While the objective of these sets of sorties has usually been couched in civilian terms, they have, in actuality, also contributed to military capabilities in space. Thus, the advances that have been made in the development of rocket-propulsion and other propulsion systems by NASA have also been made available to military space operations by the U.S. Air Force. The same can be said for the several space "maneuvers" that have been developed by NASA. Among the most notable of these are the rendezvous maneuvers of the Space Shuttle spacecraft with the International Space Station, and the precise calculation formulae that placed a spacecraft in the precise point in orbit that enabled the Cassini-Huygens mission to carry out its long-range voyage to the Jovian region of space.

Sometimes the target objective of a space sortie has been a particular point in an orbit around a given body in space, such as the Moon or a planet, rather than the body itself. In an example of how a "cultural geography" is being developed in space, many orbits are now being calculated in order to rendezvous with artificial bodies, such as the orbiting satellites and space stations, as well as other spacecraft. At other times, the target area may be a point along the orbit of the Earth itself, as is the case of reentry into its atmosphere; or the orbit the artificial satellites that have been placed on "near-Earth orbit" (NEO). Another specific set of orbit-point targets have been those of the Hubble Space Station and the International Space Station, which also occupy a respective orbit around the Earth. Then there are the longer-range sorties that have been performed by the deep space probes, whose target orbit points have been located near the outer planets, such as Jupiter, or one of the comets and other bodies that inhabit the Kuiper Belt and Oort cloud.

Therefore, the NASA programs of the first fifty years of its existence can be understood as a series of sorties. Their chronological order also happens to have a geographic dimension as well. As a geographer, I think of it as a map-making exercise, very much like what occurred on Earth during the European Age of Exploration, which began in 15th century and continued into the 20th century. In the current space age of exploration, the same process of discovery and addition to the map of the solar system has been occurring, practically since the first years which followed the end of

World War II. As a military geographer who specializes in military aerospace activities, I see a military aerospace region that is under construction, both consciously and unconsciously, in various areas within the solar system. With this in mind, the following is an analysis of the solar system as another class of space operations, the military operations that are occurring within the Geospace Domain.

The object with the closest natural orbit around Earth is its Moon. At this time, in the beginning decades of the 21st century, it still is believed the Moon is Earth's only natural satellite, and it also is the nearest neighbor, only about a quarter of a million miles away. However, this has to be said in a tentative sense, because in 2011, a couple of bodies that are similar in size to the Earth were discovered orbiting inside our own orbit around the Sun. Nevertheless, the Moon appears to be the brightest object in the sky, as seen from the surface of the Earth. Therefore, it has historically been the stuff of poetry and music. On a more practical level, the Moon has also been associated with natural events on Earth, such as the ocean tides. Its predictable movement around our planet has led to the development of lunar-based calendars since the earliest days of human history. Then, in 1969, the romance was gone; human astronauts landed on the moon and it became simply the stuff of scientific observation and analysis. Geologists, for example, now began to see the moon as simply another "dig" to be analyzed like any region on Earth. Following the visit of the astronauts on the surface of the moon, and after many years of repeated visits by both man and machine, we now have a much more scientific understanding of this heavenly body.

To the NASA explorers, however, the Moon also was viewed as a stepping stone for future space missions to other regions of the solar system. For one thing, its relatively accessible location with respect to the Earth, and its much lower gravitational attractive force, made this body a prime candidate for what might be seen as a strategic base for further operations in space; a "stepping stone" for further excursions into the next "unknown," as it were. These characteristics of the Moon have also been invoked as a reason for establishing a military space base on the Moon to control the lines of communication between the Earth and other points within the solar system. Also, the Moon might be seen as an interplanetary "coaling station" for refueling and replenishing space missions to the outer regions of the solar system. The Moon already has served as a "gravity assistance" node for space missions whose ultimate objective is the planets that lie beyond its orbit. The first of these included Mercury, Venus, and Mars. However, the Moon and the inner planets also have been used to provide gravity assistance to the outer planets that lie beyond the Asteroid Belt. These include the outer gas giant planets: Jupiter, Saturn, Uranus, and Neptune. Now, they too, are being used to provide gravity assists to space missions whose ultimate objectives are the regions of the Milky Way galaxy that lie beyond the heliopause.

———

Contemporaneously with the early attempts to launch artificial satellites into Earth orbit, NASA embarked on a program to deploy orbiting satellites to explore our own planet. One of the main reasons for this effort was a consequence of the International Geophysical Year (1957-1958). I was vaguely aware of the IGY during my high-school years; and I only understood that it was an effort to gain a more comprehensive "global" view of our planet from the external perspective of orbiting satellites. Later, I was to learn that the IGY was one of the most important geographic exploration missions of the Earth that would occur in my lifetime. I also learned that the program actually encompassed a constellation of scientific observation and research efforts in several fields related to the anatomy and physiology of our planet. The studies encompassed such aspects of our planet as: auroras and related phenomena; cosmic rays; geomagnetism; glaciology; gravity; ionosphere physics; latitude and longitude determination; meteorology; seismology; and solar activity. Ultimately, this matrix of geo-scientific studies would become the paradigm for planetary scientific studies throughout the solar system.

One of these geocentric studies was carried out by the NASA Explorer missions. At least in the initial phase, these involved a series of geo-orbiting satellites which began to be deployed, beginning in 1958 with the launching and deployment of the Explorer 1 spacecraft. This was to be the first of 70 Explorer missions, six of which are still active in the second decade of the 21st century. All but four have been successful in accomplishing their designed mission, but even the "failures" have added to the base of knowledge and skill set that would form the foundation of space operations. Among the highlights of this program has been the solid, disciplined scientific study of the Earth's magnetosphere and the associated Van Allen Radiation belt, as well as of the ionosphere. Later missions were designed to expand on the properties and extent of these zones of radiation. As such, they were the foundation for studying magnetospheres throughout the solar system. Explorer 3 was deployed as part of the International Geophysical Year program, and its main contribution to the overall Explorer program was that it included the first attempt to record a complete radiation history for each orbit. More detailed measurements of charged particles that are trapped in the terrestrial radiation belts were accomplished by Explorer 4, from October 1958 and October 23, 1959.

Another contribution to the overall endeavor of space exploration was derived from "failures," the so-called mission-ending problems. These actually served to further the stepwise improvement of the particular technology that is related to batteries and solar arrays for powering the communications and payload operations. The Explorer 4, for example, ended when the high-power transmitters ceased sending signals on October 5, 1958. The probable cause was the exhaustion of the power batteries. The upshot was that it decayed out of orbit later in the month, on October 23, 1959. According to the mission report by NASA, contact was lost when the solar cell charging current fell below the required level to maintain the satellite systems. NASA learned from this, and evidence of significant improvement in the technology of batteries was seen with the Explorer 7 mission, which was assigned much more extensive and complex objectives than any earlier missions. These included the task of measuring "solar X-ray and Lyman alpha flux and the trapped energetic particles, and heavy primary cosmic rays" in what geographers might call "charged particle regions." The secondary objective of this mission "included collecting data on micrometeoroid penetration and the study of the earth-atmosphere heat balance." The payload and operating systems of this spacecraft were powered by some 3000 solar cells, as well as a pair of nickel-cadmium batteries. The tools for doing scientific measurement of these charged particles also progressed in their sophistication by this time. The payload of Explorer 7 included an X-ray chamber, and a cosmic-ray Geiger counter.

The Explorer 8 mission was even more advanced in terms of its onboard suite of measurement tools and techniques. It was designed to obtain measurements of such phenomena as: "the electron density, the electron temperature, the ion mass, the micrometeorite distribution, and the micrometeorite mass in the ionosphere," ... at median altitudes of about 500 miles. Armed with this payload, it was intended to study the temporal and spatial distribution of these properties and their variation from "full sunlight conditions to full shadow, or nighttime conditions." The toolkit included an "RF impendence probe, an ion current monitor, a retarding potential probe, an electron current monitor, a photomultiplier-type, and a microphone-type micrometeorite detector, as well as an electric field meter, a solar horizon sensor, and thermister temperature probes." Unfortunately, the advanced battery power failed on December 27, 1960 (USA in Space).

The analysis of the hundreds of manned and unmanned space sorties, which have been conducted by NASA, shows the wide range and variety of objectives and functions of these space operations. Speaking only in terms of functions that correspond to the traditional aerospace kinds of operations, one can readily see the development of an early warning and strategic reconnaissance function in space by NASA, such as the establishment of the Great Observatory Program. Within the overall program, there is the Chandra X-ray Observatory, the Compton Gamma Ray Observatory, the Hubble Space Telescope, and the Spitzer Space Telescope. These all can be seen as early detection and tracking systems that could easily be applied in a military context. Although they are currently being oriented towards the outer reaches of the solar system, they could presumably be oriented specifically toward the Geospace Domain too. Therefore, it can be said that much of a potential "early warning" system that might be needed for the conduct of military operations in space has been in development already (nasa.gov).

Consider, also, the great advances that are being made in the "strategic reconnaissance" operations that are being performed by telescopes of all types, some of which are located on the surface of the Earth; and others in outer space. It seems that almost every few months some astronomer is discovering a new galaxy, planet, or other such phenomenon in the skies. Some of these "passive" observation sites are extremely large, complex, receptor systems that can detect visible-light radiation from great distances in the solar system, and beyond. Then there are the long-range sensors that operate in the radio wavelengths. These consist, essentially, of a radio receiver and an antenna system which is able to detect radio-frequency radiation that is emitted by distant objects. Because radio wavelengths are much longer than visible light, radio telescopes have to be extremely large, in order to attain the resolution of optical telescopes. However, there now are the so-called "virtual telescopes" which consist of arrays of smaller telescopes that are interconnected by supercomputers.

Another set of NASA space systems that could be applied to military space operations are seen in the laboratory space stations, such as the NASA Skylab. This orbiting science laboratory, which was launched in 1973, initially was designed as an orbiting platform to carry out long-term scientific experiments in the micro-gravity environment of space. But it also represented an effort to gain practical experience in maintaining long-term human habitats in space. The various science experiments were performed by crews of astronaut-scientists, who carried out various scientific tasks pertaining to their scientific specialty. One of the main objectives which were assigned to the

Skylab mission was to conduct prolonged studies of the Sun, from a point of observation that lies outside the Earth's atmosphere. From this perspective, the Skylab was able to carry out more detailed and longer-term studies of the Sun than ever before. One of its main tools for performing its scientific mission was the "The Apollo Telescope Mount," which incorporated a set of component telescopes and other devices for observing the Sun over a broad range of the electromagnetic spectrum. On the other hand, the Skylab also was given the task of conducting extensive survey exploration of the resources on Earth itself, from this global perspective. In addition to the benefits of its vantage point, the Skylab was able to conduct wide-ranging spectral analyses of mineral structures and elemental composition of soil samples, as well as possible organic materials on other celestial bodies.

Another painful, but valuable learning experience for conducting space operations was derived from a system malfunction that almost caused the complete failure of the entire Skylab program. In this case, a thermal-meteoroid shield was severely damaged during Skylab's ascent. This also caused the jamming of one of the solar power arrays that was supposed to provide electrical power to the station. The initial three-man crew managed to provide an improvised fix by deploying a "parasol" sunshade, which was later reinforced by an overlying sun shield. They also were able to release the jammed solar array. I am sure that the minute-by-minute details of the incident and the fix procedure became a permanent part of the maintenance and operations manuals for that type of space station. Ultimately, the mission was carried out successfully and ended when increased solar activity caused the space station's orbit to degrade. It reentered Earth's atmosphere on July 11, 1979 and its debris was scattered throughout the Indian Ocean.

Following the completion of the Apollo Project, NASA turned its attention and its resources to doing more science work with respect to other bodies in the solar system. The planet Mars was to be the first target for such scientific study. This involved the use of more sophisticated techniques for data acquisition and sampling, as well as protocols for scientific experimentation and analysis throughout the solar system. Furthermore, even while doing the design and planning for sustained operations in orbit and on the surface of Mars, NASA already was planning for the exploration of the rest of the solar system. Eventually, virtually all of the major planets and some of their moons and rings would be studied by generations of space orbiters, landers and rovers. The general modus operandi has been to: (1) conduct global-mapping of planets, moons, asteroids and comets (which usually is done by multiple-orbiting, standoff orbiters); (2) utilize unmanned probes and landers vehicles to descend through the atmospheres and, sometimes, onto the surfaces of planets and moons, as well as an occasional asteroid and comet; and (3) conduct more detailed and longer-term observation and scientific analyses of certain geographic phenomena, such as the magnetic fields and plasmas that occur around the Sun and other planetary nodes in the solar system.

———

Speaking of lessons learned: I would like to take special note of another, perhaps most fruitful, decision that has been taken by NASA: namely, to make available virtually all of the data that has been accumulated and the archives of the experiences and lessons learned to the overall scientific community of the Earth. Such a policy of sharing of information has been a boon to the overall study of the solar system and to the development of technologies that is seen today. This policy of openness also has been reflected in the space agency's willingness to not only share information and skills, but also to utilize the knowledge and skills of the private-sector "space-industrial complex" and of the academic clusters of scientific work to further their mission objectives. A happy coincidence which has powered the effect of these policies is the computer internet, which has provided "interested persons," such as me, with the information and material to write this book.

The effect on the science of military geography also has been significant. So, during this new phase of human space exploration, the dimension of the "land" side of the man-land equation of military geography has grown by orders of magnitude. What began as a study of the terrain and weather on predominantly surface-based military operations has been expanded on several dimensions. Thus, with the development of military atmospheric operations during the World War I period, military geography broadened definition of the "natural environment." That is to say, the terrain and the atmosphere were seen as the new definition of "land." Then, when military aerospace operations continued into the post-WWII era, the definition was broadened again to encompass the transitional zone between Earth's atmosphere and near outer space (i.e., geospace). Today, the map of the military area of operations is continuing to expand as human exploration of the Jovian subregion of the solar system is being carried out more intensively. Perhaps by the end of the 21st century, space exploration will have been extended outwards into the realm of the exoplanets of our Milky Way galaxy.

————

If we understand the term "reconnaissance" as a synonym for learning, then we can appreciate how the telescope has been a learning tool for gaining knowledge about the solar system for the last 400 years or so. The telescope observatories have provided "reconnaissance" of the solar system since the 17th century, but they came into their own during the 19th century when major advances in the science of optics were made. At the very beginnings of the Space Age, it was the surface-based telescopes that provided much of the context for mission planning. Today, these original telescopic systems are being supplemented by similar systems that are located in space itself. They now are working together with the space probes and other space platforms to form an overall system of reconnaissance of the new, expanded geography of the solar system. So, once again, the Earth-based space observatories have taken the lead in this new endeavor, especially those that are designed to detect and record the most distant and faintest bits of signals within the radio phase of the EM spectrum. In the language of electronic warfare, these are "passive" systems which are capable of detecting and processing signals that originate from various areas of the Cosmos. Also, the overall space reconnaissance system has been made more versatile through the development of specialized more versatile observatories that can operate in many of the wavelengths of the electromagnetic spectrum. And, the output "imagery" from these systems has achieved ever greater resolution and clarity through the development of computer-based enhancements.

ISR (intelligence-surveillance-reconnaissance) operations also are being performed by space-borne sensor platforms, even though they are not necessarily designed specifically for military purposes. This situation reminds me of the way in which major powers of the post-WWI era utilized "civilian" research and development in the areas of aircraft and naval vessels to circumvent the limits prescribed by the treaty that ended the "war to end all wars." The practical upshot of that experience turned out to be, the ultimate design and construction of more advanced planes, tanks, and aircraft carriers, as well as rockets, for military uses in WWII. Therefore, it can be seen that all of these manifestations of advances in "civilian" science, technology, engineering – and just the growing "body of knowledge and knowhow – will contribute to military space operations in the future. All of these remote sensor systems, as well as the in situ sensors, have served to develop a first-generation early warning and detection system which can also be utilized for military space reconnaissance and surveillance purposes.

Another phase of the exploration and study of the solar system began in 1993, when the United States and Russia agreed to merge their separate space station programs to produce a single facility which would "loiter" in orbit around the Earth, and serve as a platform for conducting various low-gravity scientific experiments. The early design of the basic architecture of the ISS was based on the concept that modules were to be individually built, sent into orbit, and then integrated onto the existing space station. It also came to be a living model for the development of future habitats in space. Ultimately, it also would come to serve as a base for the continuing exploration of the solar system and the Milky Way galaxy. Perhaps most significant to the development of military space operations in the future, however, was this first example of pragmatic cooperation among the nations of the Earth in space.

According to the overall construction plan, the segments would be integrated after they had rendezvoused with the space station. The first phase of assembly began with the launching of the Russian control module Zarya in 1998. Then, the U.S.-built Unity connecting nodule was launched in mid-2000 to begin the modular construction of the ISS. At this point, the ISS would be manned by a crew of two Russians and one American astronaut. A NASA microgravity laboratory, which was named Destiny, and other elements destined for the ISS were subsequently joined to the station. This method of "Tinker-Toy" construction called for the assembly, over a period of several years, of a complex of operating systems, laboratories and habitats, which would be linked by trusses. These would be powered by four large solar-power arrays, battery packs, and thermal radiators. The Russian modules were ferried into space by expendable launch vehicles, which then were joined to the ISS. Other elements were ferried by NASA Space Shuttles and Russian Soyuz spacecraft. The shuttle spacecraft then continued to carry out a sustained program of maintenance and logistical support, by ferrying astronauts and equipment. Then there was the Soyuz spacecraft that was permanently docked to the ISS, serving as a "lifeboat." It actually was used on at least one occasion to evacuate crew when a solar storm threatened the integrity of the ISS and the lives of its crew.

As was indicated earlier, there is another area of space exploration which has been focused on our own planet, Earth. In a sense, these space missions have been exploring our home planet from a new perspective, and armed with tools of observation and measurement that were never previously available. One such inward-looking space mission was the CLARREO (Climate Absolute Radiance and Refractory) mission, which was a joint-venture of NASA and NOAA (National Oceanic and Atmospheric Agency). Among other missions, the project was designed to carry out an ongoing series of measurements of incident solar irradiance on the surface of the Earth's and to measure its planetary energy budget. The NASA portion of the mission involved the measurement of spectrally-resolved infrared and reflected solar radiation to a high degree of accuracy. The results of these scientific studies have provided a data record which now is being used to study long-term changes in the climate system of the Earth. This mission was also designed to provide a source of absolute calibration for a wide range of visible and infrared observing sensors, which should greatly increase their value for climatic monitoring.

The Cluster II mission, which was launched in July of 2000, is an example of an in situ investigation of the Earth's magnetosphere, through the use of four identical spacecraft which worked in concert. One of the primary outcomes of the mission that was expected was to make it possible for scientists to distinguish between spatial and temporal variations within the magnetospheres, on a global scale. Then there was the Aeronomy of Ice in the Mesosphere (AIM) mission (2007), whose objective was to determine the nature and causes of the highest altitude clouds in the Earth's atmosphere. Another mission, CALIPSO (2006), was developed to provide scientists with information regarding the effect of clouds and airborne particles (aerosols) on changes in the Earth's climate. Similarly, the 2012 Carbon in Arctic Reservoirs Vulnerability Experiment (CARVE) is designed to measure the carbon budget of Arctic ecosystems, which is still not well understood.

Farther out into our solar system, a planned joint NASA-ESA Europa Jupiter System Mission, which is planned to launch in 2020, will send two robotic orbiters to conduct a three-year study of Jupiter and some of its moons. One moon in particular is Europa, which is of interest because it is theorized that beneath its icy surface, there exists an ocean which might contain enough oxygen to support life. Ganymede is all of interest because it is thought to be the only moon that has an internally generated magnetic field. For its part, Io is considered, at this time, to be the most volcanically active body in our solar system. And, Callisto is of interest because it presents a heavily cratered icy crust that might be hiding an ocean deep in its interior. On another track, NASA is conducting early studies for a launch of still another probe mission which is projected for some time in the 2020s. This mission will be designed to visit Saturn's moon, Titan, which is the only moon in the solar system that is known to have an extensive atmosphere. In this case, the methodology that is planned centers on the deployment of a balloon that would hover within Titan's nitrogen-rich clouds. Plans also call for the development of a lander that would splash into Titan's methane seas; as well as an orbiter craft that would be capable of relaying data from the lander back to Earth. Meanwhile, the ongoing NASA-ESA Cassini-Huygens mission (launched in 1997) continues to explore Saturn and its moons, and the other planetary domains of the Jovian subsystem.

Several other space missions are now taking a more detailed and sustained look at the nearest celestial bodies in our inner-space neighborhood of the solar system. One of the most prominent of these bodies is the planet Mars, and it is the object of an ongoing mission by the robotic Mars Science Laboratory, was launched in 2011. An essential component of this mission is a rover vehicle whose mission is to analyze soil and rock samples and to look for organic materials. The latter may help to answer a key question of whether Mars is capable of supporting microbial life. Later in the decade, the joint NASA-ESA ExoMars Trace Gas Orbiter (estimated to launch in 2016) will study the Martian atmosphere, paying special attention to the methane gas, which was first detected in 2003. Because one source of methane is the result of biological activity, it is conceivable that life might currently exist on Mars. Also, NASA's Gravity Recovery and Interior Laboratory (GRAIL) is now utilizing twin spacecraft in tandem orbits around the Moon. These orbits are designed to work together to make highly sensitive gravitational field measurements. Such data will allow scientists to map the lunar interior, from crust to core.

There are many reasons for focusing on Mars. Some are part of an overall program of continuing exploration of the solar system, but there are more specific considerations as well. Among the latter is the probability that Mars may contain many accessible mineral resources and water resources. For one thing, it seems that there is considerable empirical and scientific evidence that there is iron ore on Mars. Indeed, the Spirit and the Opportunity rover vehicles that were deployed by the Mars Exploration Program of NASA have returned soil samples that show promise of economically-viable mineral resources in several of the areas of the planet. In fact, by utilizing sampling and statistical analyses, scientists have hypothesized that there are sufficient quantities of iron deposits on the planet to make them economically accessible with current technologies. However, there are still many issues that must be dealt with before mining operations can begin there.

For example, at the present time, it is not economically feasible to actually send robotic rovers out to search for the mineral on a global level on Mars, mainly due to the prohibitive costs associated with such a venture, at our present level of technology. Instead, scientists are using analog and statistical models to develop a preliminary map with the data that is available today. So, within these constraints, the Spirit and Opportunity rovers of the Mars Exploration Program have already done some very preliminary sampling of the geological data – the result of which shows a high correlation between the many craters on Mars and the presence of iron. However, even if an adequate map of iron-ore deposits were to be produced, the economics of location analysis would still have to be studied to determine the feasibility of mining operations on Mars, as opposed to the comparative costs of such operations on Earth, on the Moon, or some asteroid. This is where the several dimensions of the problem are encountered: that is to say, all of the "costs" (both economic and non-economic) must be factored into the equation in order to determine the "net feasibility" of such a venture.

However, even if we assume that conducting mining activities on Mars is economically feasible, then the military dimension of the problem still must be considered. This particular trajectory also would need to be considered to be the "military" cost of conducting such operations on a planet in the solar system. But, even before determining what the function of the military would be in such an endeavor, the question again would be one of "economic geography." By this, I mean to say that the preliminary issue again would be one of location analysis, including the net "cost" analysis. So, assuming we have developed a map of the craters on Mars, and that a certain percentage of them contain enough iron ore to justify the cost of developing mining operations in some of those locations, the problem now develops into one of location analysis. From the perspective of the geographer, at this point in the overall feasibility study, the focus would turn to a particular area for conducting mining operations. Then, when the feasibility of a particular site has been established, the next issue would be how the site would interact with the other functional nodes the planned functioning regional economic system. Of course, functional linkages would have to be established in order to form the spatial collection of nodes into an interactive and interrelating system which is working toward a defined objective.

As a case study, consider how an economic geographer would seek to determine the optimum location for each of the respective processing points in a steel-production regional system. Such a site also would depend on the relative location of the other essential ingredients for steel-making. The process of making steel itself requires inputs of other natural ingredient resources, such as "coke" (a particular state of coal) and silicon, among others. The developing steel-making node then would be incorporated into another, larger-scale regional system, which includes the logistical and marketing nodes that are necessary for getting the finished product to the ultimate end-users, or customers of the steel products.

However, let us assume for the present, that at sometime in the future, it is decided that it is feasible to develop iron-mining operations on the planet Mars. Now the problem assumes geo-political and military dimensions, as well as geographic and economic dimensions. A fundamental reality that underlies any asset is that it eventually also becomes a "liability" which must be defended – either by diplomatic or military means. We are seeing this reality becoming ever more apparent with respect to the orbiting satellite systems that have been deployed by nation-states and others during the past sixty years or so. This network of communications, navigation aids, and observer nodes has suddenly become an asset that must be protected against attack from a variety of malevolent forces. And, given the relative ease with which militarized rocket technology can be purchased – either from countries like Iran and North Korea, or on the black market – it is within the level of significant probability that some non-governmental organizations and "rogue" nation-states will presumably have the wherewithal to launch significant attacks on the artificial satellites that are orbiting the Earth.

————

Another node in this hypothetical economic region in space could be an asteroid. For one thing, the asteroids also are considered to be probable sources of natural resources, including iron and water. Second, due to its low surface gravity, future space operations may come to utilize the certain optimally-located asteroids as nodes within shipping lanes to link resource extraction sites with other auxiliary raw material nodes and processing nodes. The market nodes for such a system might include the Earth, human colonies on Mars, or even forward bases for deep space operations. Asteroids also might be utilized as military bases which would provide security for the economic system. For these reasons, there also is a trajectory of exploration and scientific inquiry into the movement of the asteroids, and of their intrinsic properties and characteristics.

One aspect of these studies has to do with the fact that there also is a dark side to the asteroids. That is, much attention is being given to asteroids as possible threats to the Earth itself. Indeed, such is the heightened perception of the threat from asteroids (as well as comets and "cultural debris") that the threat population has been given its own acronym: Near-Earth Objects (NEOs). NEOs are defined as being either an asteroid or a comet with an orbit that brings them to within 100 million miles of our planet. In response to the increased sense of urgency with respect to the threat of collision events in geospace, the NASA Jet Propulsion Laboratory has implemented an automated collision-monitoring system, which has been named "Sentry." Its focus of attention is on those NEOs which have orbits that bring them within 1 AU of Earth; of these, the ones that have diameters of 500 feet or more are considered to be the most significant threats. Given these criteria, approximately 8,400 NEOs have been identified as threats of the highest order, and 1,200 asteroids among them are classified as posing a significant threat to Earth itself. Aside from the existential threat to our own planet, the NEOs now pose a higher probability than ever before of negatively affecting humans, as we venture more extensively and intensively into the interplanetary medium. This is an example of how natural and cultural geography interacts.

As a result, the probabilities of an asteroid strike, and the various levels of damage that they might cause in any given strike event, are now the subject of the most intense scientific observation and paramilitary "threat analyses." The nature of the threat that is posed by NEOs on Earth already has been manifested by the so-called "Tunguska Event." In that case, it is thought that an asteroid or a comet, or their meteorites entered the atmosphere and exploded at about 30,000 feet above the surface of Siberia, Russia in 1908. This atmospheric blast managed to flatten about a 1000 square kilometers of pine forest, and its force has been calculated to have been the equivalent of a thousand atomic bombs, of the sort that exploded over Hiroshima during WWII. One can only imagine what this kind of strike event would do to an inner solar system that is populated by space stations, satellites, and human settlements that might be located on nearby bodies. On another dimension, consider the possibility of an adversary being able to deliberately "cause" such a disastrous event to occur. Such a capability already has been discussed in serious studies about how to deflect the vector of an incoming asteroid, comet, or even an artificial satellite whose orbit has been degraded.

Nonetheless, asteroids also appear to have many positive attributes in terms of military operations in space. For example, they are perfect for use as "stepping stones" towards ultimate targets like Mars, or as space bases for operations within Earth's atmosphere – thus constituting the ultimate "high ground" over the atmospheric area of operations. Also, thousands of asteroids are distributed throughout the gap between Earth and Mars, thus providing many possible refueling and replenishment bases for missions into deep space. As a practical matter, the use of an asteroid as a secondary base would reduce or eliminate the need to expend precious fuel simply to touch down on the surface of an intermediate planet, and then to takeoff again. Because an asteroid's gravitational attraction is relatively weak, launching from one takes less energy than the surface of the Moon or Mars. For these reasons and others, an asteroid offers many options in terms of military mission planning throughout the solar system, and beyond.

In practical terms, however, in order to be able to take advantage of these potential "asteroid-bases," there is much technology that still needs to be developed. Fortunately, much of this new technology can be derived from current technologies. For example, solar arrays, battery packs, and life-support systems can be adapted from designs that have been implemented within operating space stations, such as the International Space Station. Another area of research and development that still must be done has to do with the propulsion technology that will be needed within a sortie. For example, there will probably be a need to be able to move a complete spacecraft-payload unit about the surface of the asteroid, without having to utilize the heavy propulsion systems. The most promising area of work on this problem appears to be on the development of the solar-powered ion propulsion system. This form of electric propulsion is one in which ions are bombarded within an electrostatic field in order to produce a high-speed exhaust. One such possible operations plan would be to utilize ion engines to ferry spacecraft along the surface of an asteroid, and between asteroids and other objects. Other ideas include the use of portable ion engine units that could be sent to provide additional power to robotic rovers on Mars. There also is consideration of a logistical system whereby pre-constructed bases for human activities would be placed into Mars orbit, and then be towed by smaller "tugboats" to a site on the planet. Still other ideas that are being considered include the pre-placing of chemical rocket boosters along an interplanetary trajectory in advance, so astronauts could pick them up along the way; or even towing a very small asteroid to the vicinity of a space station to provide water to the astronauts.

In all of these cases, the proper balances among the size of the platform, the quantity of integral fuel supplies, as well as the quantity of payload, will be of paramount importance for any such operations. In this regard, NASA Jet Propulsion Laboratory scientists have proposed a strategic approach to achieving the systems that will achieve the desired balance. In the December 2011 edition of the <u>Scientific American,</u> they report on the outcome of "brainstorm" sessions for maximizing the propulsion resources that might be available to space rockets and other vehicles in the coming years. After considerable research and thinking on the subject, Damon Landau and Nathan J. Strange have formulated the outlines of such a system for maximizing resources and mission objectives for space operations in outer space.

One of the main nodes in this proposed system of operations in outer space has to do with the technology of electric thrusters and light-weight, high-efficiency solar arrays to achieve the optimum balance of thrust and range in this new medium. Another major suggestion in the proposal put forth by the NASA JPL scientists is that any kind of deep space operations should rely on the "stepping stone," incremental approach (like the island-hopping American campaign towards the Japanese homeland during World War II in the Pacific). In this case, however, the islands of the Pacific Ocean are replaced by the string of asteroids that lie between the Earth and the outer solar system. Such a strategy would also suggest the model for a military space force configuration in the future. In short, many "islands" in the solar system, the Moon and the asteroids, for instance, could eventually serve the same purpose as did the islands of the Pacific during WWII. That is, as stepping stones towards the achievement of a grander, longer-range objective, such as the establishment of a permanent military base on Mars.

To summarize, we have seen how the experience of NASA in space, and the knowledge and skills that have been developed from it, has been propagated throughout the scientific community and the military-industrial complex. One of the many consequences of NASA's policy of open sharing of information has been a significant advance in the quest to understand our solar system. It also has furthered the technologies that will be needed to conduct military operations in space. Moreover, the imperative for conducting military operations in space has derived from the quest to develop economic systems, especially those having to do with the extraction and processing of the raw materials that are thought to exist on the other planets and bodies of the solar system. Assuming that such activities are actuated, it is likely that a military presence in the heliosphere also will ensue. At this point in the analysis of the military geography of the solar system, the question that follows has to do with the nature of the military systems in space. We will explore this matter in the next chapters, beginning with the electromagnetic battlefield that will occur in the 21st century.

THE ELECTROMAGNETIC BATTLEFIELD

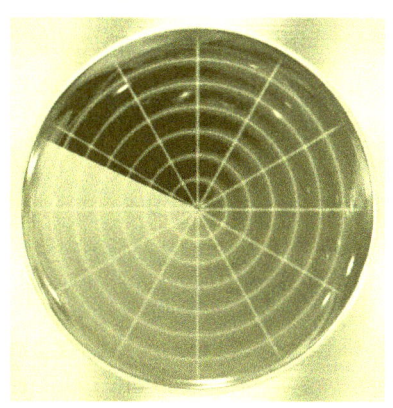

I think it is reasonable to assert that the electromagnetic spectrum, which permeates military aerospace operations today, will continue to do the same in space warfare. This phenomenon began with the utilization of the visible-light portion of the electromagnetic spectrum. This phase of the spectrum is mainly utilized by optical enhancement devices, such as the binocular and the optical camera. However, since the middle of the 20th century, the visible light spectrum also has been utilized by electro-optical devices, which employ an electrical charge to amplify the available visible light. One of these technologies involves the utilization of visible light in so-called L.A.S.E.R. (Light Amplification by Stimulated Emission of Radiation) applications. These "directed energy" beams of visible light are useful for "pointing" to a particular target, for transmitting stream of data for communications, and for navigation purposes. Other portions of the EM spectrum, including the microwave and infrared phases, are also being used as grist for the amplification by stimulated emission of radiation devices. In one particular case in space, NASA is using lasers used to pulverize rocks on Mars, so they can be chemically analyzed. Laser technology also is now being explored for possible use in generating energy that can be used as a substitute for fuel in space. One such application that has been discussed is the use of laser energy to charge photovoltaic arrays to provide power for space stations and satellites.

The practical use of the electromagnetic spectrum for aerospace operations began in 1932, when radio engineers found a source of cosmic "static" which was attributed to the interaction between the electrons and ions (charged atoms) in interstellar space. Later, it was discovered that radio waves are able to penetrate much of the gas and dust in space, as well as the clouds of the planetary atmospheres. Meanwhile, in aerial navigation, a system of radio transmitting stations (each of which sends a signal that carries identification, direction and distance information about the emitter) also has become a basic tool for a navigator to fix his position in the atmosphere. The radio spectrum itself includes the part of the electromagnetic spectrum that corresponds to frequencies that are lower than 300GHz. This is significant, because above 300GHz, the absorption of electromagnetic radiation by the Earth's atmosphere is so great that it becomes effectively opaque – until it becomes transparent again in the near infrared and optical frequency ranges. So, one of the most significant the applications of the EM spectrum in air and space military operations has been the harnessing of the various spectra to "see" things that can't otherwise be detected by the human eye, even with the enhancements of the optical technology.

However, the utilization of the visible light portion of the electromagnetic spectrum remains essential for use in the developing military space operations of the 21st century. Consider that most of what we humans know about the bodies of the solar system has been derived from data which has been captured by optical telescopes. Even though some telescopes have been deployed in space, most are still located on the surface of the Earth. At this point in time, these massive observatories are deployed on the surface of the Earth, mainly because of the prohibitive cost of launching and deploying such platforms into space. However, as launch and assembly techniques in orbit have improved, the launch procedure is becoming more cost-effective and, therefore we can expect that an increasing number of larger space-based telescopes will be deployed in space, thus adding the benefits of the array and the constellation to their other capabilities.

It is interesting to note that both terrestrial and space-based telescopes have gone through a pattern of modification and upgrading that is similar to that which has occurred to the original B-52 Stratofortress of the 1950s, which has greatly extended its service life. In the same manner, the optical telescope of the early 20th century has now been given greater visual acuity with the addition of a suite of sensors. These multi-spectral capabilities now enable the original optical telescope to "see" in the radio, radar, ultraviolet, infrared, and X-ray portions of the electromagnetic spectrum. The advances in optics technologies also have made telescopes more powerful. Thus, the latest generations of telescopes now are able to peer farther into the Cosmos, and to do so with greater clarity and resolution, than previous versions. Also, the recent advances in computer-generated, three-dimensional graphics have resulted in much more "information" being embedded within each image. More specifically, the power of digitalization and the pixel, along with the magic of computer-based animation software, is producing imagery of amazing resolution and clarity. Overall, the advances in computer technologies also have made it possible to store vast amounts of imagery and to manipulate and produce vast quantities of data for output as imagery. The quality output from the processors also is enhancing the ability of scientists and planners to develop models for scientific analysis and prediction. At the same time, continuing advances in the science of telemetry have made it possible to transport these images at almost the speed of light, from greater distances in space.

The early balloon-flights into the stratosphere of the Earth, which took place in the middle of the 20th century, can be seen as precursors to the modern field of remote sensing. Both the balloons and the orbiting satellites can be seen as manifestations of the ongoing effort to "gain the high ground" in the reconnaissance battle that occurs within the electromagnetic battlefield. For its part, the Soviet Union, rather than attempting to develop its own high-altitude aerial reconnaissance aircraft platforms, chose to follow a different path – the Earth-orbiting artificial satellite during the Cold War period. The first of these was the Soviet "Sputnik I," which was successfully placed into orbit in 1957, only to immediately be followed by the U.S. effort to develop its own artificial satellite program, again presumably to regain the "high ground" in the reconnaissance wars. The upshot of this competition was that, by the end of the Cold War in 1991, both the United States and the nascent Russian Federation had placed more than a thousand artificial satellites into orbit around the Earth.

As a derivative of the satellite contest between the Cold War adversaries, the advent of space satellite technology in the late 1960s has enabled scientists, including geographers, to view our planet from the perspective of an extraterrestrial observer. Indeed, our understanding of our home planet was no longer limited by the panoramic air photography produced by such high-altitude "spy planes" as the U2. We now had a new, vertical perspective, of not just a portion of the surface of the Earth, but a virtually global view of the planet that was generated by a constellation of satellites in geosynchronous orbits over the Earth. As a result, we now have a half-century archive of imagery of the surface of our home planet. Also during this time, we have been monitoring the dynamic geographic processes as they are occurring on Earth as never before. The upshot is that now we can "see" our planet in its totality and the operation of the various systemic processes in real-time fashion as well. The orbiting artificial satellites also enable us to monitor the course of recurring natural disruptions, including: land movements, volcanic eruptions, earthquakes, hurricanes and typhoons. On another dimension, we can comprehend the continuing changes that are brought on by human activities on the surface of the Earth, such as the development of huge urban islands of human settlements, agriculture, deforestation, depletion of water areas, as well as the pollution of the atmosphere and the oceans, and so on. And we can visualize the depletion of the ozone layer as well. Again, with the availability of modern remote sensing technologies, scientists and others now have the capability to observe the state of the climate and weather processes on a global scale of any planet. An example of this capability occurred when probes were inserted into the atmosphere of Venus to study that forbidding environment. Finally, we can comprehend the state of the surface of a planet at any given time, or over a period of cycles, through the use of constellations of orbiting satellites (especially those that are in geosynchronous orbit).

———

So, it can be said that the modern telescopic system is now capable of functioning throughout the solar system, and from anywhere within it. One reason for this is the continuing development of devices that are equipped with all sorts of optical technologies (such as mirrors and directed energy) to amplify and give coherence to the visible light and other EM radiation. Also, the increasingly sophisticated filtering and manipulation of the electromagnetic wavelength is providing these modern telescopic systems with new "penetration" capabilities through such obstructions as gas, clouds, dust, and even the obfuscating light of the Sun. Moreover, even man-made attempts at "jamming" signals and other "stealth" measures can be overcome by these new technologies. Finally, the modern generation telescopes are being given the power of "nimbleness." Thus, in just the past two decades, the optical telescope has been modified to be able to operate dynamically across most of the wavelength portions of the spectrum – from radio, x-ray and microwave, through infrared and x-ray – in order elicit the desired "information."

As we have learned on Earth, a planet's atmosphere can obscure a surface-based view of the celestial objects by absorbing or distorting light rays from them. However, a telescope stationed in the interplanetary medium is able to receive images of greater clarity and detail than do the planet-based telescopes with comparable optics technology. An example of such a space-based telescope is the Hubble Space Telescope (HST). It is basically a very large telescope whose mirror optics are able to gather more light, from the most distant celestial objects, and to direct it in a very coherent way to two cameras and an integral spectrograph device. This space telescope is still considered to be among the most sophisticated optical observatories ever placed into orbit around the Earth. The platform that bears this highly sophisticated optical observatory has been placed into orbit around the Earth, where it can view the skies from a vantage point that lies beyond the absorbing and distorting veil of clouds, dust, and water vapor of the Earth's atmosphere. From its moving orbital locations, the HST has been providing a "real-time" stream of imagery of the ever-changing solar system (and the nearest galaxies) to scientists and space explorers. To do this particular job, the HST circles the Earth every 90 minutes, moving at a speed of about five miles per second.

The HST began its operations in 1990, when it was launched aboard the space-shuttle Discovery. Since then, it has been providing unprecedented clear and high-resolution images for analysis. In reality, it has been presenting an unfolding history of the evolution of the Universe since its beginning. One advantage of having such a complete record of the history of the Universe and especially our solar system is that the so-called Big Bang theory can be continuously tested against new empirical data. According to this theory, it the Universe almost immediately began expanded rapidly from its original, highly-compressed, primordial state of being. In addition to being able to peer back through cosmic history, the HST ideally located to provide ongoing monitoring of the atmospheres and surfaces of the planets and other bodies. Because these observations have been archived in digital form, there is a comprehensive database related to the evolution of these bodies in more recent times. Such archives enable the planetary scientist to develop hypotheses and to make predictions about the subject. This is very similar to what the historical geographers have been doing on Earth for the last few centuries.

The Hubble Space Telescope is an excellent tool for studying our the planets, moon, and other bodies of the solar system; especially in conjunction with the other sensor systems that are placed on other types of platforms, such as space probes. In terms of military space applications, the capability for continuous observation that the HST provides, would allow commanders to detect any significant occurrences in space that might contribute to the situational awareness anywhere in the solar system. An illustration of this capability can be seen in an event which occurred in space, in July of 1994, when the Comet Shoemaker-Levy 9 was actually observed striking the atmosphere of the planet Jupiter. The data from the event itself and the consequent effects have provided much valuable "intelligence" about the planet and its environs as well. Had there been any military space operations occurring in the vicinity, this kind of "real-time" detection and warning could have proven useful for making the decision to activate necessary counter-measures to ensure the survivability and the operational integrity of the mission. Even though the observation in this case was made by a space probe's sensors, the point remains that the telescope, in all its configurations, can be very useful in performing the same functions as the long-range and short-range radar systems that are now being utilized in atmospheric operations.

To do its job, the HST has been equipped with an assortment of highly-sophisticated tools, the main one being its large reflecting telescope. With its mirror optics, it is able to gather light in the visible, ultraviolet, and infrared wavelengths from celestial objects. In addition to the primary and secondary mirrors, the observatory is also equipped with a system of panoramic, high-resolution cameras with which it is able to see deeply and widely into the solar system, and beyond. Thus, it produces both panoramic and detailed imagery of whatever it sees. Also, the array of multi-spectral sensors and multi-faceted optics with which it is equipped, enables the HST to send out a continuous and integrated stream of images; and these can be sent via many wave-lengths of the electromagnetic spectrum. Another element of the platform's advanced equipment is the "faint-object spectrograph," which is able to receive ultraviolet light that otherwise cannot reach the Earth because of the absorption of such energy by the atmosphere of the planet.

A continuing series of updates, as well as a program of both routine and exceptional maintenance, is expected to keep the Hubble Space Telescope in operation well into the next decade. These steps will continue to provide this proven platform with the latest scientific instruments, as well as replacement parts to maintain optimum operational status. The latter includes batteries, gyroscopes, and other equipment to keep the system operating at peak effectiveness. To date, these continuing maintenance and modification operations have been done by astronauts who are ferried to the site by space shuttles. However, with the termination of the space shuttle program, it is likely such future missions will be done by private-sector contractors which are developing similar space transport capabilities. Indeed, the use of the space-industrial complex might become the model the development of future military space systems. Historically, it essentially has been the model for the exceptional longevity successful systems like the Boeing B-52 Stratofortress, which is now in its sixth decade of active military service. Working with private contractors, there has been an excellent record of generations of modifications and upgrades. The upshot is that this model has enabled the bomber to operate effectively in many different environments, and to respond to a variety of mission requirements. It can be reasonably expected that the same "military-industrial" complex will continue to be an effective and efficient option in military space operations.

————

Another device, the spectroscope, also can be seen as another artificial enhancer of the human senses; it is an instrument which detects the spectra of elements, thus making them visible to the human eye. Spectroscopes are now being used to detect phenomena which cannot be seen by other means, even by the microscope or other similar optical devices. The resolution that is provided by the mass spectroscope is a measure of its ability to separate adjacent masses that are displayed as peaks on a detector. During the period following WWII, the techniques for manipulating short pulses of electricity allowed the construction of the "time-of-flight" mass spectrometer. This device utilizes short emissions of ions that are emitted from a source, and whose arrival times then are recorded after having traversed a sufficient distance to allow for the measurement of discrete speeds. Similarly, the microscope can see objects at the micro-level too, but it is used to look for atomic structures of objects, rather than the spectra of their signal profile. Both of these instruments are now being used to study space in many different locations, and from many perspectives. In the case of space studies, they are now being used both on the surface of the Earth to study objects from space that have been brought back to our planet, and in space itself, as remote sensors.

Throughout the Universe, there are basic elements which exist in many forms and which are codified within the so-called periodic table; a classification scheme in which the chemical elements are arranged according to the order of their atomic number. The atomic number refers to the particular configuration of an atom's electrons around its nucleus, and the number of protons that lie within the nucleus itself. The spectrometer enables one to see the nature of the atom and to determine the geography (or configuration) of its nucleus and electrons. With the aid of a computer, it is now possible to determine the atomic number of an atom by searching the periodic table for a match with the particular atom in question. With computer technology, one can also compare and contrast atomic numbers and otherwise manipulate the atomic data to suit one's purpose. Such a capability is especially important in the military context because it provides the power to detect, track, and attack only the target, with no collateral damage.

The spectroscope also has given rise to the science of spectro-photometry. This is the branch of spectroscopy that deals with the measurement of the quantity of radiant energy that is transmitted or reflected by any object in the Universe. Usually, the intensity of the energy that is being transmitted by any given body is compared to the transmitted energy of a base (known) body which serves as a standard. Different types of modern spectrophotometers cover wide ranges of the electromagnetic spectrum, including X-rays, ultra-violet, visible, infra-red, or microwave. These insights into the dynamic construction of our own planet's structure are now also being applied to the exploration of other bodies in the solar system. Thus, all sorts of probes and other exploratory technologies that have been developed on Earth are now being applied to the study of other materials in space.

———

Since all bodies in the Universe emit some kind of electromagnetic energy at various wavelengths, a sensor which is able to operate in various and multiple wavelengths is useful in the conduct of military operations, whether in our atmosphere or in space. However, as has been learned over the past century of military aerospace operations, sometimes a system that has multiple capabilities is not always the optimum solution to a particular problem. An example of this can be seen in the various attempts to produce an all-purpose bomber platform during the years of the Cold War. The upshot of the experimentation of the day was that, for example, sometimes a versatile B-52 heavy bomber might be the most optimum platform; but at other times, only a B-2 would do. In the same sense, there are some sensors that only "listen" passively for the energy signals, while other sensors are able to actively transmit a signal, and then wait for the reflected signal for input. In similar fashion, the multiple-wavelength capabilities of the modern telescopic systems may not be the optimum in certain circumstances. Ultimately, the best telescopic system, in any given situation, will depend on the information that is being sought.

Consider that in the recent history of the Space Age, there has developed a perceived imperative to detect, catalog, and track threatening extra-terrestrial objects (both natural and artificial) in order to give the alarm and to relay vital information to "defense" systems. As a response, there has been the development of specialized telescopes that likely will become an integral part of modern electronic warfare in space. These telescopic systems and their associated imagery-producing subsystems are examples of the technique of "remote sensing." In military parlance, this would be expressed as remote detection, and tracking of a target. This is usually defined as the acquisition of raw data about an object which is located at a distance from the sensor itself. In other words, it does not involve direct physical contact for purposes of empirical observation. Now, in the 21st century, the term "remote" has come to refer to both airborne and space-borne platforms, and it also can refer to telescopes and other kinds of sensors. So, the term "remote sensing" can generally refer to any of the technologies that have been developed to enhance or facilitate this input phase of the process. In the end, the one thing that all of these sensors have in common is that they all operate within the electromagnetic spectrum. My experiences in the air war in Laos led to my first realization that electromagnetic energy is a natural geographic phenomenon, which has become the most important geographic environment for the conduct of military operations.

So great is the dependence on the electromagnetic spectrum in modern warfare that, in each of the major air campaigns that have occurred since the end of World War II, the first priority has been to "blind" the enemy's air defense radar and other sensor systems. That is, by disabling the enemy's use of the electromagnetic spectrum in modern air warfare, one effectively degrades the capabilities of the ground air defense artillery and missiles, to the point where they are no longer effective against modern attack aircraft. Similarly, an attack on their computer electronics degrades the enemy's command and control systems, so that there can be no coordinated response to air attacks. On the other hand, electromagnetic radiation has increasingly been utilized to detect ground targets, and to guide the air-to-surface missiles to attack them. If this paradigm continues into the future decades of the 21st century, it is almost certain that electronic warfare will extend the geospace transition zone.

Electronic warfare is a contest of point and counterpoint. Perhaps the classic scenario in the electronic warfare battle situation is the duel between attacking aircraft and ground-based air defenses. On one side there is the ground-based electromagnetic-based, detection and guidance system – most often radar – but now electro-optical (visible light) and infrared too. Opposing these ground-based air defense systems is the enemy aircraft, which may be either a reconnaissance or an attack platform, or even a specialized electronic "pathfinder" platform which is used to provide an electromagnetic path for another attack platform. Another version of the normal reconnaissance platform is the aircraft that is specifically designed to detect ground-based radar or infrared emissions which are seen as indications of malicious intent.

Sometimes the electronic duel involves the use of materials that either camouflage or otherwise distort the "view" of the opponent's electromagnetic-based sensors. Alternatively, one side may resort to the so-called low-tech (but nevertheless effective) measures to confound the opponent. This can include the introduction of electronic "noise" or "jamming" techniques to distorting the electronic signal that is being emitted by an opposing emitter; sometimes something as crude as ejecting strips of aluminum, called "chaff," has been found to be effective in creating "noise" on the scopes of the operators. By the same token, flares ejected from an aircraft or by an enemy on the ground have been shown to be effective in distorting one's own infrared signature, thereby effectively confusing or blinding the attacking system. Other electronic warfare countermeasures rely more on materials that are designed to either camouflage one's true signature, or to shield the platform from some of the harmful effects of explosive ordnance. As an example, the material called Teflon effectively changes the surface heat "signature" of materials.

Another high-tech countermeasure, which is a response to the development of stealth technologies, involves the use of electro-optics, so as to actually reshape the signal that is being emitted by a platform. This highly-sophisticated countermeasure technique incorporates many of the earlier utilizations of the electromagnetic spectrum itself, as well as the use of angular design and specialized materials in the construction of stealth aircraft. Stealth technology itself is used to avoid detection by enemy radar systems by employing a combination of features, such as the incorporation of angular construction of the platform's skin. This effectively serves to reduce or "scramble" a platform's signature on a radar scope. The evolving response to stealth technology is referred to as a "complex design philosophy," which involves a combination of low-observance (LO) strategies, as well the utilization of active emitters, such as the Low Probability of Intercept Radars, as well as radios and laser designators. So, one might say that the response to the "mixed and distorted" messages that are emitted through stealth technologies, is to develop multiple channels and filtering techniques for detection.

Therefore, it may be said that the current advantage of stealth technology in military aerospace operations may be eroding. This is due to the recent advances in sensors which are associated with surface-to-air missile systems or other weaponry. Ongoing advances in radar detection and tracking are also serving to degrade the stealth effectiveness of all-aspect, low-observable aircraft, such as the B-2, F-22, and the F-35 Joint Strike Fighter. Stealth-killing advances in technology now include more sophisticated and discriminating VHF and UHF radars, such as those that are being developed by Russia and China. Then there is a "passive-detection" system devised by Czech researchers as well. The latter uses any available "signals of opportunity" that are reflected off stealth aircraft, and then amplifies them for detection and tracking of stealth platforms. These new detection systems could reverse a 30-year trend that has seen the U.S. Air Force gain an increasing advantage over enemy air defenses through the use of stealth technology.

The Ho Chi Minh Trail: A Case Study

The history of the interdiction campaign that was conducted by of the U.S. Air Force against the supply line that is commonly known as the Ho Chi Minh Trail provides a useful example of electronic warfare. The main objective of the U.S. Air Force interdiction campaign along this important line of supply was to destroy as many trucks as possible, somewhat in the manner of the U-boat campaign against American convoys during World War II, in which the main objective also was to destroy cargo vessels. In Laos, the interdiction strategy was based on the fact that all objects on the surface of the Earth emit a unique "signature" of infrared-frequency radiation which can be detected and measured by sophisticated "heat" sensors. To accomplish this interdiction at the "retail" level meant that we would need to see in wavelengths other than the visible-light to effectively detect and attack individual convoys of trucks, which were camouflaged by the triple-canopy vegetation and hidden by the darkness of the night.

This is when the Ho Chi Minh Trail became more of an electromagnetic battlefield. Now the "eyes" that were used to detect, track, and attack the convoys of trucks shifted to the portions of the electromagnetic spectrum that lay beyond the visible light phase. For one thing, the detection and weapons delivery systems became more dependent on the infrared, acoustic, and even the chemical signatures of the targets. Interestingly, the electronic package that was carried onboard the air platform became as important as the weaponry of the platform itself. Among the most notable of these platforms was the WWI era A-26 (which was later replaced by the jet-propelled A-10). These airborne platforms (which might be seen as analogous to the submarines of WWII) hunted enemy trucks with sensors that operated in various phases of the electromagnetic spectrum, but especially those that utilized electro-optical and infrared sensor systems. At the same time, during the course of the Vietnam War, the electro-optical sight, the television and thermal-imaging technology led to the re-incarnation of such vintage platforms as the AC-130 gunship, which also was used to carry out the attack operations along the Ho Chi Minh Trail. However, as can be seen throughout the history of electromagnetic warfare, the enemy on the ground also began to augment the visible light portion of the spectrum with the infrared and radar-based weapons. These included the Soviet ZSU-23mm "quad 4" which consisted of four 23mm guns, and the 37mm anti-aircraft gun.

Electro-optical technology also was basic to the so-called "smart bomb" aerial tactics that matured during the Vietnam War. An example of this is the electro-optical sensor system that was used for guiding ordnance. Once again, the electromagnetic-based package that was carried aboard the air platforms became as important to the mission as the platform itself. This particular utilization of the electromagnetic spectrum included a laser illuminator and designator sensor package onboard the platform. Sometimes one platform would carry the laser payload only and would act as a "pathfinder" for other attack platforms with specialized detectors. The system was designed so that a laser beam would be used to "illuminate" targets, and therefore provide an electromagnetic "path" to the target for the specially-equipped attack platforms. Later, the pathfinder equipment and the ordnance package was was placed onboard the same platform. The essence of the system was the sensor-actuator subsystem that enabled the "smart" bomb to continuously detect and be guided by the laser beam all the way to the target. Later, during the wars in the Persian Gulf region, the airborne electro-optical laser sensor would also be used to detect improvised explosive devices. In that campaign, a multi-spectral electro-optical sensor system that was sensitive to such properties as texture was used to spot recently buried objects which were suspected of being mines or IEDs.

CYBERWARFARE

Generally speaking, all computer systems are designed to perform three main tasks: (1) to receive data inputs from external sources, (2) to process the data and convert it to usable information, and (3) to output (transmit) information. The data itself can be permanently stored in archive files, or it can be deleted once it has served its purpose. The "permanent" files of data are referred to as databases or archives of data; these can be manipulated by computer programs to provide other configurations of usable "information" to a user. It is also important to note that data files that are stored in a computer system are only logical constructs which are made up of electronic ones and zeros. Because the contents of these computer "regions" are largely ephemeral in their natural state, they must be imprinted on a variety of magnetic media if they are to become "permanent." In the same vein, the data that are stored in a computer system resemble pools of water within a hydrological system. Thus, data can be located within temporary pools known as "memory," or in more permanent pools on "disk" storage location. The data that reside within "virtual libraries" are accessible to anyone who knows (or can deduce) the correct "password" to enter the library. In a sense, these computer-based virtual libraries also are like geographic regions whose areas are constantly growing and shrinking.

The important point to be made, in terms of military operations, is that all the pools of data that exist in the global network of repositories are theoretically available to anyone on the internet. Some of this data is restricted, but history has shown that any database can be accessed through the process known as "hacking." Within the context of cyberwarfare, the unauthorized person who gains access to a data repository uses specialized programming techniques for various purposes. These may include the outright theft of data, or it may involve the corruption of the data in order to degrade the quality of the information that is produced. In one sense, such activity can be compared to the work of the traditional spy, who either steals a document or simply changes its content to "scramble" the information it contains. Or, it can be viewed in terms of electronic warfare, where a transmitted signal is captured and "reconfigured" by the receiver to confuse or "scramble" the return signal to the original transmitter. Therefore, it can be said that what we are dealing with in cyberwarfare is the use of remote spying techniques versus the more traditional in situ methods of the classic spy. Although the function of the computer system is to produce a desired outcome to the particular user, the user can be the entity for whom the information is intended, or it can be a hacker entity. Thus, it is the very purpose for which the computer system is designed that makes it so vulnerable to being hacked.

Stated in its most basic terms, one can say that all information is transmitted in the form of packets of electricity. These contain "information" which can be transmitted to a human entity or to a sensor-actuator component of an electro-mechanical system. At this level of abstraction, it can be said that the output of a computer system is the impulse which is either received by a human brain or an artificial processor. Thus, another main function of computer systems is to interact with other computer processors as well as human end users. This is the world of distributed micro-processors which send electrical impulses to mechanical control-actuators in order to perform a particular task (such as moving the guidance elements of a missile). In the military aerospace environment of today, this function can be seen in the control and guidance of such diverse systems as the orbiting satellites, the space vehicles, and unmanned aerospace weapon platforms (such as the Predator unmanned aerial vehicle of the war in Afghanistan). Many of these weapon systems are still only partially autonomous; that is, they still require some degree of intervention by humans. But the trend seems to be going towards military aerospace platforms which can operate completely independent of direct human intervention; under the complete control of an onboard computer system, not unlike the "HAL" of the well-known science fiction movie "2001."

———

In addition to the processing function, a computer system also has an input and output component. The job of the first is to "acquire" elements from the computer system's external environment. In doing this, the input subsystem of a computer utilizes its sensors in the same way that humans use theirs. Indeed, the latest input technologies offer the capability to not only see, hear, and feel, but even to smell and taste through chemical sensors. This total environmental awareness is the key to the growing ability of computer systems to "comprehend" almost to the level of a human organism. It is at this point in the IPO cycle, that there are many opportunities for an adversary to engage in cyberwarfare. An instance of this would be one where the enemy deliberately places faulty or inaccurate data into the environment to introduce "bad" data into the computer system. Thus, consider how the North Vietnamese purposely hung bags filled with urine in the trees along parts of the Ho Chi Minh Trail, in order to introduce "bad data" to the chemical "sniffer" sensors which had been air dropped by the U.S. Air Force. The purpose of these olfactory devices was to detect the movement of humans (and their trucks) along the trail for possible air attacks. The upshot was that the high-technology of Igloo White sometimes was being countered by the introduction of spurious data to the computers. This case also points out how relatively low-technology methods can counter high-technology methods when it comes to electronic warfare, including that which is carried out by computer systems.

The input function can occur via many devices, including a keyboard and electromagnetic signals from sensors. The electromagnetic signals, themselves, can be emitted from the radio, radar, visible-light, infrared, or ultraviolet phases of the wavelength/frequency spectrum. Or they can be emitted in the form of sonic or seismic signals as well. However, no matter what the nature of the signal is, or how the input occurs, it is then passed onto the processing units of the computer. This is where the received data is edited, evaluated, manipulated, and otherwise converted into information which can be of use to the decision-making functions. The latter can reside within the computer system itself or as part of another system. The point to be made here is that a cyber-attack can occur within any of these phases of the computer cycle. In the input phase, an unauthorized "hacker" can interdict or disrupt the signal flow between the sensor and the processing unit. One particular measure of the seriousness of such unauthorized access is the degree of dependence of today's weapon systems on electronic signals from satellites.

There are several tactics that a hacker can utilize to degrade or corrupt the effectiveness of the input process. One of these falls under the rubric of "identity theft," where the hacker assumes the identity of an authorized user. Then there is the gambit in which "Trojan Horse" software is implanted in a computer, with the intent of remaining inactive until a certain triggering event occurs. This is an example of the sabotage aspect of cyberwarfare. In this case, such an action might occur in the form of software code which can remain as a "sleeper" until activated by some external circumstance. Such a circumstance could be a time-delay algorithm or some other variable, such as the implementation of a later batch of software code by the enemy computer system. Unlike the tactics which concentrate on the input function itself; this type of attack involves the software subsystems of the computer system. So it can be seen how the tactics that were used by the folks at Bletchley Park in England during World War II. In this case, the attacking computer system and the targeted computer system can engage in a battle of encoding/decoding and encrypting/decrypting, in order to disrupt or misinform the opposition's computer logic center.

Another gambit in cyberwarfare is to "pull the electrical cord" which provides power to the enemy's computer system. It does not have to be done physically through an attack on a generator. Instead, it can involve an attack on the computers that control the hardware systems of the power plant, for example. In the military aerospace context, such a tactic would cause all data-linkages to fail and thereby cause havoc with vital communications and command and control functions. Again, an attack on military computer systems, especially those that control the electrical power grids would effectively leave military aerospace operations "in the dark," and spatially disoriented as well. More global cyber attacks on an entire network might involve not just the power generating nodes, but also their linkages with command & control and relay nodes. Experts believe that China and Russia may have already infiltrated the U.S. electrical grid, and have left behind "Trojan horse" software programs which could activate at a later time to disrupt electrical grids.

––––––––

One way to understand the elements of cyberwarfare is to examine a relatively primitive "internet" such as the IBM System-38 which I utilized as a data processing manager, for a small corporation, during the 1980s. My "internet" (called a local area network), consisted of an IBM System-38 midsize computer and its 12 "workstations" (which I eventually upgraded to 24 by adding another "server-controller" to my mini-internet system). In addition to the monitor screens of the workstations, there was another output component, which was the printer that outputted "information" in hardcopy form. However, to convert this assortment of hardware into a true computer system, there also was a library of computer programs, each of which had its own particular function. These "instruction books" for the hardware side of the system are the means by which the command and control function (also known as the "operating system") manages all the hardware and the "applications software" subsystems.

The various components of my local internet communicated with each other via electrical impulses, which are represented mathematically by series of zeros or a ones. These bits are then packaged in the form of discrete series of six or eight bits that are called bytes (similar to photons). The bytes then are transmitted to the various components of the system. Like all computer internet systems, mine had several methods of input into the computer. The main input device was the "qwerty" keyboard, by which operators and "users" inputted data and program instructions. In reality, what was being keyed into the system at that point was considered to be only queues of alpha-numeric raw data, which could either be valid or not – as in "garbage in…garbage out."

In that state of being, the data cannot be used for any kind of decision-making or action; it still has to be filtered, validated and otherwise prepared for the processor function to do its job. In this phase of the input cycle, the "raw data" is converted into useful data for processing. That is to say, only then is the refined data suitable for processing by the operating algorithms of the processor function of the computer system. At this point, my computer system is an isolated system, which can "communicate" with other similar loners, but only if the appropriate software is written to handle the communications and interconnection protocols to make it a part of a larger internet. This is when my local net of computers and workstations would become a "node" within a larger "internet" of computers.

An understanding of my little computer "internet" enables one to begin to draw parallels between it and the classic global internet of the 21st century (which includes "dedicated" networks like that of the armed forces). The essence of the matter is that the modern global internet is simply a computer system, just like the various "computers" and their progeny – which includes the population of handheld "smart phones" of today. In the most essential terms, what makes all computers, "computers" – are their processors – both the main processor and the secondary processors. Thus, we may see the global internet as being a giant computer system. I imagine this worldwide system to be a network of nodes and linkages, like those that give form to a spatial region. The inputs into this system can come from a myriad of sources, including electrical impulses from keyboards, electronic scanners, electromagnetic energy emitters, sonic signals, and even by a kind of "touch" sensor. Regardless of how the data is inputted into the internet, once it is "captured," it becomes grist for the millions of processors that exist in the internet. The internet then is able to output usable information to the variety of "users" that utilize its processing power and storage capacities. The output product can take many forms, such as the visual image or the alphanumerical set of characters on paper or a scope of some kind. It also can be an electrical signal which can be inputted to another processor for further processing and output; or it can be an electrical pulse can be used to cause a mechanical subsystem or actuator to perform certain kind of "work," to achieve a desired outcome.

As can be seen intuitively, there are many points or nodes, within even the most complex computer system, which can be targeted by some "malevolent," agent who possesses the requisite knowledge and skills. During the 1980s and 1990s, the attack on a computer system would most likely have been done on an "up close and personal" basis. That is to say, by the use of a compact disk, flash drive, or some other such device to introduce programming code or data into the targeted computer system. Now, however, in the 21st century, it is more often done remotely, through the hard cables of the system or via a wireless "cloud" environment. In short, the keyboard and the "floppy disk" have been partially supplanted by more remote techniques means for attacking a computer system.

What should become crystal clear from the complexity of even my relatively simple computer system is that the computer-based global internet, or even a subset of it, must be extraordinarily difficult to keep secure from malevolent agents who wish to access it. The main element of this problem can be characterized by the well-known paradox of the "keepers of information." That is, how to safeguard information, even when the reason for its existence is to shared it with others.

Beyond that basic truth, it is well-understood that, like all systems, the computer-based internet will invariably contain certain vulnerabilities which can be exploited by the bad guys. This reality is the very essence of cyberwarfare within the military area of operations that is known as cyberspace. Another basic fact is that the very essential and vital role of electricity in the physiology of the internet provides a very lucrative target for an attack on it by a cyber-enemy. The information which is cached and stored throughout the internet is another place where the cyber attacker can wreak havoc or simply misinform within the communications systems of an internet. I liken the latter to the Wizards of Bletchley Park, the code-breakers of WWII, who "hacked" into the German Enigma program, and thereby engaged in cyberwarfare long before the development of the global computer internet.

Cyberspace also can be seen as a virtual military battlefield. And, and like all battlefields it can be treated as a conceptual "region." Several things can be said about the nature of this cyberspace region. First, it has "boundaries," but these exist primarily in the minds of the actors which operate in the cyberspace. Second, the military geography of cyber warfare is dominated by what can be called its physical geography; that is to say, electromagnetic phenomena. And third, computer operations within the global internet are like aerospace operations in the Earth's atmosphere. With this in mind, we also can then visualize the cyberspace region as a military system; one in which electricity is the force that drives operations within it. Within this region, the linkages are like a ganglion of nerves in an organism. In the human physiology, a ganglion refers to a dense group of nerve fibers which carry impulses from one cell to another. The same type electronic physiology occurs in cyberspace. This analogy is manifested in the language that describes the workings of the internet. Thus, "virus" is a term that is used to describe an "infection" of software within a computer system, including the internet. Another characteristic of both human physiology and cyberspace is they consist of a myriad of individual cells (nodes) which are continuously recycling; some dying off, while new ones are being created. Similarly, in the computer systems, the electronic nerve fibers form feedback loops, in which the effects of a given process can return to cause changes in other points in the system. In summary, the use of organic physiology to describe phenomena in cyberspace, especially with respect to the internet and single computer systems, now has been embraced by those who study and analyze the internet and cyberspace.

In terms of its architecture, cyberwarfare has been designed by many contributors. Two of the more influential ones have been IBM and Microsoft. The former has established many generally-accepted standards in the area of hardware, while the latter has developed into the most influential force for setting the ground rules and nomenclature for software, particularly that which is known as the operating system. Operating systems are that set of computer programs which direct the operation of the computer itself, thus establishing a "platform" on which the "application" software (e.g., Microsoft Word) and computer games can be developed. The nature of the architecture of both the hardware and software aspects of the internet that has evolved now defines the strategies and tactics for attacking it. This is another example of the history of "militarization" of civilian technology.

On another dimension, cyberwarfare is now recognized by the U.S. Department of Defense as a having created a whole new dimension of conventional warfare, including aerospace warfare. In this type of warfare, there is the absolute need for electricity; it is as important as petroleum-oil-lubricants are to "real" military aerospace operations. Thus, at the present time, this reliance of our military systems on electricity renders it as their greatest vulnerability. This is manifested in terms of maintaining the integrity of all types of command & control functions, and the efficacy of every weapons system which relies on computers. One such system is a synchronous network of Earth-orbiting satellites which the U.S. Air Force operates. As of 2012, the U.S. Air Force Space Command operated more than fifty earth-orbiting satellites, under the direct control of the Air Force Satellite Control Network. Some of major functions of the AFSC (Air Force Space Command) include: "...the operation of the space control and warning system, test launchings of ICBMs, cyberspace operations, ongoing space surveillance and warning, command and control warfare operations, and expeditionary communications/air traffic control for space operations..." (Air Force Magazine, May, 2012). This provides us some measure for appreciating the vital nature of a secure supply of electricity to military aerospace operations today.

Other agencies of the U.S. federal government, such as the NSA (National Security Agency), the CIA (Central Intelligence Agency), and the Department of Homeland Security also have begun to organize their resources in order to be able to fight in this new conflict environment. And, other countries also are developing their capabilities for what they see as a new battlefield arena. The most outstanding of these include China, Russia, the European Union, and Japan, at this point in time. However, other operatives in this growing domain of cyberwarfare include the North American Electric Reliability Corporation (NERC). In any case, it now is universally understood that a nation's electric power grid, which is linked to the internet, must be constantly monitored and managed, so as to be able to respond to both natural and man-made imminent threats. Such are the kinds of decisions that have required the organization of the USCYBERCOM (U.S. Cyber Command) whose sole mission is to operate on the battlefield of cyberspace, beyond the traditional battlefields of land, sea, air, and space.

I see cyberwarfare as being analogous to traditional electronic warfare. In cyberwarfare, the main weapon is the computer. In electronic warfare, however, the major weapons are the transmitters and receivers. Nevertheless, there are many similarities in the tactics of cyberwarfare and electromagnetic warfare. In both cases, the basic transaction is between a transmitter system and a receiver system. Like with all battlefield dynamics, there is usually an initial aggressive action by one opponent; which is then countered by the opponent which has been attacked. However, unlike the EM attack, which is usually instantaneous, the cyber attack can remain hidden for extended periods of time, and only activated in response to some future event. Thus, the computer system is both the weapon of choice and the most lucrative target to be attacked. In both cases, the "battle" can be reduced to the classic measure-countermeasure dynamic.

In the final analysis, it appears that cyberspace will be the dominant battlefield of the future. Its two main elements, electrical power and computer hardware, are to modern warfare what petroleum and the heavy platforms (e.g., heavy bombers, aircraft carriers, and tanks), have been to 20th century warfare. Therefore, it is likely that cyberwarfare will be the essence of military aerospace operations of the future.

BUILDING A MILITARY SPACE INFRASTRUCTURE

The architecture of the military space bases will likely take the form of a network of artificial nodes that are interconnected and interacting. One node that already is in existence is the International Space Station (ISS), which is a creation of many nations, including the United States and Russia. This proto-type for "permanent" space bases has accumulated a great amount of practical knowledge and skill-sets that can be applied to the development of military bases outside the Earth. More generally, the ISS is but one example of how the overall experience that has been amassed by NASA in space exploration and scientific study can be adapted to military space operations. Another source of practical experience for constructing bases in an environment that is hostile for humans is the long-range, long-duration nuclear submarine that operates in the oceans of our planet. Like the ISS, the nuclear submarine presents many similarities to a hypothetical future military base in outer space and, therefore, offers many proven solutions to the general problem of conducting military operations in the interplanetary medium.

Consider that the NASA historical experience is replete with accounts of trial and error efforts; occasional "magnificent failures" and bursts of inspirational successes. In the net, however, all of these experiences have produced the likes of the International Space Station and Skylab Project. At the same time, the Hubble Space Telescope, the Cassini and Voyager spacecraft, and numerous other long-duration space vehicles have developed a wealth of scientific knowledge and practical experience with respect to the operation of hardware systems in the space environment. On another level, the NASA space missions have provided geologists and other planetary scientists with the samples and data with which to construct models that will prove useful to those who will plan and design the military bases in the interplanetary medium, and on the various celestial bodies of the solar system.

From this cumulative experience have come proposals for the construction of a military space base that resembles the International Space Station in its architecture. One important area of practical experience in the development of such a large and complex space station by NASA has to do with the cosmic and solar radiation environment in space. One response to this problem has been to develop a shield system to counteract the of radiation nuclei that is constantly traversing a spacecraft or a space base. The usual types of shielding consist of heavy metals such as lead, or a lighter metal such as aluminum. However, even though the flux of cosmic ray particles is readily attenuated by such shields, the particles split the nuclei in the shield, which produces equally hazardous secondary radiation. Even with effective external shielding, the longer the time spent in space, the more human cells are affected. Generally, the effectiveness of the shielding is subject to the law of diminishing returns over time. And, an even a more acute situation in a military space base would be a spontaneous health event which would necessitate the evacuation of the crew back to Earth for specialized medical attention.

As far as the space base itself is concerned, in addition to the potential hazard to the material structure that is posed by cosmic radiation and electromagnetic radiation, there also is a potential risk from space debris. Regardless of the source, the degree of damage to the base can be somewhat ameliorated by the provision of pressure covering or other shielding systems to protect vital exterior components of a base. Within the space base, another hazard which has to be addressed is the potential damage to the electronic components of all systems by both types of radiation. The countermeasures that have been developed to date include the hardening of the electronic components and systems to ameliorate the damage or malfunctions that are caused by ionizing radiation and high-energy electromagnetic radiation. The use of materials, such as polyethylene, whose high density resists the penetration by such radiation, is often mentioned in the NASA literature. Water also has been suggested as a high-density barrier, perhaps by circulating it around the inner shell of the space base. One suggestion that I find particularly intriguing is to utilize the magma tubes on Mars as locales for a military space base, somewhat akin to the Cheyenne Mountain complex in the United States.

Consider also the need to protect electrical and electronic systems from cosmic radiation and fields of charged particles, in order to prevent damage to delicate space flight systems, subsystems, and components. In the case of the static electric fields, the systems can be protected by a surrounding shield that is made of a good conductor, such as copper. However, the shielding of systems from a steady magnetic field, like those found in space, is more problematic. It is possible that certain "rare earths" can be used as shielding materials. In addition to these passive measures, there other methods for countering the hazards that are posed by radiation in space, including those that are potentially disastrous. One technique is to implement backup and redundant systems to minimize the likelihood of the most catastrophic scenarios.

In any event, an effective general approach toward minimizing the hazards of cosmic and electromagnetic radiation in space seems to be the development of protective and resistant materials. In this regard, nanotechnologies at the molecular level are being applied to improve the tensile strength and density of such materials. These technologies are being developed to construct the shields and other protective structures of the space bases and space vehicles, by the manipulation of the essential atoms of materials. Nanotechnology refers to the controlled manipulation of the physical, chemical, and biological properties of materials on an atomic scale. Now, in the 21st century, it appears that these technologies and techniques are essentially formulated and ready to be put into practice. In short, with these technologies, it will be possible to utilize the atoms that exist in any environment to create the materials that will enhance the efficacy and survivability of military bases in space.

However, even with all of these engineering steps, it must be said that the establishment and maintenance of military bases in space will continue to be a risky proposition; and it will therefore require continual risk management to maximize the probability of success of the mission. Another kind of threat hazard that would be faced by military space bases results from the continual impacts from space debris and micrometeorites; these are considered to pose the most potential for causing serious harm a spacecraft or a space base. These kinds of events could even force an evacuation from an orbiting or surface-based military space base. A more likely threat; is that of collisions with other spacecraft or with the externally-protruding equipment of a base, such as a crane or a robotic arm, could also cause harm to the base and its crew. Other potential hazards include onboard fires and toxic spills. More lower-probability events that could pose risks would be such events as a deliberate attack on the bases' systems from Earth, or an accidental critical command from a mission control facility. In all these cases, one of the most important facets of minimizing the potential danger from these elements of the geographic environment in space will involve training and strict safety protocols.

———

There now is a very large and comprehensive database that is being developed with respect to the conduct of military operations in hostile environments. One aspect of such operations has to do with might be seen as a "cultural geographic" factor. I am referring to the boredom factor that would be inherent to long-term military space operations. Of course, we have the detailed "black box" readouts of long-duration space missions that have been carried out by NASA. These can be tapped for dealing with the problem of boredom and "wasting" that typically occurs during long periods in space. On Earth, one notable series of experiments were carried out on-board a deep-sea submarines. In one particular case, one of the mission objectives was to study the behavior of aquanauts in a sealed, self-contained, self-sufficient capsule, over a period of a month. During the course of the dive, NASA conducted an exhaustive series of analyses on virtually every aspect of onboard life. These included measurements of sleep quality and patterns, and even the sense of humor and mood shifts of the crew members. There also were measurements of physical reflexes and the effect of a long-term routine on the crew readiness. The experience that was gained from that submarine dive proved to be very helpful in the design of Apollo and Skylab missions, and has provided continued guidance to NASA scientists as they are designing future manned space flight missions.

Perhaps even more has been learned from space mission disasters, which have included launch site disasters, explosions in route to a designated orbit location, and problems that have occurred during reentry into the Earth's atmosphere. There was even a near disaster during splashdown into the ocean in one of the earlier missions. The most spectacular and horrifying such event was the Columbia disaster of 2004. The bitter lessons of that deadly event, and others, have been incorporated in NASA space body of knowledge and procedural expertise, and have been used in the design of the space missions that have followed. Other "learning events," included ammonia gas leaks from a solar radiator due to a damaged panel, damaged solar panels, and structural failures also have been assimilated. Then there were those problems that stemmed from natural forces, such as near-collision with space debris. Many mishaps resulted from human error too. And, finally, there have been errors in system rebooting program and other software "glitches." All of these were investigated and documented as "lessons learned" for future missions.

As an example of this, consider the set of conclusions that was developed after analysis of all the explorations of Venus by NASA and the Russians. One of these was the realization that the only way to maintain a long-term military base on that planet was to find a way to maintain operational bases at an altitude of at least 30 miles above the surface of Venus. This is because only at these levels would the Venetian atmosphere be cool enough to enable the construction and maintenance of large, permanent manned bases. Some space scientists have suggested that one approach that might prove viable would be to deploy large "balloons," on the scale of Earth dirigibles. This is thought to be feasible if the carbon dioxide gas regime of Venus itself could be used to inflate the balloons and cause them to maintain altitude at levels above 30 miles above the surface of the planet. One problem that has to be overcome is the corrosive nature of the atmosphere and the possibility of electrical phenomena similar to lightning on Earth. This means that the materials to construct the balloons would have to be somewhat like Titanium to counter the corrosive effects of the carbon dioxide regime. The latter phenomenon created the requirement for special "coating" of electronic systems to protect them from being "fried" by the electric discharges.

In terms of military space operations, the above discussion of military space bases can be placed under the rubric of potential warfighting multipliers. The reason why this is so can be seen in the warfighting experience of the U.S. Air Force. Thus, it is intuitively obvious that neither the strategic aerospace bombers of the Cold War, nor the tactical fighter-bombers of the post-Cold War period could have been efficacious in achieving their objectives without the support and multiplying effect of the "location" of their bases. Or, consider if it would have be possible to mount a bombing campaigns in World War II in the Pacific Theater without the strategically-located airfield systems that provided, not just the real-estate, but also the fuel, meteorological, command and control, navigational and other support to the warfighter platforms. A classic example of this was the World War II bombing of the Japanese home islands by the B-25s under the command of Jimmy Doolittle. In that case, their military base happened to be a naval carrier which was customized to support that particular mission. Their dropping of bombs on Tokyo and Yokohama would have been practically impossible without the support of the U.S. Navy carrier system, as well as the navigational beacons along the route, and the recovery sites in China, to name just a few nodes.

———

Leaving aside the issues of survivability and self-sustainability, we next consider the classic geographic question of where to locate the military bases. That is, which location will be optimum, and what is the set of multiple variables that affect such a decision. As has been the case on Earth during any major military conflict, but particularly with respect to the Cold War, the location decision for any military base has invariably been tied to the mission objective that it will support. The primary derivative in the Cold War analysis was that a military aerospace base had to be optimally located with respect to the "net effective time" it would take to react to a mission targeting imperative. This was the reason why it was decided to locate the SAC B-52 and Minuteman II bases as close to the Arctic Circle as possible. In other words, the least net effective time from launch to a target in the Soviet Union would be one whose trajectory would occur over the north pole of our planetary sphere.

A secondary derivative had to do with the "net effective accessibility" to the resources or backup force that would be needed to accomplish the main mission. In space, one possible factor that might particularly affect the net effective accessibility to resources might devolve from a hypothetical competition for natural resources in space, such as fuel or water. These are but a few of the reasons why control of space domain will be so important. The present-day problem of providing fuel supplies to our aircraft and ships on Earth is an apt example, and it provides us with a useful model for analyzing the problem of military base location in the solar system. Consider that effective accessibility to petroleum is more than just having adequate fuel supplies for one mission; rather, it involves having control of oil resource regions, in one way or another, as well. Thus, one way to approach the problem of providing fuel for military space operations is to focus on the geography of hydrogen distribution in the Geospace Domain, which incorporates Earth, the Moon, Mars, and the asteroids which also reside there.

A model which may be useful for projecting what a military space base system might look like is the U.S. military base structure that has been constructed to secure access and maintain control over oil resources in the Middle East. Some of these bases are located in Europe. Other air bases which are assigned the same mission are located in places like Uzbekistan, Kyrgyzstan, Afghanistan, and Pakistan, as well as Turkey, Iraq, and Kuwait. Their overall mission, essentially, is to secure the access to the oil resources of countries like Saudi Arabia, Bahrain, Iraq, Oman, Qatar, and the United Arab Emirate. More generally, the mission objective for all these military bases is the protection of oil supply for the United States and Europe. The threat environment that is the cause for all this military capability comes from such sources as the hostile and unstable nations, but also terrorist organizations, in the region. A less direct threat is posed by the increasing need for oil by rising nations such as China and India, which introduces competition for these scarce resources. Then there are military bases that are located on the seas. These include naval fleets that the U.S. Navy has deployed. Especially the 11 to 15 carriers that are deployed throughout the globe, including the waters near East Asia, the littoral of the Mediterranean region, and near the Middle-East. At least two of these have been constantly on station in the Persian Gulf.

A review of the exploration of the solar system that has taken place to date indicates that there are fuel resources, metals and other minerals, and water to be found throughout the solar system. More to the point in the short term is that these resources can be found on some of the terrestrial planets and on the Moon. Additionally, many bodies in the outer solar system may also be rich in ices and volatiles, and also may be a source for cold-temperature extraction of hydrogen from hydrocarbons. Consider also that there may be methane in Titan's atmosphere, which could be occurring via processes related to cryovolcanism. Then there are indications that metals and volatiles may be found within asteroids and comets. Water ice and other volatiles would not only be vital in sustaining military bases in the solar system, but the hydrogen and oxygen which is extracted from water, produces hydrogen-oxide reactions, which also could be used as a propellant.

———

At the present time, the most likely candidate for a surface-based military base in the solar system, outside the Earth, is the Moon. There are several reasons for this: (1) it already has been landed upon (and launched from) by humans; (2) it has been studied extensively by geologists and other planetary scientists; and (3) it is strategically located relative to both the Earth and Mars. Equally important, it has a relatively low gravitational attractive force, and it contains many strategic materials for maintaining human systems, possibly including water. But what would be the imperative for constructing a military aerospace base on the Moon in the first place? The answer to this might lie in the fact that the first explorations and probes of the planets that lie within the lunar neighborhood, such as Mars and other celestial bodies, have produced indications of significant iron and other mineral resources that could be used by earthlings to provide resources for a hypothetical military-industrial complex in the Geospace Domain. In the journal Science (2011), it is suggested that the Earth's trove of such heavy elements, such as gold, has resulted from repeated large impacts from extraterrestrial bodies during the early development of Earth. It is logical to assume that this hypothesis also applies to other rocky bodies in the inner solar system, such as the Moon and some asteroids. All of this conjures up images of the 18th century European nations, in the midst of the first Industrial Revolution, competing for resources such as bauxite and bat-guano in the "underdeveloped" portions of the globe, in Africa, Asia, and South America, as well as the oil-rich region in the Middle-East. So, it would appear that a similar paradigm is developing in space today.

As far as the architecture of a permanent military base on the Moon is concerned, there is considerable precedent on Earth for such an undertaking. One such precursor project that is occurring on our planet right now is the oil and gas drilling that is being conducted by Chevron Oil at a point along the Arctic Circle, in Canada. The Science Channel (2012) depicts the layout and operation of a base camp in this forbidding environment located near the North Pole. There, humans are relying on cutting-edge 21st-century engineering, logistics, and communications to perform their mission at the top of the world. In this scenario inflatable buildings, which resemble modular space stations, are being equipped with life-support systems to enable the personnel to live and work for extended periods of time in that hostile environment. All systems, including the drilling operation itself, are computer-controlled, and communications are heavily dependent on orbiting communication satellites. Replacement personnel, some equipment, and supplies are delivered by the huge Russian Mammoth helicopter, snow cats, and snow mobiles. I believe that this is a close approximation of what a future Moon-base would look like. The striking thing to me is how similar the environmental conditions on that spot on the Moon's surface are to the ones on some places on Earth. Having served through three winters in North Dakota during my U.S. Air Force career, I can personally attest to the hazards of the sub-zero temperatures. These include freezing winds that sap the ability of both man and machine to function without appropriate technologies. I remember how the wind-chilled cold quickly could freeze the human mind and body, and how it could freeze man's machines as well. On the other hand, there was plenty of breathable oxygen and water on Earth. The same would not be true on the Moon without considerable human engineering.

There are other plausible candidates for military bases in the solar system. The most feasible of these are the Near-Earth asteroids, whose orbits are easily accessible with today's science and technology. Their accessibility and relative location within any hypothetical military area of operations in this region of the solar system also make them prime candidates as "stepping stone" military bases to the more distant Jovian bodies and the Main Asteroid Belt – very much like the series of Pacific Ocean islands that were used to eventually project American air power near the Japanese homeland during WWII. At the present time, the most notable candidates, aside to the Near-Earth asteroids are the more massive Ceres, Vesta, Pallas, and Hygeia. Each of these is a rocky-icy body which contains a mixture of water ice and various minerals, such as carbonates and clays. They also may harbor an ocean of liquid water under their surfaces, according to preliminary spectroscopic analyses that have been done. This composition profile and their relatively low gravity pull also make them good candidates for being space bases in support of military operations.

We already have discussed the advantages of utilizing natural bodies, like the Moon or an asteroid, as a site for the first generation of surface-based military space bases. But it is also likely that artificial platforms may be used as bases within the interplanetary medium, in a manner similar to the naval carriers and other vessels on the oceans of the Earth. The feasibility of developing such military bases in space is closely tied to the problem of how to create a permanent habitat for humans in space. The solutions to this problem would enable humans to live and work in space for months at a time, and with rotating shifts of replacements, almost indefinitely. I can imagine that they would be very much like the permanent scientific stations that have been constructed in the deeper environs of the oceans on Earth, or like the nuclear submarines which provide a long-term human habitat under the surface of the oceans, or even like the naval aircraft carriers which spend months at a time at sea. Of course, much of what has been learned about maintaining permanent space bases has come from the practical experience of the International Space Station.

When completed in 2010, the ISS included living quarters and the laboratories where the scientific research was to be conducted. Since that time, many members of the consortium of nations have contributed to the enterprise. All of these participants have furthered the technology that will be useful in the development of future military space bases. Other contributions also have come from non-governmental organizations that now are engaging in space enterprises, which include the construction of space stations. An example of this is the Bigelow Aerospace BA 330 is an inflatable space station module which is longer and wider than any part of the International Space Station. Another private firm, SpaceX, has already announced plans to build a heavy-lift rocket called Falcon Heavy. Using this launch rocket (which is said to be the largest rocket to be used in space operations so far), the company has plans to launch Bigalow Aerospace's BA 330 space station as well as the additional supplies and rocket engines that would be needed to take the module beyond Earth orbit and to the Moon. The timeframe for such an enterprise is for sometime within the second decade of the 21st century.

———

One thing that can be said about these hypothetical military space bases of the future is that routine operations there probably will be comparable to the surface military air bases on Earth, but on a smaller scale. This assessment is made on the basis of my three years experience as an assistant to the Base Commander of a major Strategic Aerospace Command base, and as the Base Director for Administration for the combat support group on that installation. From this particular experience and my overall experiences in the U.S. Air Force, I have gained a comprehensive understanding of the workings of a typical military air base; especially the entire range of support systems for combat operations. These included: the immediate support functions that provide ordnance and fuel to a military aircraft; the air control and communications facilities that exert command and control over air operations; the fire-fighter personnel and the medical corpsmen that stand ready to respond in case of an emergency; and the maintenance, logistics, and munitions functions that are in place to ensure that the combat aircraft are able to fly and fight as needed.

Later, while serving as an instructor at the Armed Forces Air Intelligence Center in Colorado, I had many opportunities to gain information about the similar support activities that occur onboard a U.S. Navy aircraft carrier. In those discussions with my naval counterparts at that school, I learned that a naval carrier was essentially a U.S. Air Force base, except that it could move about the oceans at will. Otherwise, the surface air base and the naval carrier had many features in common, at least in terms of support functions. There is one other significant difference between the land and ocean bases, however. That is, in the scale and complexity of the flight line. Thus, on the carrier the takeoff and landing strip is about the length of five football fields; while land surface bases can have takeoff and landing strips that range in length from several hundred feet (in developing nations with mountainous terrain, for example) to those that provide ample thousand-foot runways, and multiple takeoff and landing strips as well.

Given the above, one can imagine a scenario in which space stations (mother ships) will operate in much the same manner as aircraft carriers today. As part of an overall military space system, these mother ships would carry a complement of space warfighters to ensure that the space shipping lanes of the future remain open and free of any disruption. In such a scenario, a space carrier would be able to both launch and recover these space warfighter platforms in order to project power in space, and to react quickly to any threat situations, as needed. To give more substance to this scenario of the future, consider the following actual mission: an unmanned U.S.A.F. spacecraft lands at a military airbase in California, thus capping a 15-month clandestine mission; it was launched from Cape Canaveral, Florida in March 2011 to carry out its assigned mission of conducting in-orbit experiments. This was actually the second such autonomous landing at Vandenberg AFB, California; it was preceded in 2010, by an identical unmanned spacecraft, which returned to Earth after 7 months in Earth orbit. In both cases, the platform was part of a military program of testing of robotically-controlled, reusable spacecraft technologies. Both missions were designed to test the geospace craft itself, and to deploy a classified payload into orbit. This experimental platform, I believe, is the precursor to the warfighter of the future in geospace, because it has shown the capability to operate seamlessly between the atmosphere of Earth and the near outer space beyond it. Another possible precursor to the military spacecraft of the future can be seen in the NASA and Russian space shuttles, which are now making supply and maintenance visits to the space stations in order to keep them functioning in orbit (nasa.gov).

It is interesting to note that the Air Force Space Command already has developed a parallel military space program, which formalizes the U.S. commitment to military space operations; there is even the preliminary definition of missions and responsibilities in the geospace domain. And, there is the sense of a steady movement toward the development of the actual capabilities for conducting such operations in space. Thus, one can visualize a scenario in which there will be a network of planetary surface military space bases and a fleet of interplanetary carriers. Both would be able maintain a complement space warcraft and support spacecraft. At the same time, I foresee a network of military space nodes operating on the surface of the Earth and exerting command and control over all of the space forces. Some of these would be analogous to the Headquarters of the Strategic Aerospace Command (SAC) at Offutt Air Force Base in Nebraska and the North American Aerospace Defense (NORAD) center in Colorado's Cheyenne Mountain.

Given the nature of the interplanetary medium in the solar system, it is conceivable that the military space fleet of the future would probably operate under the same policies and in the same manner as the naval fleets of today. That is, they would be assigned, as their main mission, the maintenance of security and freedom of movement within the space lanes of the developing economic region in geospace. Such a space economic region would have as its major nodes the mines and other primary resource sites which would be located on the inner planets and bodies of the solar system. These would include lines of communication between Earth, the Moon, and Mars – as well as the relevant asteroids to be sure. However, there also would be other nodes whose integrity would have to be guarded, such as the artificial space bases and other spacecraft throughout the system. Of course, the technical and economic details of the actual processes for mining in the solar system that lies beyond Earth still have to be worked out, but given the history of the human species, there is little doubt that it will become reality sometime during the 21st century. Whatever forms the specifics ultimately take, these economic nodes of the new mercantile system in space will need security, so there might well have to be a space "navy" to defend the resource and trading nodes of these mercantile systems.

Another aspect of naval warfare on Earth that may be replicated in space warfare is that of amphibious operations. However, in the case of military space operations, the "amphibious" landing would likely occur on a planet, a moon, or an asteroid. So, like the famous amphibious operations on the beaches of Normandy on D-Day in World War II, there conceivably would be such operations on the surfaces of the various planets and moons in the solar system. In fact, NASA continues to test lander systems that could be prototypes for similar systems in hypothetical amphibious operations in space. Consider the Morpheus project which is described in NASA on-line literature (morpheuslander.jsc.nasa.gov). Morpheus is described as a platform that is designed for testing future autonomous landing systems. It is capable of carrying about a half-ton of cargo, including a "humanoid robot, a small rover, or a small laboratory to convert moon dust into oxygen…" According to NASA, this test is to "…demonstrate the feasibility of developing an integrated propulsion and guidance, and navigation and control system that can fly a lunar descent profile." Therefore, one of its main tasks would be "…to test the Autonomous Landing and Hazard Avoidance Technology (ALHAT) safe landing sensors and closed-loop flight control…"

Again, within the context of space warfare, the space force would likely be called upon to conduct military "off planet" operations. One precursor to this type of capability might be the orbiter spacecraft that NASA and other countries have deployed around various planets and even the Sun. One, in particular, is the Mars Reconnaissance Orbiter, which has maintained an orbit around Mars to gather long-term intelligence about the planet. Its mission, in this case, is to conduct scientific studies on or about the planet Mars. The important point to be taken from this discussion of the capabilities of the Mars Reconnaissance Orbiter is the relative ease with which civilian space technology to can be adapted to military purposes. Consider the basic measurement techniques that have been developed to plan the precise orbital profile of the MRO mission. Among these, are the calculations that have enabled NASA mission engineers to determine precisely how far away the initial orbit is, and how fast the spacecraft is moving, and through the application of the principle of the Doppler Effect in that mission. This technique measured how fast the orbiter was moving away from the Earth, and it did this by measuring the magnitude of the Doppler shift occurring in the radio signal that was received from the home planet. Another method of positional analysis is one that determines the distance or range from Earth to the orbiter by measuring how long it takes for radio signals that are sent from the home planet to reach the orbiter.

Mars is another prime candidate for a central base for military operations in the geospace domain of the solar system. It has the added advantage of having an orbit that sometimes will take it to the outer edge of the hypothetical military area of operations in the solar system; toward the Main Asteroid Belt. The distance from Earth to Mars is approximately 0.5 AU (about 47 million miles). We know, based on the Viking probe missions and the Mars Global Surveyor mission that the "effective distances" (fuel and load factors) that would be involved in establishing and maintaining a permanent base on Mars are feasible, even with today's technologies. In fact, as early as 1980, a group of scientists at Martin-Marietta had come up with a working plan for establishing such a human colony on the Red Planet (nasa.gov). Moreover, the Viking Orbiter has returned imagery and chemical evidence of ample quantities of water in the region of both poles. There even have been proposals for a way of utilizing the water which is found on Mars by breaking it down into its hydrogen and oxygen components; the hydrogen could then be combined with atmospheric carbon dioxide on Mars to begin the process of "green-house warming," and thereby conceivably begin to make the planet's atmosphere actually livable for humans in the long term.

It seems that most of the NASA space missions that have been conducted following the completion of the Apollo program have been oriented toward the ultimate establishment of permanent human settlements Mars. Meanwhile, the researchers aboard the International Space Station have been working diligently to identify new processes for the potential large-scale manufacture of a wide range of products; everything from computer chips to paints and pharmaceuticals – possibly in the hopes of a future development of such industries in the low-gravity environment of an Earth orbit. Such orbiting-communities also could serve as way-stations for travelers on the way to Mars. In the same vein, ISS has served as a testing laboratory for new materials and technologies (such as nanotechnology) that could make it easier and less costly to send humans on interplanetary voyages to Mars. Another area of scientific testing is being done to find ways to maintain the health of humans on long space voyages to Mars, and the eventual residence on permanent space bases on that planet. And, on another track, the space agency is preparing for the challenges of renewing the human presence on the Moon, as well.

Planners hope that by the year 2018, when the relative positions of Earth and Mars are especially favorable, an expedition to begin the construction of bases on other celestial bodies will be ready. This begs the question of how to conduct regular missions between the two planets, for regular operations, once a base on Mars becomes a reality in the future. It is at this point that geographers would begin to focus on the universal problem of location and relative location. The first thing that would be realized is that the nodes of the relevant functional region in this case are constantly moving around the central node, which is the Sun. If we proceed on the assumption that contact with a military space base on Mars would be primarily originating from Earth, then it seems that the contact will be intermittent, depending on the alignment of the secondary nodes of the perceived functional region. What this means, I believe, is that the final configuration of the functional region which will support a military space base on Mars will be have to be multi-nodal. That is, there also will be a need for multiple launch-points within the Geospace Domain, including the Moon and some asteroids, rather than a bi-nodal (Earth-Mars) region. It may also include multiple orbiting space stations which will function as transshipment points.

In fact, the practical construction of just such a multi-nodal system in space is occurring at the present time, primarily through the efforts of the NASA space agency, but also through those of the European and Russian space agencies. Furthermore, Mars is only one target in the continuing series of robotic missions of exploration of the solar system. These initially were complex, instrument-loaded probes that were sent out primarily to perform space reconnaissance and mapping of the bodies of the solar system. However, as time has passed, these have evolved into smaller and more focused probes which are designed to deliver specific scientific results. Some of these spacecraft have already been sent to study the nearby asteroids and comets, while others have focused on the "mini solar system" that we refer to as the Jovian system. Indeed, the moons of Jupiter, along with those of Saturn, Uranus, and Neptune have been the subject of at least the same intensity of exploration and study as the parent planets themselves. Most of what we know about the Jovian system has come from data that has been returned by the Pioneers 10 and 11 (1973-74); Voyagers 1 and 2 (1979); and the Galileo orbiter and probe, which arrived at Jupiter in 1995, although we still rely on the long-term presence of the surface-based telescopes as well.

———

In any case, it is Mars that the U.S. is considering as the main base of operations in the geospace domain. This assessment comes after decades of telescopic and spacecraft "reconnaissance" and study of Mars. During this period of time, we have developed a significant body of knowledge and an expanding skill-set that will be relevant to any colonizing enterprise on the Red Planet. At the same time, there have been significant advances in the science and technologies that will be required to establish a human colony or a military base on the planet. For instance, NASA has generated a huge archive of space mission records relating to the design and planning, not only of mission to Mars itself, but also of the early experimental missions, such as those of the Mercury and Gemini projects that prepared us for the lunar landing. Further, every other space mission that has been conducted by NASA (and other space agencies) has added to our knowledge and honed the skills that will be needed to establish a human settlement or military base on Mars.

Prior to the age of space exploration, most of what humans knew about Mars came from centuries of ground-based telescopic observations and logical inference from the data that was gathered. However, since the middle of the 20th century, the knowledge-base about this planet has been steadily augmented by the spacecraft which orbit and probe the globe. So, for the past half-century, we have focused our efforts on gaining a detailed understanding of the orbital mechanics of the planet, its composition, and its structure. And, as we do so, it is as though we begin to realize that this is not a "cold, dead" planet, but rather a less vibrant body that still presents evidence of planetary activity. Indeed, the closer we look, the more we are learning about the anatomy and physiology of an organism, rather than a cold, dry object. So, now we are detecting evidence of planetary viability in certain environmental niches, similar to what we find in the Atacama Desert on Earth. Another indication of planetary viability can be seen in tundra-like deposits of frozen ground water on Mars make it an attractive prospect for human settlement.

Consider that a trip to Mars requires approximately nine months in space, given our present level of technology. However, continuing advances in technology might make it possible to shorten the travel time sometime during this century. Among these are advances in propulsion technologies, there is the VASIMIR (Variable Specific Impulse Magnetoplasma Rocket) and nuclear rockets. Such a system could bring the travel time down to 40 days. Then there are the constant-acceleration technologies, such as the solar sail and the ion drive, both of which have been proven to be feasible in practice. So, the first step in such a mission-planning algorithm would be to determine the best initial trajectory and the various orbital maneuvers that would be need to reach the final destination. If we simply were to aim our space ship toward Mars, and keep it headed that way till we got there, we would very likely miss the target. This is because a trip to Mars from Earth will take many months (Mars is never less than 35 million miles from Earth), but while we are moving toward Mars, that planet also is moving at a fast rate. The "Red Planet" moves along its orbit at a speed of more than 50,000 miles per hour.

Therefore, when we actually send space ships to orbit and probe the planet, our attention must be focused on the specific mechanics of the orbit to determine the most efficacious techniques and routes in our travels to Mars. Thus optimum initial trajectory would be one that will ultimately intercept a future point on the orbit of Mars. Therefore, instead of heading directly for the point where one actually sees Mars, one must plan a trajectory that initially will place the spacecraft on an orbit of its own around the Earth. Further, this orbit must be calculated so that it will touch the orbit of Mars at a certain point. From that point of the orbit, the spacecraft will "deorbit" and plot a transfer trajectory that will intersect the desired target point of Mar's orbit. To achieve this rendezvous event, the initial launch must be planned so that the space ship will arrive at that point of rendezvous, precisely at the moment when Mars will be there too.

More generally, the above discussion of the navigation problem in traveling from a point on Earth to a point in orbit off Mars is important, because it is representative of the dominant navigation problem that would face the conduct of military space operations. It seems that the movement of an object (including a spaceship) about the solar system is almost always in the form of some kind of an orbit; only on very short trips is it in the form of a curved line, much less a straight line. And, that is why the constant perfection and cataloging of rendezvous and insertion maneuvers is so important. Fortunately, it is this type of practical experience and knowledge that has been accumulated by the space exploration operations to date that will be easily convertible to military purposes in space. An example of such practical experience, that is available to astro-navigators, is the Mars Global Surveyor robotic probe mission that operated during the years between 1996 and 2009. This program has provided data about the surface, atmosphere, and some aspects of the interior of Mars.

Another of the important preparatory missions for the overall objective to colonize Mars actually began when NASA launched the Mars Science Laboratory (2011). Its rover vehicle (named "Curiosity") has been busy doing the advanced geological studies on the Red Planet even to the present day. It is digging up rocks and soil sample, doing some of the preliminary sifting out of extraneous material, processing some of the rest, and then relaying the results to both space and Earth-based laboratories for more sophisticated and long-term analysis. Currently, it is focusing its work on the various layers of debris that has accumulated around a crater which is thought to be the oldest section of that planet. One thing is true: the propulsion, materials, electronics and many other technologies that have gone into the design and application of this mission have all been known on Earth – even those that may have been reformulated or otherwise tweaked just for this mission.

———

The Jovian moons

Consider how each of the Jovian moons would be evaluated as possible sites for future military space bases. Unlike Mars, which is relatively close to Earth, and which can be reached within a matter of days, the planets and moons of the Jupiter-centric solar subsystem present an entirely different level of complexity and challenge in the development of military bases. Actually, it is the moons of these so-called "Gas Giants" that offer the best opportunities for the establishment of a military base in that region. (Again, we must take into consideration our present level of technologies). For one thing, they have a lower gravitational pull and, therefore, the launching of spacecraft from these bodies requires less propulsion energy than the major planets. Another factor that works in favor of utilizing the moons, such as Io and Europa, is the fact that they both appear to be endowed with large quantities of water, and that their temperature and pressure regimes are more like that of the Earth. The chemical environments on the Jovian moons also appear to be favorable to the establishment of a military base on one or more of them.

Recent discoveries have led to the identification of a total of 63 moons that are part of the gravitational hegemony of Jupiter. These, as well as the asteroids and other small bodies which the giant planet captures as they pass by, make up what is called the Jovian System. Four of these moons also are known collectively as the Galilean satellites, because they were first discovered by the Italian observer of the skies in the 17th century. Jupiter's other moons are much smaller than our own Moon, and may simply be asteroids that have been captured by the giant planet's strong gravitational field. It has been said that if Jupiter had been about 80 times more massive, it would have become a second star in orbit around the Sun. Because of its massive size, Jupiter has a huge impact on all of the objects around it. So, it might be said that this natural system of celestial bodies (nodes) and the force that determines their relative locations and orbits (gravity) has all the earmarks of a cultural region that is ready for "development" by humans.

A total of eight spacecraft – beginning with Pioneer 10 (1973) and Pioneer 11 (1974) – have provided close-up views of Jupiter and discovered its magnetic field and radiation belts. They also gathered valuable information about the planet's moons. Later missions, including Voyager 1 and Voyager 2 (1979) gathered more new information, as well as providing more detailed data about the earlier missions to the gas giant. Once again, moons were discovered, and many images were relayed back to Earth, including evidence of volcanic eruptions on Io. Another moon, Europa, displayed signs of a possible ocean of water beneath the ice. The Ulysses probe (1979) also sent back new images and new moons. The New Horizons probe (2007) passed by on its way to study Pluto and the Kuiper belt. But, perhaps the greatest contribution to our understanding of the Jovian system as an entity has been provided by the Galileo mission (1989) and the Juno mission (2011), which orbited Jupiter, and examined its atmosphere and magnetic field in more detail. The Galileo mission persisted on station until September 21, 2003 after having made 34 orbits of the planet.

One of Jupiter's moons (Io) actually is stretched by gravitational tidal waves as its orbit brings it near to the parent planet. It rotates at the same rate that it revolves around Jupiter, so it is said to be in synchronous orbit around its parent planet. The nearly circular orbit of the satellite has an inclination of only 0.04 degrees to Jupiter's equatorial plane and its orbital trajectory is also somewhat eccentric due to a gravitational resonance between it and another Jovian moon – Europa. As a result of all these perturbations, there is intense internal heating of Io which is also caused by the continual flexing of the satellite by Jupiter's powerful gravitational field. This flexing motion is also produces the powerful volcanoes that have been found on the moon; the volcanic flows that are so extensive and frequent that they appear to be resurfacing the entire satellite; and a surface that presents a dramatic landscape of erupting volcanic vents, pools and solidified flows of lava, as well as deposits of sulfur. There also are geysers of fine particles that are hurled several hundred miles into space. However, there is no evidence of impact craters.

Europa is the smallest and second nearest of the four large moons which have been discovered around Jupiter. Europa's diameter puts it in the same class as Earth's moon, at the 2,000 mile range. This moon presents many of the characteristics of a small planet. That is, it is thought to consist mainly of rock and smaller proportions of frozen or liquid water. And, like many of the planets, it has an inner metallic core which is about 800 miles in diameter. This core is surrounded by a rocky mantle, which is overlaid by an icy crust, some 90 miles in thickness. The moon has a magnetic field which is produced by internal forces, as well as being externally induced by Jupiter's powerful field. And, modeling also suggests that there may be a liquid ocean hidden beneath or within the icy crust. The surface itself is crisscrossed by an intricate array of curvilinear grooves and ridges that present an image unlike any other seen in the solar system. There also appear to be fractures that are thought to be caused by Jupiter's gravitational pull. Imagery has shown that some huge blocks of ice which apparently have shifted and rotated from their original positions; it is thought that this might be due to alternate thawing and freezing. Europa also presents evidence of water and dry ice in many of its craters. Like the planet Earth, there are few impact craters on Europa, which indicates that the actual surface on that satellite of Jupiter is also relatively recent. This is another indication of resurfacing that might be occurring, possibly through a continuing outflow of water from the interior of the body, and thus forming an instant ocean of ice, somewhat reminiscent of the ice-cover over Antarctica on Earth.

Saturn's moon, Titan, also presents many characteristics of a small (some would say, immature) planet. We have learned much about this moon thanks to the imagery and chemical data that has been acquired by the Cassini-Huygens orbiter. This mission was designed to extensively explore Saturn and its rings, as well as its magnetosphere and its moons. Cassini also delivered the European Space Agency's Huygens probe to its historic landing on Titan in 2005. With its multi-spectral and versatile sensors, the probe has been able to acquire data about Titan from both ends of the scale of observation; that is, both high-resolution detailed observations and larger-scale, panoramic observations, which provide great contextual background perception to scientists.

The twin Voyager probes also have acquired and returned a plethora of imagery related to the moons of Uranus. This new data also added to what had previously been discovered by Earth-based telescopes and previous flyby missions. From these sources, we had discovered five moons (Ariel, Miranda, Oberon, Titania, and Umbriel). This legacy data and the new imagery which has been sent back by the Voyager probes have provided grist for many years of scientific study of these moons. Among the things we have learned from these is that there are a total of 27 moons that have been confirmed, and many of them are accompanied by a series of narrow rings. Water ice has been detected on the surface of the major moons, whose appearance can be described as "dirty ice balls," as is confirmed by the fact that the reflectivities of the moons are lower than that of pure ice. The measurements of the ice-to-rock ratios of the five major moons, indicates that Miranda has a higher ratio of ice-to-rock than the average 60-40 proportion. In any case, the important point is that water ice also appears on the surface spectra of all the moons.

Imagery that was captured by Voyager 2 also shows that Oberon and Umbriel contain dense populations of craters, very much like those that are found on some portions of surface of the Earth's Moon. Many of these also exhibit some of the oldest terrains that have been seen in the solar system, thus providing geological insights into the early development of these objects and the solar system. According to one of the main theories of how the solar system was formed, the large craters point to a history that refers back to the earliest years of the solar system's existence; when the planets emerged after a period of condensation and coalescence of diffuse matter. Thus, out at the distances from the Sun where Uranus and Jupiter orbit, planetesimals with a different composition from that of their peers which formed nearer the Sun evidently coalesced at temperatures where water and other volatiles can freeze.

————

The Asteroids

Another space-base deployment strategy might be to utilize the asteroids as military space bases. One of the main reasons why these wandering asteroids are seen as being favorable as a site for a military space base is the relative ease with which spacecraft could land and launch from there, due to the lower gravitational pull of these smaller bodies. Also, there is the fact that some of these smaller bodies are known to contain water ice and minerals which could be utilized to refuel and replenish military spacecraft that would be operating within the geospace domain, or be in transit to another far-off celestial destination. Then there is the relative location factor which makes many of the asteroids valuable as strategic interconnection nodes in any military space domain. An example of this is the fact that the Asteroid Belt lies between the inner and outer regions of the solar system, thus offering optimal transshipment points in any inter-domain line of communication.

Generally, the asteroids are also known as "minor planets" or "planetoids." These include any of a class of small, rocky or metallic bodies which orbit the Sun, primarily between the orbits of Mars and Jupiter. Most of these bodies can be found within an orbital region that is known as the Main Asteroid Belt, which lies in a zone that lies more than twice as far from the Sun as Earth does. Within this zone, there are estimated billions of these bodies, whose sizes range from relatively small boulder-sized aggregates of rock, dirt and ice, to those would-be planets whose diameter can be as much as a few thousand feet. They generally are small in comparison with the planets and their moons, but two asteroids – Ceres and Vesta – have come close to achieving the status of planet. Ceres is the largest of these asteroids, having a diameter of about 600 miles; the second largest is Vesta, which has a diameter of about 300 miles. Both of these asteroids share many of the geological qualities that define Earth, Mars, and the other terrestrial planets. Indeed, the Main Belt once contained enough aggregate material to form a planet nearly 4 times as large as Earth, but Jupiter's gravitational dominance not only stopped the creation of such a planet, but also swept most of the material clear, thus leaving an insufficient amount for a planet of any size to form.

In general terms, the asteroids are thought to be fragments of planets that have been destroyed by collisions with other developing planets, or are primordial remnants of the formation of the solar system following the Big Bang some 14 billion years ago. The thinking is that these fragments continue to be influenced by the gravitational pull of Jupiter, which possibly has kept them from coalescing into a planet-sized body. Occasionally, however, some of these asteroids break out of their Jovian orbit and begin to wander through the solar system, sometimes into the gravitational influence of other celestial bodies, including Earth. This fact is relevant to military space operations because occasionally one of these wandering asteroids may follow an orbit that could threaten Earth. Moreover, as the "cultural geography" throughout the geospace domain grows and develops, there will many more "targets of opportunity" for the asteroids and their meteorites to strike. On the other hand, the growing awareness of this threat has spurred more intensive studies on how to counter the general threat that is presented by external objects to Earth and, by extension, the new assets that are being developed outside our atmosphere.

FUTURE MILITARY SPACE WEAPONS AND TACTICS

A BLEND OF THE OLD AND THE NEW

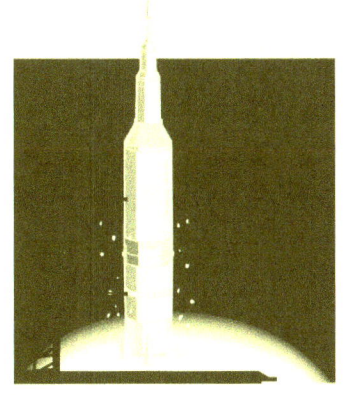 

There already is a formal U.S. Air Force statement of doctrine and strategic objectives which has been posted on the internet as of July 2012. It describes, or at least implies, what the weapons and tactics in space operations will be like in geospace. This document states that the achievement of space superiority in any military space campaign is the essential objective of all operations. This basic imperative then directs the U.S. space force to "maintain the freedom to conduct operations without interference from an adversary." Thus, the main concern is to deny any adversary the ability to, in effect, spy on our space-related operations. Our assets which need to be protected from spying include "critical command nodes, elements which we are deploying for combat, or the debarkation of supplies and forces." The statement of doctrine also mentions the need to achieve control of any battlefield in space (or along the surface littoral on Earth).

A short-hand description of the military strategic missions in geospace might include the following: (1) gathering intelligence, maintaining surveillance, and performing reconnaissance; (2) maintaining situational awareness at both a strategic and tactical level; and (3) projecting the precise level of power whenever and wherever it is needed. In order to accomplish these missions, a nested system of subsystems will likely characterize the architecture of future military space operations; initially this will consist of what we are calling the geospace domain. At the highest level of nesting, there will have to be a system to ensure situational awareness with respect to the entire area of operations. There also must be a capability for reacting in a timely fashion to any given situation. Thus, there must be in place a robust readiness and sustainment system to enable the fielding of adequate and appropriate space-oriented weapon systems for the assigned mission. The weapon systems themselves will probably consist of smaller integrated systems; composed of a variety of platforms and their "payloads" (i.e., weapon systems and their sensors-control-actuator subsystems). In the end, there must also be an omnipresent system of external platforms (such as satellites) to provide navigation, surveillance, acquisition, and tracking/battle assessment support to the warfighters.

In order to maintain a geospace domain space-strike capability by the middle of the 21st century, for instance, the optimum capability-set appears to be one which includes a capability for prompt response to any threat or actual event requiring a military action. One such response-system that has been discussed is an Earth-based system of laser devices. Such a system would be able to bounce high-powered, directed energy off a constellation of space-based mirrors. At least in theory, such a precise, megawatt-class and light-speed weapon system could theoretically act within seconds or minutes to resolve a rapidly developing threat from an adversary in the geospace domain. Like all directed-energy weaponry, the beam of light would be able to deliver concentrated and heightened energy to either disable an attacking platform, or to disrupt the energy-force of an adversary. Another example of such a flexible response also might be provided by a fleet of tactical robotic platforms that would be equipped with a variety of payloads, including kinetic-energy weapons and compact laser weapons. On the other hand, given current trends, it may be that satellites will no longer need to be large, heavy structures overloaded with redundant systems. Instead, small and microsatellites will be able to perform all the functions carried out by today's large "one of a kind" satellites. This would be an example of using the strategy of redundancy and large numbers to counter any threat or attack on the satellite system as a whole. Each of these multiple, redundant satellites would possess ever greater onboard processing power and suites of advanced visible, infrared, and radar sensors to enhance their individual and overall survivability; and thus their ability to complete their assigned mission. Timeliness, responsiveness, flexibility, survivability, reliability, precision, and selective lethality will be the hallmark of the directed energy, projectile, space sortie, and cyberspace weapon systems.

As far as the weapon systems themselves are concerned, we may not recognize them when we first see them, but it is quite likely that the space weapons of tomorrow will carry the DNA of many of the weapon systems that are used today in the atmosphere. This is so for two main reasons: (1) the projected strategic and tactical objectives will resemble those of the armed forces of today and (2) the tactical environment in space differs only within a spectrum of sufficiently well-known and resolved environmental challenges of the interplanetary medium. The planets and the other bodies, however, are still less known and understood, but the pace of knowledge and technologies indicates that these too will be sufficiently reconciled during the 21st century. In short, the challenges that are presented in space to the conduct of military operations appear to be within the same spectrum as in the Earth's atmosphere. Therefore, given the expected mission objectives and the military technologies of the future in geospace, we can begin to contemplate what the force structure might look like in that domain.

————

Thus, for example, given the vast expanse of just the geospace domain of the solar system, I think we can foresee that there will be some sort of "space carrier," comparable to the naval aircraft carrier of the 21st century. Sometimes referred to as a "mother ship," we can imagine that the space carrier will have an onboard complement of "space warfighters" that will carry out the functions of the F-18 platform of today, including that of space interceptor and attack spacecraft. And, assuming that the platform would be manned by humans, like its naval counterpart of today, such a space carrier would essentially be self-contained in terms of human life support and creature comforts, thus resolving the main impediment to long-duration operations in the geospace domain. The possible solution set to this requirement can be derived from the years of operational experience that has been accumulated by the International Space Station; it can be seen as a foundation for the development of a hypothetical military space platform that could perform the same strategic and tactical functions as the naval carrier of the oceans of Earth; only this one would operate throughout the interplanetary medium.

Another requirement for such a hypothetical space carrier has to do with propulsion. That is, it must be capable of long-term, steady-state propulsion. But it also must have some kind of "after burner" capabilities as well. These stem from the supposed mission of the platform, which basically is to patrol vast regions of the space domain, on the one hand, and to be able to respond with great quickness to an emergency event when necessary. The propulsion strategy that is being followed in the case of the ISS relies on ion propulsion for quick spurts of maneuver, while depending on solar panels and battery packs for steady state operations. Another possible technology that may be utilized to propel the space carrier is drawn from the experience of the Cassini rocket, which was powered by Plutonium-238. This innovation not only enhanced the load-thrust ratio factor; but the heat from the decay of the radioactive material also is being turned into electricity for all of the spacecraft (orbiter and probe) operational needs during the mission. Other power requirements such as would be needed to operate the various "avionic" subsystems of the space carrier will also be provided by some combination of nuclear reaction and enhanced batteries.

———

The intercontinental ballistic missile (ICBM), in some configuration or other, will still be a mainstay weapon system of a future space force, and it is likely that it will develop enhanced capabilities as an interplanetary ballistic missile. In this capacity, it is conceivable that Interplanetary Ballistic Missiles (IPBMs) will assume the role of its predecessor on Earth. Assuming a permissive geopolitical environment in the future, such a force of long-range missiles might be placed under the control of a multi-national force whose main mission would be to enforce "peace through power" throughout the solar system, in a manner reminiscent of the Strategic Aerospace Command during the Cold War. Such a force of IPBMs could be deployed on a smaller natural body, such as an asteroid, or on an artificial platform that would be analogous to the nuclear-armed submarines of the Earth's oceans.

However, just as the IPBM might be used as a peacekeeper system, it also will continue to present the greatest threat to the Earth-orbiting satellites and other near-Earth threats; it all depends on the motivation and intent of a potential user. In either case, the following trends can be expected to continue, for several reasons: (1) these rockets will continue be enhanced in terms of their capacity to launch large payloads into geospace; (2) the continuing advances in nanotechnology and micro-electronics will enable designers to launch ever smaller, but at least equally powerful platforms to either strike a target in geospace, or to deploy another platform, such as a weaponized satellite, for conducting intra-orbit military operations; and (3) advances in propulsion "throttling" technologies and micro-miniaturization will be translated into more accurate guidance and more precise control capabilities.

Another possible platform for use in military space operations is the space warplane. This is visualized as having the capability to operate seamlessly in both the Earth's atmosphere and in the interplanetary medium. It also would possess the stealth capabilities and precise armory that are incorporated into today's most advanced warfighter platforms. Ultimately, the development of such a space warplane will depend on the continued research and development that is being done on a rocket-powered spaceplane. However, another observation that is relevant to this discussion is that there could very well be a continued need for more than one, all-purpose space warplane platform. To understand this better, consider that the current mix of U.S. Air Force warplanes includes the B-2 bomber, the F-117 fighter-bomber, and the F-18 fighter. Each of these platforms possesses certain specialized characteristics and capabilities that contribute to the overall mission of maintaining control of the skies and projecting power where it is needed. Given these observations, we can expect that the demonstrated capabilities of the B-2 within Earth's atmosphere will probably continue to be enhanced, especially in its stealth technologies and precise munitions like the JDAM missiles. Then, there is the X-47B platform that apparently is being groomed to function like the NASA Space Shuttle; that is, being able to traverse seamlessly between the Earth's atmosphere and near space. More generally, there will be a fleet of flexible, mission-tailored, "trans-atmospheric vehicles" (TAVs) that will carry out tactical missions against defined targets in space.

———

Transcending the platforms, I think we also can expect that there will be an array of weaponry in space that can be classified in terms of the energies and forces that are employed to cause damage and destruction upon designated targets. These generally will include those that utilize directed electromagnetic energies (such as laser), kinetic energy, or explosive forces. One example of the use of a directed electromagnetic energy weapon system is one that operates in the visible-light wavelengths of the spectrum; within the sunlight portion to be precise. The most discussed technique for using direct energy in the visible light portion of the spectrum is to deploy focusing mirrors, equipped with pointing and tracking and maneuvering systems. Single, very large mirrors (on the order of kilometers in diameter); or large arrays of smaller mirrors (in the 100 meter class) working in concert, would be needed to make this concept practical. By intercepting and redirecting sunlight, orbiting platforms could focus the light from many mirrors onto a single spot, causing battlefield temperatures to be raised (a potential form of weather modification) and optical sensors could be temporarily blinded. Such systems could be deployed on the surface of the Earth; but to take advantage of the sunlight that is not filtered by the atmosphere, where the enormous flux of natural (incoherent) could be harnessed. But even in low-Earth orbit, these mirrors would need pointing and tracking accuracies of 10 to 100 nano-radians to qualify as precision aimed weapons. Even larger, but still lightweight structures could potentially be made from advanced aerogel materials, advanced ceramics, engineered composites, structurally supported optically coated plastics, suspended or spun-reflective liquids (a liquid mirror), or inflatable mirrors.

———

Among the lessons that have been drawn from the history of human warfare is that "plowshares" can be as easily converted to "swords," and vice versa. So, since the early development of the human species, there have been many instances of the utilization of agricultural implements, like the scythe and the axe, as weaponry. In the case of military aviation, many of the developments in military aircraft, as well as navigation, and communications have been borrowed from the civilian antecedents. The history of space exploration also has developed this pattern in which the science and engineering, and "know-how" first have been devoted to civilian purposes – but which are now ripe for military applications as well. Thus, the harnessing of laser technology in medicine, which has been manifested in precision surgery now is being tested for an array of military applications, including precision bombing. During the balance of the 21st century, there is every indication that laser beams (or directed energy) will also be used in space, against artificial satellites, for instance. In the same vein, the non-military applications of the electromagnetic spectrum in communications, remote sensing, and spectral analysis have been and will continue to be appropriated for military purposes in space. In other words, the electromagnetic battle of measure-countermeasure is likely to continue in any hypothetical military area of operations within the solar system.

I believe that warfare in space will be the first in which energy is paramount over matter. As we saw earlier in this book, the electromagnetic spectrum will be the virtual "battlefield" upon which military space operations will take place. In this kind of warfare, directed energy beams will be prime weapon systems, and the tactics for their use will be similar to those that characterize "Electronic Warfare (EW)" that has occurred in military operations since World War II; that is: it will be dueling measures and countermeasures along the entire spectrum of electromagnetic energy. There will jamming of emissions, at all points in the transmission-reception cycles, all sorts of "stimulation and amplification" of energy beams, and a lot of jockeying of wavelength and frequency between the adversaries. There also will be considerable encryption and decryption of energy transmissions and receptions, and engineers will develop materials and architectures that will produce the equivalents of the present-day stealth technologies.

There also will be "hacking" and other tactics that are presently associated with cyber warfare in the 21st century. This will likely increase in consonance with the increasing utilization of computer-based command and control technologies that will be integrated into all space systems. One fascinating example of these tactics that are being used today in military aerospace operations involves the reported attempts by that "non-state organizations" (terrorists) to "hack" the computer-control-actuator subsystem of the so-called UAVs (Unmanned Aerial Vehicles), such as the Predator in Afghanistan. Such tactics may very likely be used in space wars to disrupt the data-links between space-based weapon systems and the surface-based "pilot function." Or, they might be used to "capture" the processors, or even to "corrupt" the software involved in the command and control of systems in space. In short, every tactic that is being used to disrupt and shutdown the internet systems operations on Earth today, can and very likely be used to achieve the same objectives in a space war.

On another dimension, one can say that every natural hazard; or any glitch, design problem, and any other "bug" that has been encountered during the past half-century of space exploration operations will be studied and exploited by an adversary in future space wars. This may be a corollary to "Murphy's Law" ("If anything can go wrong, it probably will"), which might be stated as: "...any malfunction or glitch in space systems can be exploited to attack a military space system." One illustration of how this might operate, consider that two main weak points of a spacecraft are the solar array dishes and the ablative heat shield on the nose of the platform. Imagine that the technology could be developed to focus and heighten the flow of charged particles to attack the solar array, to ablate the protective material on the shield. If I give my imagination free reign, I can even visualize a technology that gives one the capacity to direct cosmic energy onto a target in space.

The U.S. Air Force has assigned the responsibility for conducting military operations in space by the National Authority; and to carry out this charge, the U.S. Air Force Space Command has been established. The area of operations in space is called a "space domain" in the current vernacular of U.S. Air Force doctrine. This military space domain is currently experiencing many ongoing military operations, and these are increasing in intensity, as may be measured by the number of military sorties that are occurring there. The physical infrastructure of this new space command also is becoming more complex as the number of assets and capabilities are being built into it by the U.S. Air Force and the overall military-industrial complex. This new military operations region is still only in the initial stages of development at the turn of the 21st century. However, many of the essential nodes are being already being defined and constructed, as have the linkages that make it a truly functional system.

Based on the overall national policy guidance that has been given to this space command, there has been ongoing development and testing of weapons and tactics; those that will be needed to accomplish the strategic and tactical mission objectives that develop in the space domain. At this point in time, it can be said that this process is in the "capability-building stage, in which the potential weapon systems and the tactics for their use are being developed; many of them by NASA indirectly. However, I argue here that it will not be a great leap forward to make them operational. In this particular space military region (or "space domain"), the weapons of choice, at this time, seem to be the surface-to-air missile and the space-to-space missile. In theory, the orbiting satellite also could be used as a military weapons platform, but at this point in time, there are international agreements which prohibit it. Also, it must be noted that there does not exist a clear comparative-advantage on the part the individual nation-states in doing so. Nevertheless, at least one nation-state, China, has been actually testing its ability to destroy a satellite in orbit, with some early success.

———

Earlier we dealt with the "why" and the "what" aspects of the matter when we discussed the role of the national authority function in determining which national objectives that would be facilitated through the use of military power in space. Now we turn to the "how" element, which involves the selection of the right tool for the task at hand, that is to say, the weapon systems themselves. As has been the case throughout the history of military air operations, we can expect that weapon systems in space will be developed with the idea of accomplishing a given task, in support of an assigned mission objective. As also has happened in the history of military aviation, the "best" tool sometimes will have to be designed from scratch; while at other times it can be "taken off the shelf" with some modifications, or as is. Thus, we will explore the nature of the future weapon systems in space with the realization that many, if not most, of these will represent modifications of existing systems; not of which are "civilian" tools in their current usage.

One way to approach the question of what the space weapon systems of the future will be like is to review the inventory of such platforms that are "in the pipeline," so to speak. Some of these weapon systems already are operational, while others are still are in the developmental spectrum that ranges from experimental to final testing. Among the former, are the weapon systems that are already deployed and operational within the military space domain that is active today. In rough terms, this is the area of space that extends from the surface of the Earth to altitudes of more than 250 miles. This region has been called the geospace region because it straddles the boundary between the Earth's atmosphere and surface, and the near-space area above it. Within this new military aerospace region (military space domain), there are two main military systems that are operational at the present time. One of these is the long-range ballistic missile system which, in fact, is the main weapon system in military space operations today. The other is the Intelligence, Surveillance, and Reconnaissance (ISR) capabilities that are focused on the orbiting artificial satellites.

Looking forward into the there seems to be no shortage of potential prototype platforms, both on Earth and in space, for conducting future military operations within the geospace domain. I would suggest that a useful starting point for dealing with the subject is to first abstract the essential "platform" from the myriad of aircraft, spacecraft, naval vessels, and other specific systems in place today. Consider the extensive conversion of platforms to military purposes during the Vietnam War. One of these is the iconic AC-130 gunship which evolved from the original C-130 cargo transport aircraft. Another is the Apache-style attack helicopters that were originally designed as cargo and troop ferries within an area of operations. And so the story continues with the T-28 trainer which was reborn as a lethal fighter-bomber that provided close air support to the Hmong army in the Plainne des Jarres region of northern Laos. Most recently, we have the example of the deadly Predator drone which is an erstwhile reconnaissance platform.

It also is quite possible that an integrated "satellite space force," will be deployed to establish control of the particular space domain; and to ensure the security of the space lanes. One scenario that is being discussed by military planners with respect to hypothetical space operations is based on the Predator drone program that is operating in Afghanistan and other terrorist "hot spots" around the globe. Assuming that there always will be some budgetary constraints, both the U.S. military is moving toward the development of a warfighting force in space that is based on what might be called "small, smart, and lethal weaponry." Thus, in the case of space warfare, and especially along the space lanes that interconnect with the artificial satellite zones, "clusters" of small, smart "bomblets" might be deployed on various orbit trajectories. These would then be "on alert" (like the B-52s of the Cold War) and ready to respond quickly to direction from a command center in response to a threat. The key to the effectiveness of such a strategy has already been established through the practical engineering and technology that has evolved in space by NASA and the U.S. Air Force. One example of this is the decision that was made by NASA to send smaller rocket-payload units into orbit and to assemble the larger unit in orbit. The collateral to this strategy is that NASA continues to learn how best to guide and command spacecraft remotely – from the surface of the Earth and in space itself.

On another dimension, the U.S. Air Force also is trying to place more onboard capabilities onto platforms like the unmanned and automated Predators, Reapers, and the new Global Hawks. To be able to do this, engineers are striving to make sensors lighter and less unwieldy. Also, with greater onboard processing power, the strain on communication networks and bandwidth will be partially alleviated. Ultimately, the desire is to develop more powerful data links to connect the pilot to the controller-actuator subsystems. In the long term, it is hoped that automatic sensor-actuator control subsystems will continue to improve, and thus compensate for not having human eyes in the cockpit. This trend also can be seen in the wide-array of sophisticated capabilities which have been built into the space probes by NASA. I am reminded of these when I ponder the growing capabilities of the remotely-piloted aircraft that are being used today in places like Afghanistan. As an example, the intelligence-surveillance platforms, like the Predator and Reaper platforms, are evolving into highly-effective ground-attack platforms. This is being done through the addition of a package of air-to-surface missiles, along with onboard target detection and tracking systems, and the guidance and control systems on the ground.

Another possible prototype of the space warfighters that would be deployed on a space warcraft carrier is the NASA Space Shuttle; it already has proven that it can launch from the surface of the Earth, access the orbit of a large space platform that is orbiting in space, rendezvous with it, and then deploy from it back to Earth. More recently, even similar civilian spacecraft are able to dock onto the larger platform to perform logistical or maintenance support. It is not much of a stretch to imagine future military warfighter platforms, like the X-47B, being able to dock onto a space carrier that would be similar to the ISS. The X-47B is designed to be the first aircraft to handle all maneuvers, including aerial refueling and aircraft-carrier landings, without any human assistance. The tailless, fighter-sized jet makes its own decisions using billions of pre-written code, which is derived from thousands of hours of simulated flight time. As a former computer programmer, my mind boggles at the concept of the vast quantity and the complexity of the lines of code that would have to be written by software developers in such a circumstance. One begins with the reality that any computer system is really like a human toddler, in the sense that every aspect of its "situation" must be painstakingly described and defined before any further instructions can be given.

A completely automated vehicle is fundamentally different from a remotely-piloted vehicle that has been used in Afghanistan at the turn of the 21st century; in the automated vehicle a "robot brain" has virtually replaced the human one in the sortie loop. "Automation" in this case means that an there is an onboard computer that is capable of operating even in the most complex setting in which there are multiple variables and limited response times. Further, it can process vast quantities of flight data, make almost-instantaneous decisions, and provide guidance and control to the platform, all on its own. This autonomy also implies having the ability to actually assess the situational environment in real-time, respond dynamically as necessary, and perform the immediate task – all within the imperative of accomplishing the underlying mission objective. In order to achieve this degree of autonomy, the platform is equipped with an array of sophisticated sensors, processors, and the appropriate electronic commands to the various subsystems for actualization.

One such platform is the National Aerospace Plane (NSP), which first was considered in 1986 by NASA. Initially, it was to be the prototype for a reusable launch vehicle for space operations. That is, it was designed to be a "single-stage-to-orbit" (SSTO) vehicle that could land and take off on conventional Earth airbase runways. The concept specifications called for the development of a plane which could travel at hypersonic speeds and, therefore would require the development of new materials and technologies to make it a reality. Another early prototype (the X-15) was to be the first of the so-called Military Transatmospheric Vehicles (TAVs). The X-15 platform was designed to reach speeds of Mach 6.7 and achieve "geo-space" altitudes of as much as 62 miles. Most significantly, the X-15 test-flights yielded a trove of valuable information about frictional heating from the atmosphere.

As far as current space weapon systems are concerned, the ballistic missile, which can be launched from either surface or atmospheric-based platforms, still is the mainstay of the military space force.

Today, ICBMs can be used to: attack space objects in orbit; attack targets on Earth via a suborbital trajectory; and interdict missiles that are traveling through space. The offensive weapon systems are primarily surface-to-surface and surface-to-space missiles. These now are being deployed by many nations, including not only the U.S. and Russia, but also China and India; and the list continues to grow. Further, the threat is not just hypothetical; both the U.S. and the Soviet Union have actually exploded a nuclear device in near-Earth space, and they and China have tested their capabilities by striking their own artificial satellites in orbit.

So, now in the space age, the ICBM has taken on an additional mission, which is to respond directly to any attack from the upper limits of the atmosphere and, from imminent attack by both natural and man-made objects from space. The former are the asteroids, comets, and other natural large objects, including meteoroids. The threat from man-made "near-Earth" objects mainly comes from the artificial satellites and the other spacecraft in geospace. To counter both of these threats, the ICBM has been provided a modified warhead to fit the circumstances of the space environment and the nature of the target itself. However, the original threat profile from surface-to-surface missiles remains even in the space age and must be dealt with. Thus, there is still the threat of a missile being launched from one point on the Earth's surface, then being placed in orbit, and finally deorbiting and striking a target point on the surface. This added dimension to the overall threat profile has required adjustments in order to determine the precise interception point along the target's trajectory. It may occur at the point of launch, while the target is in orbit, or during its descent toward the target. In any case, the point to be made here is that the original ICBM system is still viable in the space age, although certain modifications to the guidance and control system and the warhead may be required. I also would point out that the ICBM represents a case in which an original "military system" has been converted to "civilian" use by NASA when it adopted the Titan as its primary rocket launch vehicle.

It is likely that the ISR function in space also will continue to develop the overall system for supporting military operations, both on the surface of the Earth and in space. Indeed, an existing system has been performing both of these missions since the latter end of the 20th century. The heart of this space-based ISR capability is a constellation of orbiting artificial satellites; it currently provides early warning, surveillance, communications and navigational support to military forces throughout the geospace domain. Although these ISR capabilities are mainly oriented toward the belt of artificial satellites that orbit the Earth, there also has been continued development of these capabilities for supporting military commanders on the ground. Thus, during the course of the wars in Iraq and Afghanistan at the turn of the 21st century, the U.S. Air Force has developed a satellite-based ISR system that provides commanders with real-time battlefield data to support their operations. The emphasis, therefore, is to deal with all data as a raw material, which has to be analyzed and processed in order to convert it to usable information. That also means that the information has to be given to the ultimate "user" on the battlefield as quickly as possible, so that it can be acted upon it in real-time. Given this imperative, it appears that the future of ISR, while still relying on airborne and spaceborne platforms, and their sensor payloads, will also focus more intently on the processes and systems for delivering the product (information) to the customer or user, wherever that might occur in the network. Thus, one could say that the future of ISR lies in the ultimate utility of the information that is produced. Remarkably, that is the same conclusion that has been reached by NASA in the building of its "ISR" architecture for space exploration and scientific study. In essence, the ISR environment of the future – both on Earth and in space – will resemble the so-called "information cloud" of the worldwide internet.

The fact of the matter is that data about a commander's battlefield environment is being collected from a variety of sources, including signals, imagery, and others. Since the period of the Gulf Wars in Iraq and in Afghanistan, it has been realized that commanders, analysts and, above all, the warfighters in the field, should have real-time access to all relevant intelligence information. This means that raw intelligence data, from all sources, will have to be more quickly validated and processed into the usable intelligence "information," so that a warfighter (both human and robotic) can respond to perceived battle situations in "real-time." One model for the delivery of intelligence to the end user, "just in time," to wherever the user is located, is the United Parcel Service (UPS) method of tagging every piece of material to be transported with a barcode signature. Thereafter, the unit of (information) can be tracked throughout the delivery system (or area of operations). In the case of military ISR, it must be known where every packet of intelligence from surveillance and reconnaissance is created, where it is stored, and to whom to deliver it is to be delivered. Another aspect of the UPS model is that the information (intelligence) is made available in a variety of ways (or platforms), including hand-held devices in the field.

———

Looking into a hypothetical future in space, let's assume that there will be a need for a continuing presence of military power in a given space domain – in order to maintain "peace through strength." This implies some sort of dominant military platform that can project military power anywhere within that region. On the water surfaces of the Earth, such a platform has taken the form of an aircraft, a submarine, or the silo-based ICBM. However, if one expands the criteria horizon, it can be seen that it also could take the form of the naval aircraft carrier on the surface waters of the planet. One model for such a military space carrier platform might be the latest generation of naval aircraft carriers. An example of this type of platform is the U.S.S. Enterprise, which began its operational career in 1961. Powered by 8 nuclear reactors, it can steam three years without refueling, and 13 years before its uranium cores must be replaced. With such a propulsion system, it can reach speeds of 30 knots. This huge platform is also capable of maintaining a force of 100 jet aircraft, along with the required fuel and ordnance. Some idea of the features that might be included in a space warcraft carrier also can be inferred from an analysis of the U.S.S. Zumwalt, the U.S. Navy's newest warship. It too is massive, but it has a radar signature that is 50 times smaller than the typical naval destroyer. Packed with state-of-the-art radar, stealth, weapons and propulsion systems, it is designed to be the most technologically-advanced warship in all naval history. For example, this platform employs the latest stealth technology and its all-electric integrated power system produces 78 megawatts of electricity. More to the point, the ship's most immediate role on Earth is that of an incubator for potential technologies that can be utilized for 21st century conflict. It is intriguing to extrapolate from this state-of-art naval warship of today to a potential space warship with many of these technological attributes in the future.

———

Some potential space weapons are still in the early stages of research and development. One of these is a kinetic weapon system that has been proposed is referred to as "Project Thor." It basically postulates the so-called "tungsten telephone pole" as a weapon in space operations. Essentially an inert 20-foot long torpedo, it is the brainchild of Jerry E. Pournelle who, along with Stefan T. Possony, and Colonel Francis X. Kane (USAF, Ret.) proposed a long range, kinetic "kill missile system" in 1978. It is designed to be placed in orbit, where it eventually might deorbit, and then utilize kinetic energy to strike its target. Pournelle is credited with originating the concept while working in operations research at Boeing Corporation in the 1950s. As designed then, the material that was chosen for this kinetic energy weapon was to be made of tungsten, which is a chemical element known for its high density (19 times that of water) and exceptional hardness. These characteristics make it an attractive material for military applications which required penetrating projectiles. Aside from its intrinsic properties, such a weapon system has the advantage of not being prohibited by existing space treaties.

According to the experts, it is thought that kinetic-energy weapons would inflict damage in space because they move at orbital velocities of at least 5.4 miles per second. Smaller weapons can deliver measured amounts of energy as small as a 100 lb. conventional bomb. But a 6.1-meter by 0.3-meter tungsten cylinder impacting at Mach 10 has a kinetic energy equivalent to approximately 11.5 tons of TNT (7.2 tons of dynamite). The mass of such a cylinder is itself greater than 8 tons, so it is clear that the practical applications of such a system would be limited to those situations where its other characteristics would provide a decisive strategic advantage. On the other hand, the weapon would be hard to defend against for a variety of reasons, but mainly, because it would have a very high closing velocity and a small radar cross-section. Also, a launch of such a weapon would be difficult to detect because any infrared launch signature would be very small. Also, because it would likely be used in orbit, its position would be difficult to pinpoint. As an orbit-to-planetary surface weapon system, this type of kinetic weapon system would prove attractive to planners because the time between the deorbiting launch and the impact on target would only require a few minutes. And, depending on the orbits and positions within the orbits, the system could be applied on a global basis.

Next, imagine a "space warfighter" that is able to launch and land repeatedly on both natural and artificial surfaces in space – very much like the combat aircraft on today's naval carrier. Then imagine that this war spacecraft is equipped with a material or paint that renders it "invisible" to radar and infrared sensors. One such prototype may be the newly-developed X-47B. When it comes to stealth design, this latest-generation war-fighting platform represents the state of the art. Unlike the earlier-generation F-22, this platform does not require heavy inputs of regular and expensive low observable materials maintenance. Instead, the conductive material that is the essence of the camouflage capability actually becomes even smoother over time, thus reducing the platforms' original radar signature even further. Moreover, only serious structural damage to the skin of the platform will disturb its low-observability qualities; minor scratches and even dents won't affect the aircraft's stealth qualities enough to degrade its camouflage state in combat. Neither do the doors of the platform require special caulk materials to restore stealthiness, the application of which is a time-consuming and expensive chore in other stealth aircraft. This class of aircraft also has a serpentine inlet that makes the engine fan blades invisible from any point outside the fuselage, thus eliminating one of the biggest problems in reducing the radar cross-section of the platform. Moreover, the air intakes constitute a single piece of composite material which is devoid of seams, rivets, or fasteners. This is crucial because it is these elements of a structure that are huge RCS reflectors; and are known to have caused massive signatures on earlier-generation aircraft.

There are other systems that might be used in future military planetary operations. Some of these currently are being used in civilian space operations on some of the other planets and the Earth's Moon. Among these, are the NASA Mars Science Laboratory and its rover, named Curiosity. Both of these systems are examples of space systems that have incorporated all that has been learned by scientists and engineers since the start of the Space Age. The rover itself can be seen as a possible precursor to some future "platform" that might be developed for conducting military operations on the surface of other bodies in the solar system. Indeed, the NASA rover is as rugged as any Humvee or other military transport vehicle on Earth today. It can maneuver over all kinds of rugged terrain and has the battery range to cover the dimensions of about two football fields. With its plutonium-powered thermoelectric generator, the six-wheeler rover of today has a designed life span of about two Earth-years. Furthermore, rovers of the Curiosity class also are designed to function as a mobile laboratory, and their remote-sensing package, in addition to being integral material-retrieving systems. Together, these features enable them to gather and analyze large quantities of rocks and soil samples taken from the surface of a planet. Moreover, they have the onboard capability to conduct sophisticated analyses of the geological samples, including the ability to vaporize rocks through the use of a laser, and then analyze the resulting gases. Looking forward, all of these "intelligence-gathering" capabilities would be valuable to a future commander in a space domain.

Then there are the autonomous "robotic" systems of today that probably might be utilized in space military operations some day. The reasons for utilizing such robotic operatives include, once again, the need for handling increasingly complex experiments throughout the solar system where it is not feasible to utilize manned systems. The reason why robotic systems have become so attractive, too, is that they can be programmed to: devise a hypothesis and carry out experiments to test it; assess the results of the experiment; and cause a certain task to be done; all without human intervention. Even today on Earth, these robotic systems are becoming a standard option in military mission planning in every venue. Looking forward, the increasing power of analysis of the robotic systems could be utilized in a military situation in space in a variety of ways. In general, whether the mission involves ISR work or combat operations, a true robotic platform in deep space would help to ensure that one is "holding the high ground."

So, it is evident that in an environment of increasing complexity and time-stress, manual operations will be challenging, at the very least. This leads us to a discussion of "approximation" in the development of autonomous robotic weapon systems. The planetary rover of the future will need to interact with a multitude of dynamic environments. And they will almost certainly introduce unexpected events that must be analyzed and processed by the most comprehensive and sophisticated suite of software programs. This has been borne out by many such events throughout the history of NASA missions. Thus, we might expect that in future scenarios, such as the design and maintenance of a permanent base on the Moon, for example, robots and humans will need to interface through the use of a common machine-human natural language, very much like the "machine language" that is used to communicate directly with computer entities today. The human operators, in such a circumstance, might likely be in charge of defining high-level orders that are then to be executed by the robotic system on board. Beyond that, the on-board robot might be able to handle decision-making at both levels, and would be the final responsible entity for the generation of its own high-level goals in case of unexpected events. Once again, the onboard computer "HAL" in the movie "2001" comes to mind.

One existing example of a robotic spacecraft that may become the model for future weapon platform systems in space is the NASA space probe. These are unmanned spacecraft that are designed to do extended studies of the atmospheres of all the planets and bodies of the solar system. They carry preprogrammed instructions, but they are still subject to the directions of a surface control function. An example of this is the New Horizons probe (2006), which was the first to travel to the outer reaches of the solar region. There, for example, it did a flyby of Pluto, which now is recognized as an important member of a growing population of small icy worlds of the Kuiper Belt region. Equipped with a variety of sensors, it also detected X-rays and fields of charged particles within the various levels of these atmospheres. And, it proved to be successful in gathering data about the ultraviolet radiation that might be present there, as well as measuring the atmospheric pressure, temperature, composition and density of the atmospheric regions of that domain.

Continuing to peer into the short-term future, I can foresee see a time when militarized platforms will be operating within the zone of artificial satellites in geospace. Further, it is probable that the current international agreements that prohibit such activities then would be treated in the same way that such proscriptions on certain kinds of military operations are dealt with today. That is to say, national self-interest would invariably trump international agreements, if history is any guide on such matters. So, if there should be space battles within the artificial satellite zone, it most likely will involve electronic warfare and/or cyberwarfare. But it also is possible that some satellites will be armed with kinetic energy or directed energy weapons. Moreover, it is conceivable that there will eventually be hundreds, if not thousands, of satellites in the geospace military area of operations. The reason for all these possibilities is that the "cost of entry" would be relatively low. The upshot of this is that some of these military micro-satellites would friendly, but others might not. With this in mind, it also is possible that the threat to "our" satellites may come from other satellites, some of which may be controlled by non-governmental organizations (terrorists and pirates).

———

Much attention also is being given to what are called "directed energy" weapon systems for use in space, especially tactical laser systems. The emphasis in this area of weapons development has focused on the development of solid-state technology. One benefit of this technological approach is a smaller and lighter laser system that is more feasible for tactical purposes. The goal has been to provide a single flight of tactical spacecraft with the ability to strike many more targets during each combat sortie, much more than was possible with conventional missiles. Ultimately, it is thought that laser weapon systems would enable the combat space warrior to strike targets more quickly, and without the extended setup time of conventional bullets, rockets, and missiles. Also, direct-energy attacks would have a greater impact on enemy targets upon striking them than would be the case with conventional munitions. In the end, the magnitude of the impact energy could be manipulated with greater nuance, in accordance with the requirements of the specific tactical situation. As an example, one interesting aspect of the use of tactical directed weapon is that it can be used as a powerful "stun gun," so that targets can only be rendered defenseless, but not destroyed. And collateral damage could be limited as well. On the other hand, explosive ordnance is seen to be more problematic in the context of space operations. Most of these weapon systems employ both explosive and kinetic force energies, whose characteristics can be predicted in terms of their effect in space. However, the main constraining factor against using them, at the present time, has to do with the amount of space debris that would be created, and the concurrent desire to avoid triggering what is called the Kessler Syndrome. This refers to a chaotic phenomenon, in which cascading collisions would create unacceptable damage to all man-made systems in space. Such an event would render of space exploration and satellite operations unfeasible for many years.

On another dimension, as we have seen, one of the main developments in military space operations is a greater reliance on robotics, and the corresponding movement to cede more autonomy to them for guidance and control of the ordnance delivery systems. The reasons for this trend revolve around the heightened demands of time and space in the solar system outside Earth. Thus, as humans move their areas of operation to deep space, beyond the inner solar-system region, we find that there is less capability for direct communication between an Earth-based command and control function and the spacecraft, whether crewed or not. One of the reasons for this degradation in direct communication is that the deep space missions will be occurring within the domain of the asteroid belt and the Jovian solar subsystem, where there is a greater problem of dispersal and other interference of signals from Earth. In addition to the problem of distance, and the concomitant time delays, there will be a greater likelihood of "eclipse events," where the direct line of sight to the spacecraft is blocked by another space body. For these and other reasons, the space vehicles of the future will likely have a greater capability for "domain pattern recognition" (situational awareness) so that the robotic system can be more autonomous in every aspect of a space combat mission.

There also would be another set of limiting factors with respect to the conduct of military operations in space if manned space vehicles and stations are contemplated. These include the effects of living and working in an alien environment, and the psychological stresses resulting from long-term isolation and close living quarters. One attempt to deal with this potential problem has been continued study and experimentation with the development of artificial environments which are designed to remain in space for extended periods. At this time, the International Space Station is the main prototype for such a future long-lived manned base in space. Aside from the integral systems experience that is being gained, this particular manned space station has been providing valuable practical experience in the areas of continuing maintenance and logistical support by space shuttle sorties for several years. Thus, it appears quite likely that future space operations will include the establishment of a similar artificial space base system, at least during the 21st century.

THE TECHNOLOGIES OF MILITARY SPACE OPERATIONS

All of these proposed weapon systems for future space warfare are being made possible by the ongoing advances in various technologies in the 21st century. That is, the outlines of the technology systems that will be needed for future military space operations are now being developed in several countries by the so-called military-industrial complex, academic centers of research and development, among others. If the experience of the past century is any indication, from these clusters of innovation there will emerge the weapon systems that will be used in space. In the United States, some of the more prominent centers for military space weapons include such "brain trusts" as the Bell Laboratories, California Tech, Chicago University, and the Jet Propulsion Laboratories. Their efforts are being guided and coordinated by the various national space agencies and military establishments in a variety of nation-states. From these, there already have emerged most of the advances in science, technology and engineering that have led us to the present state of affairs in space. And, as has been the case since the beginning of military aviation, certain nodes of technological advancement have produced different weapon and other military systems, at different times; it has not been a straight-line progression, but rather a braiding of various lines to form the overall military systems structure. The following is an outline sketch of the major areas of technology that are being pursued in military space operations today.

Propulsion Technologies

As we have seen, the major phases of a sortie in space operations include: the launch event; the orbit insertion process, the transfer to the target site, the operations phase, and the return recovery event. During the launch phase, the most important technology has to do with propulsion and the development of adequate thrust to escape the Earth's gravitational pull. In this regard, NASA that has been doing a great deal of work in launch rocket technologies. Nevertheless, due to the legacy of rocket technology that was inherited from the U.S. military, the architecture of the present-day launch systems still involves huge platforms and extensive associated support facilities. Basically, the same chemical-based propellants that were developed by the Germans during WWII continue to provide the propulsion "muscle" that makes possible the launching of heavy rockets and their "payloads" beyond the gravitational pull of the Earth. However, unlike its military predecessors the typical payload of the NASA rocket is not an explosive device. Instead, the payload it can be a satellite, space station, space probe, space shuttle, laboratory, telescope, or any other type of space hardware. It can even be in the form of a spacecraft with a capsule carrying or astronauts. In any case, one of the limiting factors in terms of the size or weight of a payload is its mass, because greater mass requires greater propulsion capabilities and, ultimately, greater cost.

Today, at the turn of the 21st century, there is a trend toward the development of smaller and lighter, more nimble and nuanced, launch platforms. This is part of the overall effort to maximize the mass of the payload, while minimizing the mass of the propellant that is being used. One strategy for reducing the net cost of putting objects into space is to break down the whole payload unit into smaller parts, each of which can be launched separately, and then be reassembled in the low-gravity environment of space. Another strategy is to launch space rockets from large mother platforms, flying at stratospheric heights, in order to decrease the length of atmosphere and gravity that has to be overcome to achieve escape velocity. Ultimately, regardless of the particular launching strategy, it seems clear that the increasingly complexity of space missions will require a more efficient propulsion technology; one that will involve greater thrust per unit of mass. As a more efficient combination of materials and propulsion technology emerges, it is also likely that we will be able to launch greater units of mass at escape velocity, with the same quantity of propulsion material. Finally, as I see it, rocketry and launch techniques will continue to be refined and made more elegant through the use of micro-miniaturization technologies.

So, it can be seen that one of the major areas of research and development in space operations has to do with the mass and the composition of the payload that is being thrown into orbit. This is an area of science and technology that deals with both propulsion and materials. It also deals with the fact that the less massive the payload, the more efficient a given unit of propulsion will be. In practice, the composition of the payload may vary, but the familiar "loadmaster's" problem will be the same, regardless of the properties of the load. As far as a manned space mission is concerned, the typical payload will include a payload unit and an operations capsule, which is configuration of a transport aircraft on Earth. And, for long-term manned missions, the payload also may include life support and "creature comfort" systems.

One strategy that has been developed by NASA and the other space agencies to optimize the payload/propulsion ratio can be seen in the design and construction of the International Space Station. In that case, the optimum ratio was achieved by breaking down the overall propulsion load into a number of smaller components. In this scenario, each of these self-contained modules was launched into orbit as a separate payload, so as to make the most efficient use of a given quantity of propulsion fuel. These individual modules were reassembled in the low-gravity environment of space by crews of astronauts that were ferried to the "construction site" in space by space shuttle platforms. Today, in the early decades of the 21st century, the integrated ISS space platform, which consists of many forms of capsules and connecting "tubular" linkages, provides a model for this payload/propulsion technique. In this situation, the overall space station includes capsules that are designed to provide a "cockpit" for the onboard operation of the spacecraft, while others are designed to be spaces for conducting scientific experimentation and other mission operations. Then there are the capsules that are used for daily maintenance repair work.

Once the launch phase of a space sortie has been completed, the next event usually the insertion of the spacecraft onto a transit orbit. Such transit orbits are utilized to transport the remaining rocket and payload configuration to the next maneuver event point, or to the target point itself. In this phase the sortie the required technologies will be more complex and nuanced, as the astronaut or flight controller on the ground begin to utilize propulsion throttling for maneuvering, as well as cruising in the interplanetary medium. The calculus of energy and propulsion in this environment reminds me of an earlier situation in which naval submarines had only a limited range in the conduct of their operations; that is, they were effectively tethered to their source of diesel fuel. Consider the case of the German U-Boat operations during WWI and WWII in the Atlantic Ocean. The scope of those operations was effectively limited by the amount of diesel fuel the submarine could carry onboard. In much the same manner, the scope of the early space operations is still being limited by the amount of rocket fuel that can be carried onboard the rocket and the spacecraft. These limits are manageable in the case of NASS operations, but such constraints probably will be problematical in a military environment in geospace, where both the range of operations and maneuverability in three dimensions would be more demanding.

One potential new propulsion technology that holds some promise in fulfilling the expected requirements of future military space operations is described in a report published by NASA, NASA's Human Exploration and Development of Space Enterprise (Published by NASAexplore, May 15, 2003), Franklin Chang-Diaz. He is the Director of Advanced Space Propulsion Laboratory, as well as being an experienced astronaut. In the report he explains the potential benefits that are attached to the Variable Specific Impulse Magnetoplasma Rocket (VASMIR). Essentially, such propulsion technology would allow spacecraft to travel through space with more flexibility than is possible with current chemically-powered rockets. Such technologies are designed to deal with issue related to changes in acceleration, as opposed to the search for sheer velocity. In such situations, the VASIMR-powered spacecraft would only need to carry sufficient fuel to reach a destination the nearest refueling node for achieving a series of short-distance of intermediary legs of a longer mission. Ultimately, the development of such nuanced propulsion technologies can be seen as one of the first steps in the achievement of true "space flight" that is comparable to heavier-than-air flight within Earth's atmosphere.

———

Electrical Power Technologies

Advances also are being made in the search for better, longer-lasting integral power systems for spacecraft and the onboard systems. At the present time, systems that derive their power from the Sun are widely used. The applications of this technology have clustered around lightweight, high-efficiency solar arrays, and it is thought to be a promising solution to the problem of weight vs. thrust, maximization of payloads, and creature comfort for human crews. Currently, one area of innovative work has been centered on the solar cell (a crystalline wafer or photovoltaic cell) which can convert sunlight directly into electricity without the need for any moving parts. These cells are then are placed in array on a panel which is connected to a spacecraft in an outrigger configuration. Generally, one major limiting factor in the use of solar arrays is the distance from their source of energy, the Sun. At the present time, the operational limit of solar power extends along the orbit of Mars.

Following a hiatus of several years, there appears to be a renewed interest in the use of nuclear fission power for space operations. The main reasons for this include the fact that humans are traveling farther out toward the boundary of the solar system, and we are also "loitering" longer in orbit, and on the surfaces of the Moon and the planets. So, there is now a recognized need to develop more permanent electrical power generating systems, on the scale of the electric grids on the surface of Earth. Another area of work in this area is the search for efficacious engineering strategies for delivering power to the future human settlements and the expected industrial activities on other celestial bodies. This also would include providing electrical power to any manned and unmanned "bases" which would be deployed on the Moon, the planet Mars, or even on some asteroids and comets. Again, the main impetus for the search for alternatives to solar propulsion is the fact that solar technology is not always feasible in some situations. For example, in the outer area of the solar system, and within "shadow" regions on planets and other bodies, the number of photons per square meter falls below a certain required level. On the other hand, battery-produced electricity, at the present level of development, still is inadequate to fill the needs of larger-scale space activities.

Development of small-scale fission power systems is another area of focus in the overall pursuit of the objective of developing systems that are small and light enough to overcome the cost-of-launch barrier. In terms of spacecraft propulsion in the post-launch mission phases, attention is now turning to the possible use of nuclear electric systems, where nuclear reactors are a heat source for electric ion drives expelling plasma from a nozzle. According to the article, Nuclear Reactors in Space (world-nuclear.org), "...Power System (HPS) reactors are compact systems that can produce up to 100kWe for about 10 years to power a spacecraft or planetary surface vehicle." Such systems already have been developed since 1994, at the Los Alamos National Laboratory. One of these has been described as being a "...robust and low technical risk system, with an emphasis on high reliability and safety." This very succinctly lays out the parameters of the research and developmental work that is being done in this area. A complementary line of research is being done to increase the output of the reactors from kilo-watt to mega-watt levels. Finally, efforts are continuing on such areas of research and development with the objective of achieving greater life-service and to minimize on-going maintenance requirements.

———

Navigation and Communications Technologies

Another major constellation of technology in military space operations likely will be in the area of navigation. One trajectory of development will focus on the utilization of objects that lie outside the heliosphere as navigational orientation points. More likely in the short term, however, the bulk of the research and development work will focus on constructing a more permanent navigational infrastructure in space; perhaps by placing individual or constellations of artificial satellites in various orbits around the Earth or other planets, or even at selected LaGrange points. Such an "add-on" system would be complement the so-called near-Earth orbiting satellites which are now conducting a variety of "military" navigational and positioning operations. A third method for interplanetary space flight navigation that might prove feasible would be to use certain well-placed planets as navigational beacons whose emissions could be used as a form of triangulation. The principle behind such navigation by planetary triangulation is based on the known speed of light and, therefore, the time it takes for a beam to travel a certain distance in space.

Another trend in technological development is in the area of communications. In the short-term an effective infrastructure is planned that would take advantage of the lessons of the communication satellites that are oriented to both Earth-base and space-based communications. These are now providing communications support, intelligence, surveillance, and early warning functions with respect to Earth-based and space-based threats – both natural and man-made. However, in this era of tight budgets, it is being realized that there is a limiting cost-factor in terms of the total number of these relatively large satellites that can be deployed. One way in which this problem now is being addressed is to develop satellites that can multi-task, or that can operate as a member of an array of smaller, specialized satellites. An alternative to this first-generation of satellites – which is becoming more feasible with the advances in nanotechnology and computer micro-miniaturization, along with nano-power sources – is to deploy a new generation of micro-satellites to supersede them. The key to the efficacy of such a "swarm" strategy is the micro-miniaturization of the input-processing-output circuit. Actually, the trend toward such a strategy in space can be said to have begun in the 1980s with the age of microminiaturization in the architecture of computers. Thus, advances such as the silicon-based computer chip, enabled computer engineers to continue to build ever smaller, but more powerful processors.

Electromagnetic Technologies

The utilization of the spectrum of electromagnetic energy for military applications will continue to be a basic technological imperative in military space operations. Much of the work in the application of this natural phenomenon to space operations has been done by NASA and its partners in the private sector. The fruits of this work also have provided many opportunities for the application of this technology to military space operations. Thus, many of the advances in the space-related sciences and technologies now are being applied to military operations, both within the terrestrial atmosphere and in geospace. On Earth, the war that is fought along the spectrum of electromagnetic radiation can be seen as a type of virtual battlefield; and it is likely to continue in military space warfare. In geospace, this particular form of warfare already has begun developing its own strategic doctrine, and weapons and tactics. It began with the practical application of the radio and radar portions of the electromagnetic spectrum, mainly for navigational and detection-tracking purposes of spacecraft. Later, it advanced to applications in the area of weapon guidance and control, especially with respect to long-range ballistic missiles that can reach the orbiting artificial satellites.

The warfare that will occur in space will require an immediacy of response or counteraction. Also, notification of a threat or other impulse along an "awareness web" in space will require real-time responsiveness, and the ability to respond to a threat or event with appropriate force. This is where "smart" velocity and acceleration comes into play. Like its ancestors, the V-2 rockets of WWII, the space rocket was initially a "dumb vehicle," if you will. That is to say, its trajectory into space was pre-calculated prior to launch and the distance and velocities were essentially a function of the mass of the propellant and the rate of burning of the fuel. However, over the course of time, the unguided rocket was transformed into a remotely-piloted "guided missile." This also was accomplished through a process of "education" which not only provided more guidance information to the brain of the rocket, but also a kind of "self-awareness" that computer programming can provide. Therefore, it can be seen that the rockets that ply the interplanetary medium today have become smarter. In the same way, one of the first steps toward a "smart" rocket system for launching was the adaptation whereby the overall launch rocket was configured as multiple, staged rockets to maximize the thrust-payload problem. And then, some space rockets were provided with a "throttling" capability in order to vary the amount of thrust as needed. In summary, it can be said that the "dumb" rockets of the early age of space exploration have been superseded by a new generation of space vehicles that are conducting the next phase of space activities.

————

Sensor Technologies

Sensor technology also is closely intertwined with the technologies of the electromagnetic spectrum. As I see it, there are several ways in which sensors are being used by NASA. One of these is to use artificial devices to enhance the wavelength limitations of the human eye. There are, for instance, the various artificial sensors that are sensitive not only to the visible-light portion of the electromagnetic spectrum, but also the infrared and ultraviolet portions. Another feature of these artificial sensors is that they enhance the sight-distance and visual acuity and resolution of the human eye in the various wavelength regions. Also, when these artificial enhancers move into the radio and radar portions of the spectrum, they enable the human eye to penetrate cloud and other states of precipitation in the medium involved. Finally, the infrared sensors a unique perspective from which to view light as heat. All of these enhancement properties are currently the subject of research and development.

One trajectory in the use of sensors in space involves the tiny sensors which are being constructed with state-of-the-art materials. One use of these advanced sensors is to detect trace elements within the planetary atmospheres of the solar system. Another is used as a nano-sized detector that is designed to detect atomic oxygen; it is a low-mass, grapheme-based detector that is able to measure the amount of atomic ultraviolet radiation from the Sun which breaks apart molecule oxygen. Graphene was only recently discovered in 2004. It is an example of the carbon-based materials that are being used in space. One of its main attractive attributes is that is so strong, about 100 times stronger than steel. This makes it the strongest material ever measured. On the other hand, it is also the most sensitive material, even at extreme temperatures. For these reasons and others, graphene promises to have many applications in space operations. According to NASA scientists, one particular application will be to resist the highly corrosive effects of the upper atmosphere of Earth and in space orbit. Graphene-based sensors also are quite effective for measuring methane, carbon monoxide, and other gases that are found on other planetary bodies in the solar system (nasa.gov).

However, the graphene-based sensor is but one of the many types of chemical sensors that are being used in space. Rather, scientists use a specific material that is sensitive to each chemical they wish to detect. So when a trace of a particular substance touches the sensing material, it can trigger a chemical reaction which causes an electric current. The intensity of this current can be measured by sensors which then can activate a given electro-mechanical response. One application of this technology occurs on long-term missions in space, where harmful chemical contaminants may build up gradually in the crew's air supply. In such a case, nanosensors that are located throughout the cabin to detect even minute amounts of these contaminants can alert the crew to a dangerous buildup of the chemical. This ability to detect minute amounts of the most critical contaminants and then alert the crew of the hazard makes these tiny chemical sensors extremely vital in space operations. In 2007, NASA tested the first nanotechnology-based electronic device to fly in space. It was designed to show that the "nanosensor" could monitor trace gases inside a spaceship. The results of these first tests have since led to the development of smaller and more capable environmental monitors in astronaut crew habitats.

Military space operations also depend greatly on a variety of sensors types. In terms of flight operations, the functions of the sensors are somewhat analogous to the Instrument Flight Rules (IFR) in atmospheric operations on Earth. But the use of sensors also will be extended to such purposes as detecting, tracking, and precisely attacking of targets in the interplanetary medium and on the surface of celestial bodies in the solar system. Generally speaking, situational awareness in military space operations is greatly dependent on artificial sensors, or more particularly, the receptor cells within them. The sensory stimuli that they detect are typically classified into chemical, visual, mechanical categories. Receptor cells, such as "olfactory" receptor neurons and "taste" receptors, must come into close or direct contact with a chemical stimulus in order to generate a response. However, direct exposure makes these cells susceptible to damage from environmental toxins such as harsh chemicals. As a result, considerable research and development is being invested in the area of mechanisms designed to protect sensory systems from debilitating permanent damage.

Chemical sensors, in particular, are enhanced by nanotechnology in several ways. Adsorption of a chemical at contacts between particles produces local charge states that alter electrical properties, such as resistance and capacitance. Sensors containing sheets of a material such as grapheme, detect such changes in resistance when even low levels of vapor from certain chemicals, such as ammonia, are present. Hydrogen sensors use a layer of closely-spaced palladium nano-particles that are formed by a beading action on a surface. When hydrogen is absorbed, the palladium nano-particles swell, causing shorts between nano-particles which lower the resistance of the palladium layer. Typically, separate types of chemical sensors are used to measure different chemical parameters; but sometimes sensors are combined into multi-functional sensor platforms. In space, sensors using carbon nanotube detection elements capable of detecting a range of chemical vapors have been developed. These tiny sensors are powered by electricity generated by piezoelectric (stressed by mechanical forces or under an applied electromagnetic field) nanowires.

There are other areas of sensor development that are specifically focused on military space systems. The qualities that are being sought generally revolve around wavelength sensitivity and high resolution of output. Mass and size attributes also are a major consideration in space operations. Thus, higher-powered and smaller lasers will improve detection and warning, as well as ordnance-delivery systems. Other lines of research are centered on more efficient optics and more efficient detectors, which hopefully will result in more efficacious and efficient sensor systems. Perhaps the most important improvements will be in the areas of resolution and brightness of the display – the chief limitation of most night viewing systems.

The Artificial Intelligence Technologies

Artificial Intelligence (AI) also will be essential to future military space operations. Among the reasons for this is that human spaceflight is relatively expensive, both in terms of human lives and resource cost. Mainly, this is because the systems that are needed to support humans in space add tremendous costs to the overall space mission. And, it is for these reasons that other alternatives are being sought for application to future military space operations. One alternative that is being considered is the utilization of unmanned spacecraft, such as the NASA space probe and the lander vehicles. A space probe refers to an unmanned spacecraft that is given enough velocity to escape the Earth's gravitational attraction, and then, to operate in one or more orbits throughout the solar systems. And, of course, artificial satellites have been operating in space since 1957. Currently, research and development in the area of AI is concerned with developing "robotic astronauts" to work on spacecraft of all types in space, some of which are working even as far out as the nearest galaxies.

As I see it, one of the main concerns with respect to the unmanned operations has to do with the delegation of decision-making. That is, how far out on the decision tree one wants to go in locating the decision-making function. Generally, the closer the decision-making function is located to the origin of the decision-tree, there exists the greatest degree of certitude. This is because, as I see it, problem-solving is a systematic movement through the branches of a decision tree; and it involves continuous choosing of trajectories, through the making of a series of Boolean logic decisions. These are made with the goal of achieving an ultimate, predefined objective. One technique for navigating the decision tree is called means-end analysis. This is a step-by-step reduction in the difference between the current state of affairs and the desired final outcome. In artificial intelligence, the environment typically is scanned by means of various artificial sensory organs, and the "scene" is decomposed into separate objects having various spatial relationships. This kind of analysis is complicated by the fact that an object may appear differently, depending on the angle of view, as well as the direction and intensity of the perception field nodes, or pixels. This basic paradigm also applies to the decision of where to locate specific robotic systems in space.

On another dimension, artificial intelligence (AI) can be seen as being a metaphor for one way in which the military geographer and the intelligence officer can interface with military space operations. Consider that the established role of both these disciplines is to communicate the relevant environment, in its entirety, to the commander or other user. Further, the information about the environment must be delivered in a usable form; that is, in a form that facilitates the actions taken by the user entity. The way in which this is relevant to military space operations is that the user "entity" has assumed the role of the "command pilot" as it has developed in within the Earth's atmosphere. So, now I would argue the command entity in military space operations has been evolving along a human-computer spectrum. Now the issue of where the locus of decision-making and supervision should occur within a system has developed into an analysis of the human-machine dynamic. In other words, the decision of whether to use manned or unmanned systems has morphed into one of how best to use the given combinations of the two.

It can also be said that the increasing reliance by humans on computer-mechanical-electronic systems in space to accomplish such functions as navigation, propulsion, ordnance delivery, among others, also is driving the move to more autonomous artificial systems. Pilots and astronauts are relying more on a Global Positioning Systems and LORAN (long-range navigation) systems to fly unerringly towards the target. At the same time machines and electronics are being developed to provide the best solution to the problem of placing the ordnance on target. So now, the computer's complex of on-off binary code has become the common language of the command and control of function. The military aviator's problem variables, including one's position over the surface of the Earth, and the best "azimuth" heading for reaching the target, are now being calculated by a computer that is utilizing binary code language. Orbiting satellite, Input data and output data, advancing bionics would push the concept of bionics even further, as computers begin to write the program instructions for other computers on their own. Also, as the analytical and decision-making locus has begun to shift from the human operator to the pre-programmed computer; and evolution of the term, "military aviator" continues. The "pilot" of the 20th century has now been gradually superseded by the on-board computer, or by a "Remotely Piloted Vehicle operator". Now, the autonomous platform is beginning to make its own navigational and targeting decisions, guided only by an artificial intelligence component.

Directed-Energy Technologies

The technology of directed energy also is prominent in military space operations. At this point in time, the Achilles heel of laser energy radiation and propagation is that it can be attenuated and distorted by atmospheric conditions, especially with respect to humidity and temperature differences. Outside the Earth's protective atmosphere and ionosphere, cosmic and solar energy bursts also can degrade the cohesiveness and precision of the laser beam. Also, at the present time directed energy requires a clear line of sight to be most precise and efficacious. It also has trouble penetrating concrete walls and heavily-armored vehicles, for example.

The essence of a beam of directed energy is that it can be made "coherent" throughout the period of generation of the particular energy. The property of coherence means that a beam of directed energy can be given more accuracy and range by various methods. One of these is to find a source element whose molecules of are suitable for the task. Whether the source element is a ruby crystal, glass, carbon dioxide, neon, or other base element of a suitable nature, they make up the "ammunition" that produces the energy beam. The objective is to apply enough initial force to stimulate the molecules of the "source element" through a kind of "shock and awe," stimulation in order to shake them from their original state. These stimulated molecules are then rearranged into a tight, disciplined formation, which can then be amplified (the waves) into the ultimate "laser" energy beam.

"Functional laser weapons are just five years away..." So says an article in the Air Force Magazine (April, 2012). Both of the airborne laser systems that are being developed today use chemical lasers. These can produce high power in the megawatt class, but they are still too cumbersome to operate, mainly because of the large quantities of chemicals that are consumed. As a result, research and development attention has shifted more toward solid-state, electric, or fiber lasers, which depend only on electricity for power; power which can be generated by an aircraft's own engines. However, at the present level of technology, these electric lasers generate a beam in a lesser-watt class (only about 150 kilowatts of laser power), which is deemed insufficient for practical applications in military space operations. Nevertheless, there still exists a viable option for using the chemical laser systems, for at least ground-based deployment. Northrop Grumman has been involved with programs such as the Tactical High Energy Laser (THEL), which has demonstrated a capability for shooting down mortars and surface-to-surface rockets on Earth. The company has also demonstrated a capability for shooting lasers from moving ships, against other vessels which are bobbing dynamically in the water. Ultimately, there still remains the question about the feasibility of using chemically-based laser weapon systems in space, either from planetary bases or orbiting satellites. In addition, there still are many practical contradictions to their actual use, mainly because of the probability of collateral damage to friendly assets.

———

Imagine having the ability to hurl lightning bolts against the infrastructure of the enemy. Or, the ability to damage electrical systems on a mega-scale, so that electrical energy could not be provided to key elements of the enemy's overall infrastructure of modern civilization. This would include electronic systems and computer networks. Consider what would happen if the internet were to be degraded or otherwise be thrown into cyber chaos by a human-created mega-electrical charge. Is it possible that in the near future earthlings will conduct warfare using the ionosphere as the battlefield? All of these potential phenomena certainly will take military geography into a whole new area of inquiry. One scientist, in particular, by the name of Nikola Tesla was a pioneer in field of electricity. In 1900, he discovered the presence of terrestrial stationary waves. From that, he showed that the Earth itself could be used as a conductor of electricity. Beyond that, he proved that this energy could be made to resonate and the wave would thus be amplified repetitively to increase the power of the electrical charge.

Some of Tesla's ideas actually have been implemented. One example of this is located on a military installation near Anchorage, AK. It consists of an area equal to about three football fields, and it contains a giant array of electromagnetic transmitters. These powerful transmitters multiply single energy rays into a major pulse, which then emits a highly-cohesive, electro-magnetic ray at a specific point or points along the ionosphere. The electromagnetic energy pulses that are generated are at ELF (Extremely-Low Frequency) EM energy levels. At another such U.S. installation, located in Arecibo, Puerto Rico, there are systems which can generate powerful and focused EM energy to enrich the quantity water particles in clouds, and to also push the clouds up in the atmosphere.

Electromagnetic energy also does several other things: it literally ionizes the water particles, which could adversely affect military aerospace operations at certain altitudes of our atmosphere. It does this by causing local regions of atmosphere to heat up, thereby increases the ability of each unit of cloud to absorb water like a sponge and, therefore, creates turbulent weather. At the present time, there reportedly are five such ionic "heaters" on the surface of the Earth. At least in theory, these may one day enable humans to actually manipulate the ionosphere, either for good or evil. There are even some reports that these technologies may be able to cause all sorts of weather anomalies, including floods, hurricanes, and landslides, within local regions. Therefore, to some extent, such phenomena could well intensify the normal negative effects of weather on affected military aerospace operations. And, weather warfare allows the offending party to have "plausible deniability."

In each case, energy in the atmosphere would be manipulated by such systems. In general, we know that there is a lot of energy in the ionosphere, and some experts think it can be manipulated by science. However, one thing we know now for sure is that an electrical surge can effectively disable an automobile ignition system. It will shut it down, and thereby render the vehicle unusable. Now, what if the electric charge that hit the automobile engine could be amplified to a magnitude of 100 foot bolts of lightning, and strike the millions of engines over an extended geographic region of the Earth? Imagine the impact of such a mega-charge on the myriad of military weapons that rely on sensitive micro-chips in geospace. As another example of the threat that is posed by electrical charges, we are all familiar with the upset to our equilibrium that can be caused by extremely powerful stereo speakers in a neighboring apartment or a passing automobile. Imagine these 30 watt assaults on our human systems being amplified to a level of 3.6 million watts by an extra-low frequency installation. The medium for such an electromagnetic attack would likely be the ionosphere, which is the layer of atmosphere around the Earth which would provide a very effective "highway" for projecting artificial pulses of electro-magnetic energy from one point on earth to another point on earth "over the horizon," so to speak.

————

Even today, in the early years of the 21st century, many of the technological advances that have been achieved in space by NASA and other space agencies are being utilized by the national military establishment in the development of weapons and tactics within geospace. One of the most important media for technological innovation has been the electromagnetic spectrum. It has been the essence of modern navigation and communications to be sure, but it has also been the essence of the technologies that have developed in remote sensing and the computerization of military operations, among others. Another arena for technological innovation in space operations has been in development of new materials that have made possible space operations under extremes of temperature and pressure that are encountered within the interplanetary medium and within the magnetic and atmospheric domains of the planets. And, on another dimension, the microminiaturization of the electromagnetic circuitry has proven advantageous the efficacy of space operations, as the quest to conquer the barriers of space and time with less energy input continues.

Another area of technology which will facilitate quick-reaction tactics in space is the so-called "stealth technology." This might be seen as an example of what is called "adaptive camouflage" in modern military parlance. A more precise term might be "electro-optical camouflage technology" (fas.org). In any case, this camouflage technology is grounded on the fundamental tenet of the axiom, which says that: "if you can see the enemy, but he can't see you...it's easier to kill him, and much harder for him to kill you..." One member of the military-industrial complex, Stealth Technology Systems (STS), claims to have developed a camouflage technology that can be applied as a form of "makeup" for the human warrior, or to weave a fabric that can cover the warrior in a cloak of "invisibility" to the naked eye. Also, reportedly, the camouflage material can be applied as a form of paint to provide this same invisibility to any kind of military craft – including a spacecraft. Generally, it appears that this application of stealth technology, in essence, is one that enables this new camouflage material to actually "bend" light rays, from the visible-light portion of the spectrum to the infrared and X-ray portions as well.

It should also be noted that continuing advances in the technologies of space materials will be required in the future. Just as the advances in military power on Earth during the past 5,000 years or so have been attended by the development of "better" materials; that is, from stone to bronze, and then iron and steel, the advances in space military capabilities will depend on the development of still another generation of "better" materials that are made with nanotechnology and in a microgravity environment. It all starts with carbon; carbon atoms are ubiquitous. Nano building blocks made from gases such as propane are turned into solids such as diamonds, the hardest material found in nature. Sharing of atoms is what holds atoms together in a molecule. Covalent bonding: carbon atoms can bond together with many types of atoms, using a process called covalent bonding. There is no other element in the periodic table which bonds itself so strongly to itself and in as many ways as the carbon atom. Graphite sheets can slide across each other easily, which makes graphite useful as a lubricant molecular orbital allows delocalized electrons to move freely throughout an entire graphite sheet; that is why graphite conducts electricity.

———

Nano Technologies

In any case, the "holy grail" of space operations, whether civilian or military, is to maximize the thrust-to-payload ratios for each rocket or other launch vehicle in order to get into the IPM. There are several ways to achieve this goal. The most prevalent one at this point in time is through the development of more powerful chemical rocket systems, either by increasing the ratio of the rocket propulsion system vis a vis the payload component system. Another approach has been to minimize the mass of the payload component so as to increase the efficiency of the thrust energy per kilogram of the total launch mass. Then there is the strategy of reducing the overall mass of the launch vehicle as well as the payload. In practice, this has produced considerable research and development of lighter, but stronger, materials. Hence, there has been increasing interest in such technologies as nano-technology, which enables the engineers to manipulate the materials at the level of the molecule; to develop customized materials that serve to enhance the thrust-to-payload ratios. At the same time, the advances that are being made with respect to orbiting laboratories will potentially leverage the power of nanotechnology by adding even more control over the molecule architecture in the micro-gravity environment.

So, nanotechnology is the art of manipulating materials on an atomic or molecular scale, especially when building microscopic devices (such as micro-robots). It involves the designing and building of machines in which every atom and chemical bond is precisely specified. The imperative is the ability to image at the atomic scale with scanning probe microscopes. This is nano-scale science and technology; the manipulation of things with a scanning probe. The ultimate end of this scale of science and technology is toward the ability to build things smaller, and with greater precision.

This is the level of scale where we can design things atom by atom. There is already such a technology occurring; it is biotechnology. So, we are talking now about the processes of manipulation and manufacture of materials and devices on the scale of the atom. This so-called, nano-scale is usually measured in nanometers, or billionths of a meter. Materials that are built at this scale often are seen to exhibit distinctive physical and chemical properties that are related to quantum mechanical effects. Quantum mechanics is the science which deals with the behavior of matter and light on an atomic or sub-atomic scale. Thus, it is an attempt to describe and explain the properties of molecules and atoms, and their constituents – electrons, protons, neutrons, and the more elusive particles, such as quarks and gluons.

Molecular nanotechnology (MNT) also offers the prospect of providing significant increases in various technical performance properties, such as material strength and density. Such performance enhancement would provide the capabilities of many classes of space systems. Two of these are in the area of chemical rocket systems and for putting payloads into Earth orbit, with existing rocket architectures. Other systems that might be enhanced by the use of MNT are the synchronous and rotating skyhooks, solar sails, and solar electric ion engines. There is even serious discussion of the application of this technology to the construction of large-scale space bases, including military installations. The kinds of improvements that are considered feasible through the use of this technology include: product tensile strength and corresponding material density; mechano-synthetic device operating rates; and in marginal manufacturing costs. The essence of this technology in its application to space operations is the increased strength-to-density ratios that could be achieved. Aluminum, steel, titanium, graphite crystals, and MNT structural materials are those that are considered the most appropriate to this technology.

Equally important to such quick-reaction missions will be development of a comprehensive logistical and refueling network nearby. This would include the deployment of a logistical network in space; it would include the utilization of spent heavy rockets in space as "caches" or "coaling stations" in which fuel and other materiel could be located for future use by other space missions. The main point of this is that humans have already been developing the communications and logistics systems that could easily be converted to military purposes in space. However, there is still one other very important aspect of military space operations that still needs significant "application research," that is: weapons and technology.

It is the combination of all these technological developments that has enabled space systems to operate more efficaciously and reliably throughout the solar system. Each stream of technological advancement intertwines with all the others to form a powerful symbiosis that provides a total effect that is greater than the sum of its parts. Thus, we have seen how the technologies of propulsion, materials, sensors, and the electromagnetic spectrum all work together to produce platforms and weapon systems that will be optimum for the environment in geospace, and beyond. In the future, advances in micro-miniaturization and nano-technology will make it possible to take advantage smaller mass and weight to make possible the "swarm strategy" in which weapons and tactics will be reduced to the scale of the atom and, thereby improve the propulsion/thrust ratio form movement throughout the solar system.

Modeling Technologies

As we have seen, the solar system, outside our planet's atmosphere, is inherently hostile to human life, mainly because of radiation and extreme temperatures, among other environmental factors. So, as we continue to operate in these environments, there will be a continuing need to simulate the conditions in a controlled manner in order to develop technologies for dealing with them. The technologies of simulation offer a way to create realistic models of an environment under controlled conditions. In fact, NASA and other space agencies have been engaging in simulation of the exotic environments of the solar system since the 1960s, when the first astronauts were sent into orbit, outside our atmosphere. Initially, most of the simulation activities were focused on reproducing the elements of space operations. Thus, there were the familiar simulators which test pilots had been utilizing since the middle of the 20th century. These were designed to reproduce actual events and processes that would be encountered when flying and fighting within our atmosphere, under controlled conditions. However, with the beginning of the lunar landing missions, new simulators were devised to reproduce conditions in space and on the Moon. These were based on models of the environments that would be encountered during the various phases of the space mission. So, there were simulators that reproduced the conditions of flying in low-gravity environments. There also were simulators that prepared the astronauts for dealing with other environmental variables in space, such as techniques for oxygen management and fire safety procedures, among others. These simulators were, in effect, models that were designed to reproduce cockpit conditions in space. Simulators then began to be utilized to reproduce the environmental conditions of space themselves. This type of simulation involves the identification and measurement of the many variables that make up the reality that is being modeled. Once this phase of abstraction is completed, these variables then can be manipulated via a series of mathematical equations that test all the possible aspects of the reality system. At this level of abstraction, the various elements of the real-world system can be dealt with in a mathematical or computer model. This level of abstraction also allows for the application of an infinite iteration of events to develop a model of a future reality.

There are many benefits that are derived from simulation. The ability to compress time is one of these. In space, events that can take billions of years in real time can be simulated in a few minutes. Other benefits of simulation include: (1) it allows researchers to perform experiments without risking lives, in the case of manned space operations; (2) scarce or expensive resources do not have to consumed during the repetitive experimentation that occurs in designing and planning military space operations; (3) costly specialized equipment does not have to utilized during the experimentation and development phase of a space mission. In addition, computer simulations can be devised to study the dynamic behavior of objects or systems in response to conditions that cannot be easily or safely applied in real situations. Simulations are especially useful in enabling observers to predict how the functioning of an entire system may be affected by altering individual components within that system, in a given geographic environment. If the nature of the forces between objects is known, then a computer model that represents the behavior of those bodies by developed to solve the equations of motion that are involved. For example, astronomers using computer simulations of galactic movements can demonstrate events that take millions of years to complete.

———

Unifying Technologies

One conclusion that I hope will be drawn from this book is that the science of geography has much to offer to the understanding of the solar system, both in a general sense and for specific applications. It provides a unique perspective for studying our corner of the Milky Way galaxy, as a spatial region within a nested set of spatial regions. Such a spatial viewpoint also enables one to see the solar system as a functional region, in which a given complex of nodes and linkages interact under a set of physical laws for a specific purpose. Such a regional methodology also makes the enormity of the solar system more manageable; and it provides the framework for the formulation of predictions about the spatial behavior of the spatial system under various sets of effects and circumstances. At the same time, an understanding of the internal properties and internal spatial behaviors within each of these nodes can enable one to draw logical inferences about their nature. More generally, the knowledge of these nodes and their linkages in one spatial system can then provide insights into an analogous system through the methodology of modeling.

One of the most useful technologies that have derived from the spatial perspective of the geographer is that of the region. Like the millennia and the centuries of the historian, the region is a unifying principle that can be applied to the study of not just our own planet, but the other planets and bodies of the solar system too. Indeed, this is one of the most powerful conceptual tools of the military geographer in any environmental or operational context.

SUMMARY CONCLUSIONS

Perhaps the most sweeping conclusion that can be drawn from this book is that there is a basic "sameness" that occurs throughout the solar system. One major reason for this is that all of the matter that has created the geography of the solar system has derived from a common genesis; that is, a primal cloud of gas that formed shortly after the original event that we call the "Big Bang." It is from this cloud that all the matter and energy in the Universe, including what would become the solar system, has evolved. Since then, the process has produced the atom and the elements that are common to the Sun and all the planets and other celestial bodies throughout the solar system. Another reason for the sameness that we see throughout the solar system is that the same basic forces of nature have and continue to operate on all matter and energy throughout the solar region. The most significant of these forces is gravity, because it is so pervasive in shaping the anatomy and physiology of everything within the solar system. It also accounts for the apparent "clockwork" elegance of the distribution of the major planets and their moons. The force of electromagnetism also has been found to operate in the same predictable manner throughout the solar region. We see it in the waves of charged particles that are flowing throughout the region. These include the solar winds and the magnetic fields that develop around many of the planets and some moons.

The atom has been found to look the same, and to operate in the same fashion throughout the heliosphere. So too, do the elements that are familiar to chemists on Earth. For this reason, scientists such as geologists, geographers, physicists and chemists, among others, have found that the theories and models of their particular scientific fields are applicable on other planets and bodies throughout the region of our Sun. Geographers, in particular, can see that the traditional conceptual frameworks and methodologies that have proven to be effective in the analysis of the spatial aspects of our own planet can be applied successfully to the study of the solar system as a subregion of the Milky Way galaxy, and as a region in its own right that is made up of various subregions itself.

It is precisely because of this sameness natural environment that humans have been able to develop technologies that operate effectively both on Earth and in the rest of the solar system. We have found this to be the case during the first half-century of space exploration. So, our machines and our technical systems are being successfully applied on other planets and through the interplanetary medium of our solar neighborhood. Furthermore, the machines and the other manifestations of human technology that have proven to work well in space exploration should be useful in the development of military operations in space as well.

To be sure, there are differences within the overall context of sameness in the solar system, but these are only variations within given parameters. This is why scientists and engineers on Earth have been able to devise effective simulators – both natural and man-made – to plan and design space systems on Earth prior to deploying them in space. And, it is this quality of sameness in space exploration that military space operations can draw from the experience of the space agencies, like NASA, in order to develop efficacious military weapon systems and tactics in space as well.

SELECTED BIBLIOGRAPHY

Air Force Basic Doctrine, Organization and Command, Air Force Doctrine Document 1, 14 October 2011 (af.mil)

Atom: Journey Across The Subatomic Cosmos, Isaac Asimov, Truman Talley Books, 1991

Chemistry for changing times, John W. Hill and Doris K. Kolb, Prentice Hall, (Eight Edition), 1998

The Complete Book of Spaceflight: From Apollo 1 to Zero Gravity, David Darling, John Wiley and Sons, Inc., Hoboken, New Jersey, 2003

Electric Universe: The Shocking True Story of Electricity, David Bodamis, Crown Publishers, 2005

Exploring Mars: Chronicles From a Decade of Discovery, Scott Hubbard, The University of Arizona Press, 2011

Mass Spectrometry: Principles & Applications, Second Edition, Ian Howe, Dudley H. Williams, Richard D. Bowen, McGraw-Hill, 1981

Rolling Thunder: Jet Combat From WWII to the Gulf War, Ivan Randall, The Free Press, 1999

Geography: The Study of Location, Culture, and Environment, John F. Kolars and John D. Nystuen, McGraw-Hill Book Company, 1974

The Idea Factory: Bell Labs and the Great Age of American Innovation, John Gertner, Recorded Books

Spatial Organization: The Geographer's View of the World, Ronald Abler, John S. Adams, and Peter Gould, Prentice-Hall, Inc., 1971

Parallax: The Race to Measure the Cosmos, Alan W. Hirshfeld, W.H. Freeman and Company, 2001

Techniques of Archeological Excavation, Philip Barker, B.T. Batsford Ltd., London, 1959

NASA's Human Exploration & Development of Space Enterprise, Published by NASAexplore, May 15, 2003

USA in Space, Second Edition, Edited by Russell R. Tobias, Salem Press, Inc., 2001

Towards a General Theory of Geographic Representation, International Journal of Geographic Information Science, Vol. 21, No. 3, March 2007, pp. 239-260

A special note of acknowledgement goes to the vast treasure trove of raw data and refined information that has been provided by the NASA space agency to all scientists and researchers who are interested in the various aspects of studies of the solar system and beyond. The main venue for the transmission of such data and information is the internet, especially the website nasa.gov.

Another manifestation of the U.S. space agency's policy of openness can be seen in the willingness to share the various space mission reports. These allow interested parties to tap into a second-by-second narration of each of the missions. These provide great insights into all the skill-sets that have been developed during the past half-century of NASA space exploration activities.

www.ingramcontent.com/pod-product-compliance
Lightning Source LLC
Chambersburg PA
CBHW081104170526
45165CB00008B/2319

* 9 7 8 1 4 8 2 5 6 8 0 6 6 *